Numerical Methods for Engineers and Scientists

Numerical Methods for Engineers and Scientists

Second Edition

J.N. Sharma

Alpha Science International Ltd.
Oxford, U.K.

J.N. Sharma
Professor of Mathematics
National Institute of Technology
(Deemed University)
Hamirpur (H.P.), India

Second Edition 2007

Alpha Science International Ltd.
7200 The Quorum, Oxford Business Park North
Garsington Road, Oxford OX4 2JZ, U.K.

ISBN-13: 978-1-84265-365-4
ISBN-10: 1-84265-365-2

Printed in India

Dedicated

To my wife

Soma Sharma

and children

Nisha, Risha and Vivek

PREFACE TO THE SECOND EDITION

The rapid advances made in high-speed digital computers and softwares development have led to many need based changes and new features in this edition of the book causing it to differ substantially from the first edition. The choice and style of presentation has also been greatly influenced by the revised syllabi of undergraduate and postgraduate programs of various universities and engineering institutions as per the latest guidelines and requirements prescribed by the UGC/AICTE and other academic forums. In view of this some of the sections have been rewritten and various additions, deletions and refinements have been made in order to incorporate corrections of misprinting and errors in the first edition of this book. A new chapter 9 on features of C-language and containing C-programs of some standard numerical methods commonly used is added in this revised edition of the book. The QR method in chapter 3, Chebyshev polynomials with applications to Lanczos process of economization of power series in chapter 8, and Descartes method to solve fourth degree nonlinear equations in the appendix are also added. A number of new examples and problems have also been added based on the new additions. The text is enriched with solved examples, algorithms, C-programs and a variety of exercises at the end of each chapter so as to drill the students in problem solving both manually and through computer programming. The book is designed as a text book for undergraduate and post graduate students of engineering, mathematics, physics, BCA, MCA and competitive examinations such as NET, SLET, GATE and IAS etc. The book would be useful and of interest to the scientists and engineers engaged in research. It is hoped that the book in its present form will enjoy its ever increasing popularity.

I am extremely thankful to my colleagues, research students and teachers of various institutions in the country and abroad who took interest in this book and gracefully helped me in making this version of the book error free in addition to widening its user's sphere. I would like to express my gratitude to various authors whose books and research papers are referred to and have been listed in the bibliography. My gratitude's and appreciation are also due to the M/S Narosa Publishing House, New Delhi India, for their very careful and meticulous processing of the manuscript, both during the editorial and production stages. The author is also grateful to the authorities of National Institute of Technology (Deemed University), Hamirpur (H.P) for the help, encouragement and inspirations provided by them during this work.

Finally, I sincerely thank my wife Soma Sharma for her patience and encouragement while writing this book. I also appreciate the understanding shown by my daughters Nisha and Risha and son Vivek who had to forego my attention and care during this work.

Constructive comments and suggestions for the improvement of this book are welcomed and will be thankfully acknowledged.

Jagan Nath Sharma

PREFACE TO THE FIRST EDITION

The desire for numerical answers to applied problems has increased manifold with the advances made in various branches of science and engineering and rapid development of high-speed digital computers. Although numerical methods have always been useful, their role in the present day scientific computations and research is of fundamental importance. Any kind of application is amenable to numerical treatment, the only limitation being the reader's imagination. In fact, many practical problems in science and engineering when formulated mathematically give rise to several interesting and useful situations, for which numerical solutions can be obtained while analytical methods may fail. It is therefore clear training in applied sciences or engineering would be incomplete without an adequate understanding of numerical methods. It is also believed that students who can understand, enjoy, and successfully apply the methods of numerical computing to solve problems make better logicians and thereby better engineers and scientists.

This self-contained textbook has numerous distinguishing features over the already existing textbooks on the same topic. The contents of the book have been organized in a logical order and the topics are discussed in a system manner. In chapter 1 numerical computations and error analysis have been discussed. The chapter 2 deals with various direct and iterative numerical methods for the solution of algebraic and transcendental equations including the solutions of non-linear simultaneous equations. In chapter 3 various direct and iterative methods for the solution of system of algebraic equations, inversion of matrices and determination of eigenvalues and eigenvector have been discussed. Chapter 4 deals with methods of curve fitting function approximations and interpolation in additions to finite differences. In chapter 5 various numerical methods for the evaluation of derivatives and integrals including singular and double integrals have been discussed. Chapters 6, 7 and 8 respectively, deal with numerical methods for the solution of ordinary, partial and integral equations. The text is enriched with solved examples in order to illustrate various methods and concepts at various stages. It is also supplemented with algorithms and a variety of exercises at the end of each chapter so as to drill the students in problem solving both manually and through computer programming. The knowledge of calculus, algebra, geometry, trigonometry and differential equations will be of an additional advantage to the readers. The book is designed as a text book on computational numerical methods for undergraduate and post graduate students of engineering, mathematics, physics, BCA and MCA. The book would be of interest and useful to scientists and engineers engaged in research. In addition to an exhaustive bibliography, an appendix containing some important and useful iterative methods for the solution of nonlinear complex equations has been added at the end.

I would like to express my gratitude to various authors whose books and research papers are referred to and have been listed in the bibliography. I am thankful to Mr. Kuldeep Kumar for the excellent typing and computer setting of the manuscript. My gratitude's and appreciation are also due to the M/S Narosa Publishing House, New Delhi India, for their very careful and meticulous processing of the manuscript, both during the editorial and production stages. The author is also grateful to the authorities of National Institute of Technology, Deemed University, Hamirpur (H.P) for the help, encouragement and inspirations provided by them during this work. The fruitful discussions with my mathematics colleagues and my research students from time to time are also acknowledged. Finally, I sincerely thank my wife Soma Sharma for her patience and encouragement while writing this book. I also appreciate the understanding shown by my daughters Nisha and Risha and son Vivek who had to forego my attention and care during the preparation of this book.

Constructive comments for the improvement of contents of this book are welcomed and will be warmly appreciated.

Jagan Nath Sharma

CONTENTS

CHAPTER 1

NUMERICAL COMPUTATIONS AND ERROR ANALYSIS

1.1 Introduction

The limitations of analytical methods in practical applications have led mathematicians to evolve numerical methods. The analytic methods are unable to produce desirable results in drawing plausible inference from a given set of tabulated data, finding roots of transcendental equations, solving non-linear differential equations and in many more such situations. The aim of numerical analysis is therefore, to provide constructive methods for obtaining answers to such problems in a numerical form. When solving problems one usually starts with some initial data and then high precision computers, after some intermediate steps, the final results. The numerical data used are approximate being true to two, three or more figures. Sometimes the methods used are also approximate and therefore the error in a computed result may be due to the errors in the data, or the errors in the method or both. We shall describe in this chapter, some basic ideas concerning numerical computations, errors and their analysis.

1.2 Numbers and their accuracy

In every day life we use numbers based on the decimal system. Any integer N is expressible as a polynomial in the base 10 with integral coefficients between 0 and 9 in the from

$$N = (a_n a_{n-1} a_{n-2} ... a_0 a_{-1} a_{-2} a_{-3} ... a_{-m})_{10}$$

$$= a_n 10^n + a_{n-1} 10^{n-1} + a_{n-2} 10^{n-2} + ... + a_0 10^0 + a_{-1} 10^{-1} + a_{-2} 10^{-2} + ... + a_{-m} 10^{-m} \qquad (1.2.1)$$

Modern computers read pulses sent by electrical components. The state of an electrical impulse is either on or off. It is therefore convenient to represent numbers in computers in the binary system having base 2 and the integer coefficients may take the values 0 or 1. Thus in the binary system the number N is expressible in the form

$$N = (b_n b_{n-1} b_{n-2} .. b_0 b_{-1} b_{-2} b_{-3} .. b_{-m})_2$$

$$= b_n 2^n + b_{n-1} 2^{n-1} + b_{n-2} 2^{n-2} + ... + b_0 2^0 + b_{-1} 2^{-1} + b_{-2} 2^{-2} + b_{-3} 2^{-3} + ... b_{-m} 2^{-m} \qquad (1.2.2)$$

where coefficients b_k are either 0 or 1.

Thus if x is a real number, then its integral part x_I is the largest integer less than or equal to x while $x_F = x - x_I$, is its fractional part. The fractional part can always be written as a decimal fraction

$$x_F = \sum_{k=1}^{\infty} b_k 10^{-k} \tag{1.2.3}$$

where each b_k is non-negative integer less than 10. If $b_k = 0$ for all k greater than a certain integer, then the fraction is called terminating. For example $\frac{1}{4} = 0.25 = 2\times10^{-1} + 5\times10^{-2}$ is a terminating decimal fraction while $\frac{1}{3} = 0.333... = 3 + 10^{-1} \times 3\times10^{-2} + 3\times10^{-3} + ...$ is a non-terminating fraction.

It is customary to write the integral part and the fractional part one after the other, separated by a point, called the 'decimal point' as

$$x = (a_n a_{n-1} ... a_0 \cdot b_1 b_2 b_3 ...)_{10} \tag{1.2.4}$$

Analogously, we can write the number x in binary system as

$$x = (a_n a_{n-1} ... a_0 \cdot b_1 b_2 b_3 ...)_2 \tag{1.2.5}$$

where

$$x_I = (a_n, a_{n-1} .... a_0)_2 \text{ and } x_F = \sum_{k=1}^{\infty} b_k 2^{-k} \tag{1.2.6}$$

The binary fraction $(\cdot b_1, b_2, b_3 ...)_2$ for a given number x_F between zero and one can be calculated as follows:

if
$$x_F = \sum_{k=1}^{\infty} b_k 2^{-k}$$

then
$$2x_F = \sum_{k=1}^{\infty} b_k 2^{-k+1} = b_1 + \sum_{k=1}^{\infty} b_{k+1} 2^{-k}$$

Hence b_1 in the integral part of $2x_F$, while

$$(2x_F)_F = 2x_F - b_1 = \sum_{k=1}^{\infty} b_{k+1} 2^{-k}$$

Therefore, repeating this procedure, we find that b_2 is the integral part of $2(2x_F)_F$, b_3 is the integral part of $2(2(2x_F)_F)_F$, and so on.

For example, if $x = 0.625 = x_F$ then

$$2(0.625) = 1.25 \Rightarrow b_1 = 1$$
$$2(0.25) = 0.5 \Rightarrow b_2 = 0$$
$$2(0.5) = 1.0 \Rightarrow b_3 = 1$$

and $b_k = 0$ for all $k = 4, 5, 6 ....$

Hence $$0.625 = (\cdot 101)_2$$

It may be noted here that every terminating decimal fraction may or may not give rise to a terminating binary fraction. This is due to the fact that the binary fraction for $x_F = 10^{-1} = 0.1$ is not terminating.

Algorithm 1.2.1: Conversion of numbers from base β to the base 10.

Given the coefficients $a_n, a_{n-1}, \ldots a_0$ of the polynomial

$$p(x) = a_n x^n + a_{n-1} x^{n-1} + \ldots + a_1 x + a_0 \tag{1.2.7}$$

and a number β. Compute recursively the numbers $b_n, b_{n-1}, \ldots bo$:

$$b_n := a_n$$

$$b_{n-1} := a_{n-1} + b_n \beta$$

$$b_{n-2} := a_{n-2} + b_{n-1} \beta$$

$$b_{n-3} := a_{n-3} + b_{n-2} \beta$$

$$\vdots \qquad \vdots \qquad \vdots$$

$$b_0 := a_0 + b_1 \beta$$

then $b_0 = p(\beta)$. This computes the integral part x_I of the real number x. Now given x between 0 and 1 and an integer β greater than 1. Generate recursively $b_1, b_2, b_3 \ldots$ by

$$c_0 := x$$

$$b_1 := (\beta c_0)_I, \quad c_1 := (\beta c_0)_F$$

$$b_2 := (\beta c_1)_I, \quad c_2 := (\beta c_1)_F$$

$$b_3 := (\beta c_2)_I, \quad c_3 := (\beta c_2)_F$$

$$\vdots \qquad \vdots \qquad \vdots$$

Then

$$x = (\cdot b_1 \text{ of } b_2 \text{ of } b_3 \ldots)_\beta = \sum_{k=1}^{\infty} b_k \beta^{-k}$$

We have stated this algorithm for a general base β rather than for the specific binary base $\beta = 2$ for two reasons.

(i) If this conversion to binary is carried out with pencil and paper, it is usually faster to convert first to octal *i.e.* use $\beta = 8$ and then to convert from octal to binary.

(ii) This algorithm can be used to convert binary (or octal) to decimal, by choosing $\beta = 10$ and using binary (or octal) arithmetic.

The procedure of conversion of numbers from one base to other is explained in the following examples. Many computers used in scientific work operate internally in the binary system. Users of computers, however, prefer to work in the decimal system. It is therefore necessary to have some means of converting from decimal to binary when information is submitted to the computer, and from binary to decimal for output purposes.

Example 1.2.1: Convert the following numbers from decimal to binary

(*i*) 187.625 (*ii*) 14.875 (*iii*) 17.375 (*iv*) 634.640625

Solution: (*i*) 187.625

2	187		↑
2	93	– 1	2^0
2	46	– 1	2^1
2	23	– 0	2^2
2	11	– 1	2^3
2	5	– 1	2^4
2	2	– 1	2^5
2	1	– 0	2^6
2	0	– 1	2^7

Also we have

$.625\times 2$	$= 1.25 \rightarrow 1$	↓	2^{-1}
$.25\times 2$	$= 0.50 \rightarrow 0$		2^{-2}
$.50\times 2$	$= 1.00 \rightarrow 1$		2^{-3}

$\therefore (.625)_{10} = 1\times 2^{-1} + 0\times 2^{-1} + 1\times 2^{-3} = (.101)_2$

$$\therefore \quad (187)_{10} = 1\times 2^7 + 0\times 2^6 + 1\times 2^5 + 1\times 2^3 + 0\times 2^2 + \times 2^1 + 1\times 2^0 = (10111011)_2$$

Thus $(187.625)_{10} = (10111011.1011)_2$

(iii) 14.875

2	14		↑
2	7	– 0	2^0
2	3	– 1	2^1
2	1	– 1	2^2
	0	- 1	2^3

Also we have

$.875\times 2 = 1.75$	$\rightarrow 1$	↓	2^{-1}
$.75\times 2 = 1.50$	$\rightarrow 1$		2^{-2}
$.50\times 2 = 1.0$	$\rightarrow 1$		2^{-3}

$\therefore (14)_{10} = (1110)_2$ and $(.875)_{10} = (.111)_2$

Thus $(14.875)_{10} = (1110.111)_2$

(iv) 17.375

2	17		
2	8	– 1	2^0
2	4	– 0	2^1
2	2	– 0	2^2
2	1	– 0	2^3
	0	– 1	2^4

$.375 \times 2 = 0.75 \rightarrow 0 \quad 2^{-1}$

$.75 \times 2 = 1.50 \rightarrow 1 \quad 2^{-2}$

$.5 \times 2 = 1.0 \rightarrow 1 \quad 2^{-3}$

$\therefore (.375)_{10} = (.011)_2$

$\therefore (17)_{10} = (10001)_2$

Thus $(17.375)_{10} = (10001.011)_2$

(iv) 634.640625

2	634		
2	317	– 0	2^0
2	158	– 1	2^1
2	79	– 0	2^2
2	39	– 1	2^3
2	19	– 1	2^4
2	9	– 1	2^5
2	4	– 1	2^6
2	2	– 0	2^7
2	1	– 0	2^8
	0	– 1	2^9

$.640625 \times 2$	$= 1.28125$	$\rightarrow 1$	2^{-1}
$.28125 \times 2$	$= 0.5625$	$\rightarrow 0$	2^{-2}
$.5625 \times 2$	$= 1.125$	$\rightarrow 1$	2^{-3}
$.125 \times 2$	$= 0.250$	$\rightarrow 0$	2^{-4}
$.25 \times 2$	$= 0.50$	$\rightarrow 0$	2^{-5}
$.050 \times 2$	$= 1.00$	$\rightarrow 1$	2^{-6}

$\therefore (.640625)_{10} = (.101001)_2$

$\therefore (634)_{10} = (1001111010)_2$

Thus $(634.640625)_{10} = (1001111010.101001)_2$

Example 1.2.2: Convert the following binary number into decimal from

(*i*) $(11100111.101)_2$ (*ii*) $(100101.011)_2$ (*iii*) $(110100101101.1100)_2$

Solution:

(i) $(11100111.101)_2$

$$= 1\times 2^7 + 1\times 2^6 \times 1\times 2^5 + 0\times 2^4 + 0\times 2^3 + 1\times 2^2 + 1\times 2^1 + 1\times 2^0 + 1\times 2^{-1} + 0\times 2^{-2} + 1\times 2^{-3}$$

$$= 128 + 64 + 32 + 0 + 0 + 4 + 2 + 1 + .5 + 0 + .125 = (231.625)_{10}$$

(ii) $(100101.011)_2$

$$= 1\times 2^5 + 0\times 2^4 + 0\times 2^3 + 1\times 2^2 + 0\times 2^1 + 1\times 2^0 + 0.\times 2^{-1} + 1\times 2^{-2} + 1\times 2^{-3}$$

$$= 32 + 0 + 0 + 4 + 0 + 1 + .0 + .25 + .125 = (37.375)_{10}$$

(iii) $(110100101101.1100)_2$

$$= 1\times 2^{11} + 1\times 2^{10} + 0\times 2^{9} + 1\times 2^{8} + 0\times 2^{7} \times 0\times 2^{6} + 1\times 2^{5} + 0\times 2^{4} + 1\times 2^{3}$$

$$+ 0\times 2^{2} + 1\times 2^{1} + 0\times 2^{0} + 1\times 2^{-1} + 1\times 2^{-2}$$

$$= 2048 + 1024 + 256 + 32 + 8 + 2 + 0.5 + .25 \ = (3370.75)_{10}$$

Binary numbers and binary arithmetic though, ideally suited for today's computers, are somewhat tiresome for the people because of the number of digits necessary to represent even moderately sized numbers. The number of bits in a binary number equivalent of a decimal number is on the average equal to $(\log_2 10) = 3.32$. Thus eight the number of bits in binary digits is necessary to represent the three digit number *187*. The octal number system, using the base *8*, presents a kind of compromise between the computer preferred binary and the people–preferred decimal system. The octal system used the eight digits 0, 1, 2, 3, 4, 5, 6, 7. It is easy to covert from octal to binary and back since three binary digits make one octal digit. To convert from octal to binary, one merely replaces all octal digits by their binary equivalent and thus

$$(347)_8 = (011 \quad 100 \quad 111)_2 = (11100111)_2$$

Conversely, to convert from binary to octal, one partitions the binary digits in groups of three, starting from the right, and then replaces each group by its octal digit and thus

$$(10111011)_2 = (010\ 111\ 011)_2 = (273)_8$$

Another popular base is known as hexadecimal (base 16) in which there are sixteen symbols namely, 0 to 9, A, B, C, D, E, F. Thus a group of 4 bits can be represented by an equivalent hexadecimal digit. A table of binary bits and their octal as well as hexadecimal equivalents is given below:

Binary string	Octal	Hexadecimal	Decimal
0000	0	0	0
0001	1	1	1
0010	2	2	2
0011	3	3	3
0100	4	4	4
0101	5	5	5
0110	6	6	6
0111	7	7	7
1000	10	8	8
1001	11	9	9
1010	12	A	10
1011	13	B	11
1100	14	C	12
1101	15	D	13
1110	16	E	14
1111	17	F	16

Thus $(\underline{1101}\quad \underline{0010}\quad \underline{1010}\quad \underline{1100})_2 = 13\times 16^2 + 2\times 16^1 + 10\times 16^0 + 12\times 16^{-1} = 3370.75$

D 2 A C

1.3 Floating point arithmetic

There are two types of arithmetic operations available in a computer:

1. **Integer arithmetic:** This deals with integer operands that are numbers without fractional parts. It is mainly used in counting and as subscripts.

2. **Real or Floating point arithmetic:** It uses numbers with fractional parts as operands and is used in most computations.

Due to economic considerations computers are designed such that each location, also called a word, in memory stores only a finite number of digits. Consequently, all operands in memory stores only a finite number of digits. One method of representing real numbers in computers would be to assume a fixed position for the decimal points and store all numbers, after appropriate shifting if necessary, with an assumed decimal point. If such a convention is used, the maximum and minimum (in magnitude) numbers that may be stored in a computer having memory in which each location can store 6 digits with a provision to store one or more signs, are 9999.99 and 0000.01 respectively. This range was quite inadequate in practice and hence a different convention for representing real numbers is adopted. This convention aims to preserve the maximum number of significant digits in a real number and also increase the range of values of real number stored. This representation is called *normalized floating point* mode of representing and storing real numbers. In this mode a real number is expressed as a combination of a *mantissa* and an *exponent*. An n-digit floating-point number in base β has the form

$$x = \pm(\cdot d_1 d_2 d_3 \ldots d_n)_\beta \beta^e \qquad (1.3.1)$$

where $(\cdot d_1 d_2 d_3 \ldots d_n)_\beta$ is a β-fraction called the *mantissa*, and *e* is an integer called the *exponent*. Such a floating –point number is said to be normalized in case $d_1 \neq$ o, or else $d_1 = d_2 = \ldots d_n = 0$. For most computers $\beta = 2$, although on some, $\beta = 16$ and on most desk and pocket calculators $\beta = 10$. The precision or length n of floating-point number on any particular computer is usually determined by the word length of the computer and may therefore vary from computer to computer. Computing systems which accept FORTRAN programs are expected to provide floating-point number of two different lengths, one roughly double the other. The shorter one, called single precision, is ordinarily used unless the other called double precision is specifically asked for. The double precision usually doubles the storage requirement and more than double running time as compared with single precision. The exponent e is limited to a range

$$m < e < M \qquad (1.3.2)$$

for certain integers m and M. Usually $m = -M$, but the limits may vary widely from computer system to computer system. Let $fl(x)$ denotes a floating-point number. A real number x can be translated into an n β-digit floating -point number $fl(x)$ in two ways, rounding and chopping.

1. **Rounding:** In this $fl(x)$ is chosen as the normalized floating-point number nearest x, where symmetric rounding (rounding to an even digit) is used in case of a tie.

2. **Chopping:** In this $fl(x)$ is chosen as the nearest normalized floating point number between x and 0. For example

$$fl\left(\frac{2}{3}\right)=\begin{cases}(0.67)10^0, \text{ rounded}\\(0.66)10^0, \text{ chopped}\end{cases} \quad \text{and} \quad fl(-838)=\begin{cases}-(0.84)10^3, \text{ rounded}\\-(0.83)10^3, \text{ chopped}\end{cases}.$$

In some cases, this definition of fl (x) is modified in case $|x| \geq \beta^m$ *(overflow)* or $0 < |x| \leq \beta^{m-n}$ *(under flow)*, where *m and M* are the bounds on the exponents. When an arithmetic operation is applied to two floating-points numbers the result usually fails to be a floating-point number of the same length. If for example, we take two decimal-digit numbers

$$x=(0.20\ )10^1=2,\ y=(0.77)10^{-6}, z=(0.30)10^1\ =3$$

$$\text{then}\ \ x+y=(0.200000077)10^1$$

$$x-y=(0.154)10^{-5}$$

$$\frac{\text{x}}{\text{z}}=(0.666.......)10^6$$

Hence if ω denotes one of the arithmetic operations (addition, subtraction, multiplication. or division) and ω^* denotes the floating-point operation then usually

$$x\omega^* y \neq x\omega y$$

Although the floating-point operation ω^* corresponding to ω may vary from machine to machine, ω is usually constructed so that

$$x\omega^* y = fl(x\omega y) \tag{1.3.3}$$

Thus the floating –point sum (difference, product, or quotient) of two floating -point numbers usually equals the floating point number which represents the exact sum (difference, product, or quotient) of the two numbers. Hence (unless overflow or underflow occurs) we have

$$x\omega^* y = x\omega y(1+\delta), \text{ for some } \delta \text{ with } |\delta| \leq u \tag{1.3.4}$$

where u is the round off. In certain situations, it is more convenient to use the equivalent formula

$$x\omega^* y = (x\omega y)/(1+\delta), \text{for some } \delta \text{ with } |\delta| \leq u \tag{1.3.5}$$

Thus consequence of the floating-point representation is that the associative and the distributive laws of arithmetic are not always valid i.e.

$$(a+b)-c \neq (a-c)+b \tag{1.3.6}$$

$$a(b-c) \neq (ab-ac) \tag{1.3.7}$$

Example 1.3.1: Illustrate the law (1.3.6) with the help of examples.

Solution: Consider a = .5665 E1

b = .5556 E-1

c = .5644 E1

$a+b$ = .5665 E1 + .5556E – 1

$= .566\text{ E} - 1 + .0053\text{E}1$

$= .5720\text{E}1$

$(a + b) - c = .5720\text{E}1 - .5644\text{E}1$

$= .0076\text{E}1$

$= .7600\text{E}{-1}$

$a - c = .5665\text{E}1 - .5644\text{ E}1$

$= .0021\text{E}1$

$= .2100\text{ E} - 1$

$(a - c) + b = .2100\text{ E}{-1} + .5556\text{ E}{-1}$

$= .7656\text{E}{-1}$

Thus $(a + b) - c \neq (a - c) + b$

In fact the correct answer, if no number is truncated, is .76 56E – 1

Example 1.3.2: Explain with the help the of an example the rule (1.3.7)

Solution: Let $a = .05555\text{ E}1$

$b = .4545\text{ E }1$

$c = .4535\text{ E}1$

$b - c = .0010\text{ E}1 = .1000\text{ E} - 1$

$a(b - c) = .5555\text{ E}1 \times .1000\text{E} - 1$

$= .0555\text{ E }0 = .5550\text{ E} - 1$

$a\,b = .5555\text{E} - 1 \times .4545\text{ E}{+1}$

$= .2524\text{E}2$

$a\,b - ac = .0005\text{E}2$

$= .5000\text{E} - 1$

Thus a (b – c) $\neq$ ab – ac

In fact, if the intermediate results are not truncated, the correct answer is .5555 E –1.

Concept of zero in floating point arithmetic

The number zero has a precise meaning in arithmetic. However, while doing arithmetic with real number representation in normalized floating point mode, the exact equality of a number to zero can never be ensured. This is again due to the fact that most of the numbers in floating point are only approximations. For example, the roots of the quadratic equation $x^2 + 2x - 2 = 0$ are $-1 \pm \sqrt{3}$. These roots in floating point with 4 digit mantissa are, .7320 E0 and -.2732 E 1. If we substitute x =.7320 E0 in the expression $\left(x^2 + 2x\right)$ -2 we should ideally obtain zero. Using floating point arithmetic, however, we get

$$(.7320\text{ E}0 \times .7320\text{ E}0 + .2000\text{ E}1 \times .7320\text{E}0) - 2000\text{ E}1$$

$$= (.535\ 8\text{E}0 + .1464\text{E}1 - .2000\text{E}1$$

$$= - .1000\text{E} - 2$$

Thus main point to be noted here is that in any computation algorithm it is not advisable to give a branching instruction based on whether a floating point quantity is zero or non zero.

1.4 Errors in numbers

If the number x^* is an approximation to the exact answer x, then we call the difference $x - x^*$ as error in x^* and thus Exact= approximation + error. In numerical computation, we come across the following types of errors:

(1) **Inherent Errors:** The errors already present in the statement of a problem before its solution, are called inherent errors. Such errors arise due to the given data being approximate or due to the limitations of mathematical tables, calculations or digital computers. These types of errors can be minimized by taking better data or by using high precision computing aids.

(2) **Rounding errors:** These errors arise from the process of rounding off the number during the computation. Such errors are avoidable in most of the calculations. Due to the limitations of the computing aids the rounding errors can be reduced. (i) by changing the calculation procedure so as to avoid subtraction of nearly equal numbers or division by a small number or (ii) by retaining atleast one more significant figure at each step than that given in the data and rounding off the last step only.

(3) **Truncation errors:** These errors are caused by using approximate results or on replacing an infinite process by a finite one. If we are using a decimal computer having fixed word length 4 digits, rounding off of 43.658 gives 43.66 whereas truncation gives 43.65. Truncation error is a type of algorithm error.

(4) **Absolute, relative and percentage errors:** It X is the true value of a quantity and X' is its approximate value, and then $|X - X'|$ is called the absolute error E_a.

The relative error is defined by

$$E_r = \left|\frac{X - X'}{X}\right|$$

and the percentage error is

$$E_p = 100E_r = 100\left|\frac{X - X'}{X}\right|.$$

Let $\overline{X}$ is a number such that $|X - X'| \le \overline{X}$, then $\overline{X}$ is an upper limit for the magnitude of absolute error and it measures the absolute accuracy. The relative and percentage errors are independent of the units while absolute error is expressed in terms of these units. If the first significant figure of a number is k and the number is correct to n significant figures, then the relative error $< 1/\left(k \times 10^{n-1}\right)$. For example, let us verify this result by finding relative error in the number 864.32 correct to five significant figures.

Here $k = 8$, $n = 5$ and absolute error $0.01 \times \dfrac{1}{2} = 0.005$

$$\therefore \text{ Relative error} \le \frac{0.005}{864.32} = \frac{5}{864320} = \frac{1}{2\times 8643} < \frac{1}{2\times 80000} = \frac{1}{2\times 8\times 10^4} < \frac{1}{8\times 10^4} = \frac{1}{k\times 10^{n-1}}$$

Hence the result is verified. Once an error is committed, it contaminates subsequent results. This error propagation through subsequent calculations is conveniently studied in terms of the two related concepts of conditions and instability.

The word condition is used to describe the sensitivity of the function value *f (x)* to changes in the argument x. The condition is measured by the maximum relative change in the function value *f (x)* caused by a unit relative change in the argument. Thus, condition of *f* at *x*

$$\max\left\{\left|\frac{f(x)-f(x^*)}{f(x)}\right| \Big/ \left|\frac{x-x^*}{x}\right| : x-x^* \text{ is small}\right.$$

$$\cong \left\{\frac{f'(x)x}{f(x)}\right\}, \text{ where we have used } f(x)-f(x^*) \approx f'(x)(x-x^*)$$

The larger the condition, the more ill – conditioned the fraction is said to be. For example, we take $f(x)=\sqrt{x}$ so that $f'(x)=\dfrac{1}{2\sqrt{x}}$ and hence the condition of f is approximately,

$$\left|\frac{f'(x)x}{f(x)}\right| = \frac{\left\{\frac{1}{2\sqrt{x}}\right\}x}{\sqrt{x}} = \frac{1}{2}.$$

This says that taking square root is a well- conditioned process since it actually reduces the relative error. On the other hand, if

$$f(x)=\frac{10}{1-x^2} \text{ then } f'(x)=20x/(1-x^2)^2$$

so that

$$\left|\frac{f'(x)x}{f(x)}\right| = \left|\frac{\left[\frac{20x}{(1-x^2)^2}\right]x}{10/(1-x^2)}\right| = \frac{2x^2}{|1-x^2|}$$

and this number is quite large for $|x|$ near 1. Thus for x near 1 or -1, this function is quite ill-conditioned. It very much magnifies relative errors in the argument there. The instability describes the sensitivity of a numerical process for the calculation of $f(x)$ from x to the inevitable rounding errors committed during its execution in finite precision arithmetic. It is difficult to determine the effect of these errors on the accuracy of the computed value of $f(x)$, but it is possible to estimate these effects roughly by considering the rounding errors one at a time by looking up at the individual computational steps which make up the process. Suppose there are n such steps. Let x_i denote the output from the ith step and we take $x_0 = x$. Such a x_i serves as input to one or more step later and hence influence the final answer $x_n = f(x)$. Let f_i denote the function which describes the dependence of the final answer on the intermediate result x_i. In particular, f_0 is just f. Then the total process is unstable to the extent that one or more of their functions f_i is ill conditioned. More precisely, the process is unstable to the extent that one or more of the f_i's has a much larger condition than $f = f_0$ has. For example, consider the function,

$$f(x) = \sqrt{x+1} - \sqrt{x}$$

for "large" x, say for $x \approx 10^4$. Then condition is

$$\left|\frac{f'(x)x}{f(x)}\right| = \frac{1}{2}\frac{\left|1/\sqrt{x+1} - 1/\sqrt{x}\right||x|}{\left|\sqrt{x+1} - \sqrt{x}\right|} = \frac{1}{2}\left|\frac{x}{\sqrt{x+1}\sqrt{x}}\right| \approx \frac{1}{2}$$

which is quite good. But if we compute f (12345) in six-decimal arithmetic, we find

$$f(12345) = \sqrt{12346} - \sqrt{12345} = 111.113 - 111.108 = 0.005$$

while actually, $f(12345) = 0.0045\ 0003262627751...$ and hence we are encountered with 10% error. The computational process consists of the following steps.

$$\begin{aligned} x_0 &:= 12345 \\ x_1 &:= x_0 + 1 \\ x_2 &:= \sqrt{x_1} \\ x_3 &:= \sqrt{x_0} \\ x_4 &:= x_2 - x_3 \end{aligned} \tag{1.4.1}$$

Here the function f_3 that describes show the final answer x_4 depends on x_3, given by

$f_3(t) = x_2 - t$ has condition $\frac{f_3'(t)}{f_3(t)} = \left| t/(x_2 - t)\right|$. This number is usually near to 1 implying that f_3 is well conditioned except when t is near to x_2. Here f_3 can be quite badly conditioned. For example, in particular $t \cong 111.11$ while $x_2 - t = 0.005$, so that condition is ~ 22,222 or more than 40.000 times as big as the condition of f itself. Thus the process described in (1.4.1.) is an unstable way to evaluate f.

Now consider an equivalent formula,

$$f(x) = \frac{1}{\sqrt{x+1} + \sqrt{x}}$$

In six-decimal arithmetic, this gives

$$f(12345) = \frac{1}{\sqrt{12346} + \sqrt{12345}} = \frac{1}{222.221} = .00450002$$

which is in error by only 0.0003 percent. The computational process is

$$\begin{aligned} x_0 &:= 12345 \\ x_1 &:= x_0 + 1 \\ x_2 &:= \sqrt{x_1} \\ x_3 &:= \sqrt{x_0} \\ x_4 &:= x_2 + x_3 \\ x_5 &:= 1/x_4 \end{aligned} \tag{1.4.2}$$

Here, $f_3(t) = 1/(x_2 + t)$, and the condition of this function is, approximately,

$$\left|\frac{f'(t)}{f_3(t)}\right| = \left|\frac{t}{x_2 + t}\right| \approx 2 \text{ for } t \approx x_2 .$$

Thus, the condition of f_3 is quite good; it is as good as that of f itself.

1.5 Computational methods for error estimation

The realistic estimates of the total error are difficult to make in a practical problem and adequate mathematical theories still lacking. Therefore, various methods to make use of computers to provide us with such error estimates have been proposed. We shall briefly outline five of them.

(i) **Double precision method:** In this method one simply solves the same problem twice once in single precision and once in double precision. An estimate of the total round off error can then be obtained by subtracting both the results. This method is quite costly in machine time because double – precision arithmetic increases computer time by a factor of 8 on some machines and in addition, it is not always possible to isolate other errors.

(ii) **Interval arithmetic method:** In this method each number is represented by two machine numbers, the maximum and the minimum values that it might have. Whenever an operation is performed, one computes its maximum and minimum values there by obtains two solutions at every step. The true solution is necessarily contained with in the range determined by the maximum and minimum values. This method requires more than twice the amount of computer time and about twice the storage of a standard run. Also the usual assumption that the true solution lies about midway with in the range, is not in general, valid.

(iii) **Significant digit arithmetic method:** In significant –digit arithmetic an attempt is made to keep track of digits loss due to the subtraction of two nearly equal machine numbers. In one version only the significant digits in an number are retained and all others being discarded. So at the end of a computation we will thus be assured that all digits retained are significant. The main draw back of this method is that some information is lost whenever the digits are discarded and the results obtained are likely to be much conservative. Experimentation with this method is still going on, although the experience to date is not too promising.

(iv) **Statistical approach:** This method begins with the assumption that round off error is independent. Although this assumption is not valid because if the same problem is run on the same machine several times, the answer will always be the same, however we can adopt a stochastic model of the propagation of round – off errors in which the local errors are treated as random variables. Thus we can assume that the local round – off errors are either uniformly or normally distributed between their extreme values. Using statistical methods, we can then obtain the standard deviation, the variance of distribution, and estimates of the accumulated round-off error This method does involve substantial analysis and additional computer time, but in the experiments conducted to date it has obtained error estimates which are in remarkable agreement with experimentally available evidence.

(v) **Backward error Analysis:** It reduces the analysis of rounding error effects to a study of perturbations in exact arithmetic and ultimately to a question of condition as was seen and illustrated in section 1.3.

1.6 General error formulae

1.6.1 Approximation of a function

Consider the function u of several variables $x_i(i = 1,2,3,.....n)$ as

$$u = f(x_1, x_2 .x_3 ....x_n)$$

Let Δx_i denotes error in each x_i so that the error in u is given by

$$u + \Delta u = f(x_1 + \Delta x_1,\ x_2 + \Delta x_2, ..., x_n + \Delta x_n)$$

Expanding right hand side by Taylor's series, we obtain

$$u + \Delta u = f(x_1, x_{2,} x_3, ..., x_n) + \sum_{i=1}^{n} \frac{\partial f}{\partial x_i} \Delta x_i + o(\Delta x_i^2)$$

Assuming that errors in x_i are small and that $\frac{\Delta x_i}{x_i} << 1$, so that the squares and higher powers of Δx_i can be neglected, the above relation leads to

$$\Delta u = \sum_{n=1}^{n} \frac{\partial f}{\partial x_i} \Delta x_i$$

$$= \frac{\partial u}{\partial x_i} \Delta x_1 + \frac{\partial u}{\partial x_i} \Delta x_2 + ... \frac{\partial u}{\partial x_n} \Delta x_n \qquad (1.6.1)$$

The relative error formula is given by

$$E_R = \frac{\Delta u}{u} = \frac{\partial u}{\partial x_1} \frac{\Delta x_1}{u} + \frac{\partial u}{\partial x_2} \frac{\Delta x_2}{u} + ... + \frac{\partial u}{\partial x_n} \frac{\Delta x_n}{u} \qquad (1.6.2)$$

Order of approximations

We often replace a function $f(h)$ with its approximation $\phi(h)$ and the error bound is known to be $\mu(h^n)$, *n* being a positive integer, so that

$$|f(h) - \phi(h)| \le \mu |h^n|$$

for sufficiently small h. Then we say that $\phi(h)$ approximates $f(h)$ with order of approximation $o(h^n)$ and we write

$$f(h) = \phi(h) + o(h^n).$$

Example 1.6.1 Consider the function $u = 4x^2y^3/z^4$.

$$\frac{\partial u}{\partial x} = \frac{8xy^3}{z^4}, \frac{\partial u}{\partial y} = \frac{12x^2y^2}{z^4}, \frac{\partial u}{\partial z} = -\frac{16x^2y^3}{z^5}$$

Therefore, $$\Delta u = \frac{8xy^3}{z^4}\Delta x + \frac{12x^2y^2}{z^4}\Delta y - \frac{16x^2y^3}{z^5}\Delta z$$

In general, the errors $\Delta x, \Delta y$ and Δz may be positive or negative; we take the absolute values of the terms on the right side. This gives

$$(\Delta u)_{\max} \approx \left|\frac{8xy^3}{z^4}\Delta x\right| + \left|\frac{12x^2y^2}{z^4}\Delta y\right| + \left|\frac{16x^2y^3}{z^5}\Delta z\right|.$$

Suppose $\Delta x = \Delta y = \Delta z = 0.001$ and $x = y = z = 1$. Then the maximum relative error is given by

$$(E_R)_{\max} = \frac{(\Delta u)_{\max}}{u} \approx \frac{0.036}{4} = 0.009$$

1.6.2 Series approximations

The error committed in a series approximation can be evaluated by using the remainder after n terms. Taylor's series for $f(x)$ at $x = a$ is given by

$$f(x) = f(a) + (x-a)f'(a) + \frac{(x-a)^2}{2!}f''(a) + ... + \frac{(x-a)^{n-1}}{(n-1)!}f^{(n-1)}(a) + R_n(x)$$

where $R_n(x) = \frac{(x-a)^n}{n!}f^{(n)}(\xi), \quad a < \xi < x.$

The last term $R_n(x)$ is called remainder term and $R_n(x) \to 0$ as $n \to \infty$, if the series is convergent. Thus if we approximate $f(x)$ by the first n-terms of a series, the maximum error committed in this approximation is given by the remainder term. Conversely, if the accuracy required is specified in advance, it would be possible to find n, the number of terms, such that the finite series yields the required accuracy.

Example 1.6.2 Consider the Maclaurins expansion for e^x given by

$$e^x = 1 + x + \frac{x^2}{2!} + \frac{x^3}{3!} + ... + \frac{x^{n-1}}{(n-1)!} + \frac{x^n}{n!}e^{\xi}, \quad 0 < \xi < x$$

We shall find n, the number of the terms, such that their sum yields the value of e^x correct to 8 decimal places at $x = 1$. Clearly, the error term, here is $R_n(x) = \frac{x^n}{n!} e^\xi$, so that at $\xi = x$ this gives the maximum absolute error, and hence the maximum relative error is $\frac{x^n}{n!}$.

For an 8 decimal accuracy at $x = 1$, we must have $\frac{1}{n!} < \frac{1}{2} \cdot 10^{-8} \Rightarrow n = 12$

Thus, we need to take 12 terms of the exponential series in order that its sum is correct to 8 decimal places.

1.6.3 Error propagation in computation

Definition 1.6.1: Let $e(n)$ represents the growth of error after n steps of a computation process and $\in$ be an arbitrarily small number. If $|e(n)| \approx n \in$, we say that the growth of error is linear and if $|e(n)| \approx \delta^n \in$, we say the growth of error is exponential.

If $\delta > 1$ the exponential error grows indefinitely as $n \to \infty$ and for $0 \angle \delta \angle 1$, the exponential error decreases to zero as $n \to \infty$. Consider the addition of two numbers p and q (the true values) with the approximate values $\bar{p}$ and $\bar{q}$ which contains errors $\in_p$ and $\in_q$ respectively. Then

$p = \bar{p} + \in_p$ and $q = \bar{q} + \in_q$. Now $p + q = (\bar{p} + \in_p) + (\bar{q} + \in_q) = (\bar{p} + \bar{q}) + (\in_p + \in_q)$

Hence, the error in sum of number is the sum of errors of addends. The propagation of error in multiplication is more complicated. The product is

$$pq = (\bar{p} + \in_p)(\bar{q} + \in_q) = \bar{p}\bar{q} + \bar{p} \in_q + \bar{q} \in_p + \in_p \in_q$$

If $\bar{p}$ and $\bar{q}$ are larger than 1 in absolute value, the terms $\bar{p} \in_q$ and $\bar{q} \in_p$ show that there is possibility of magnification of the original errors $\in_q$ and $\in_p$. Rearranging the terms in above equation we get

$$pq - \bar{p}\bar{q} = \bar{p} \in_q + \bar{q} \in_p + \in_p \in_q$$

If $p \neq 0, q \neq 0$, then we divide it by pq and obtain

$$\frac{pq - \bar{p}\bar{q}}{pq} = \frac{\bar{p} \in_q + \bar{q} \in_p + \in_p \in_q}{pq}$$

$$= \frac{\bar{p} \in_q}{pq} + \frac{\bar{q} \in_p}{pq} + \frac{\in_p \in_q}{pq}$$

Further more, suppose that $\bar{p}/p \approx 1 = \frac{\bar{q}}{q}$ and $\frac{\in_p}{p} \cdot \frac{\in_q}{q} = R_p R_q \approx 0$, we get

$$\frac{pq - \bar{p}\bar{q}}{pq} \approx \frac{\in_q}{q} + \frac{\in_p}{p} + 0 = R_q + R_p$$

This shows that the relative error in the product pq is approximately the sum of the relative error in the approximations $\bar{q}$ and $\bar{p}$. Oftenly an initial error will be propagated in a sequence of calculations. A quantity which is desirable for any numerical process is that a small error in the initial conditions will produce small changes in the final result. An algorithm with this feature is called stable; otherwise it is called unstable. Whenever possible we shall choose methods that are stable. The definition 1.6.1 is used to describe the propagation of error.

Example 1.6.2: Show that the following three schemes can be used with infinite-precision rational arithmetic to recursively generate the sequence

$$\left\{1, \frac{1}{3}, \frac{1}{9}, \frac{1}{27}, \frac{1}{81}, \cdots\right\}: \; r_0 = 1 \text{ and } r_n = \frac{1}{3} r_{n-1}, \text{ for } n = 1,2,3, \cdots$$

$$p_0 = 1; p = \frac{1}{3} \; and \; p_n = \frac{4}{3} p_{n-1} - \frac{1}{3} p_{n-2}, \text{for } n = 2,3,4, \cdots$$

$$q_0 = 1, q_1 = \frac{1}{3} \text{ and } q_n = \frac{10}{3} q_{n-1} - q_{n-2}, \text{ for } n = 2,3,4, \cdots$$

Solution: The first formula is obvious. In the second formula the difference equation $p_n = \frac{4}{3} p_{n-1} - \frac{1}{3} p_{n-2}$ has the solution $p_n = A\left(\frac{1}{3^n}\right) + B$, which can be verified by direct substitution.

$$\frac{4}{3} p_{n-1} - \frac{1}{3} p_{n-2} = \frac{4}{3}\left(A.\frac{1}{3^{n-1}} + B\right) - \frac{1}{3}\left(A\frac{1}{3^{n-2}} + B\right)$$

$$= \left(\frac{4}{3^n} - \frac{3}{3^n}\right)A - \left(\frac{4}{3} - \frac{1}{3}\right)B$$

$$= A.\frac{1}{3^n} + B = p_n.$$

Setting A=1 and B=0 will generate the desired sequence. In the third sequence, the difference equation $q_n = \frac{10}{3} q_{n-1} - q_{n-2}$ has the general solution $q_n = \frac{A}{3^n} + B.3^n$. This can also be verified as follows

$$\frac{10}{3} q_{n-1} - q_{n-2} = \frac{10}{3}\left(A\frac{1}{3^{n-1}} + B3^{n-1}\right) - \left(A\frac{1}{3^{n-2}} + B3^{n-2}\right)$$

$$= \left(\frac{10}{3^n} - \frac{9}{3^n}\right)A - (10-1)3^{n-2}B$$

$$= A.\frac{1}{3^n} + B3^n = q_n$$

Setting A=1 and B=0 generates the required sequence.

Uncertainty in Data

The data from real-world problems contain uncertainty or error. This type of error is called a noise. It will affect the accuracy of any numerical computation that is based on the data. An improvement of precisions is not accomplished by performing successive computations on noisy data. Hence, if we start with data having d significant digits of accuracy, the result of a computation should be reported in d significant digits of accuracy. For example, $p_1 = 4.152$ and $p_2 = 0.07931$ both have four d significant digits of accuracy. Then $p_1 + p_2 = 4.23131$ is tempted and report as it appeared on the calculator due to oversight. Therefore we should not report conclusions from the noisy data that have more significant digits than the original one. The proper answer in this situation is $p_1 + p_2 = 4.231$.

Example 1.6.3: Generate approximations to the sequence $\{x_n\} = \left\{\dfrac{1}{3^n}\right\}$ using the schemes

$$r_0 = 0.99996 \text{ and } r_n = \frac{1}{3} r_{n-1}, \text{ for } n = 1,2,3,\cdots$$

$$p_0 = 1,\ p_1 = 0.33332 \text{ and } p_n = \frac{4}{3} p_{n-1} - \frac{1}{3} p_{n-2}, \text{ for } n = 2,3,4,\cdots$$

$$q_0 = 1, q_1 = 0.33332, \text{ and } q_n = \frac{4}{3} q_{n-1} - q_{n-2}, \text{ for } n = 2,3,\cdots$$

Investigate the propagation of error for each scheme when the initial errors ϵ_r, ϵ_p and ϵ_q in r_0, p_1 and q_1, respectively, are 0.00004, 0.0000 133333 and 0.0000 133333.

Solution: The table 1.6.1and table 1.6.2 gives numerical approximations for each sequence and the error in each formula, respectively. The error for $\{r_n\}$ is stable and decreases in an exponential manner. Although the error for $\{p_n\}$ is stable, the terms $p_n \to 0$ as $n \to \infty$, so that the error eventually dominates and the terms past p_8 have no significant digits. The error for $\{q_n\}$ is unstable and grows at an exponential rate.

Table 1.6.1 The sequence $\{x_n\} = \left\{\dfrac{1}{3^n}\right\}$ and the approximations $\{r_n\}$and $\{q_n\}$.

N	x_n	r_n	p_n	q_n
0	1.0000000000	0.9999600000	1.0000000000	1.0000000000
1	0.3333333333	0.3333200000	0.3333200000	0.3333200000
2	0.1111111111	0.3333200000	0.1110933330	0.1110666667
3	0.0370370370	0.1111066667	0.0370177778	0.0369022222
4	0.0123456790	0.0370355556	0.0123259259	0.0119407407
5	0.0041152263	0.0123451852	0.0040953086	0.0029002469
6	0.0013717421	0.0041150617	0.0013517695	-0.0022732510

Table 1.6.2: The error sequences $\{x_n - r_n\}$, $\{x_n - p_n\}$ and $\{x_n - q_n\}$

N			
0	0.0000400000	0.0000000000	0.0000000000
1	0.0000133333	0.0000133333	0.0000133333
2	0.0000044444	0.0000177778	0.0000444444
3	0.0000014815	0.0000192593	0.0001348148
4	0.0000004938	0.0000197531	0.0004049383
5	0.0000001646	0.0000199177	0.0012149794
6	0.0000000549	0.0000199726	0.0036449931

Exercises

1. Round-off the following numbers correct to four significant figures:

   38.46235, 3.26425, 4985561, 0.00032217, 19.235101

2. Find the percentage error if 625.483 is approximated to three significant figures.

3. Calculate the value of $\sqrt{102} - \sqrt{101}$, correct to four significant figures.

4. Define the term absolute error. Given that a=10.00±0.05, b=0.0356±0.0002, c=15300±100, d=62000±500. Find the maximum value of the absolute error in

   (a) a + b + c + d (b) a + 5c - d and (c) c^3

5. $\sqrt{29} = 5.385$ and $\sqrt{\pi} = 3.317$, correct to 4 significant figures. Find the relative errors in their sum and difference.

6. Find the smaller root of the equation $x^2 - 400x + 1 = 0$, correct to 4 decimal places.

7. Calculate the value of $(x^2 - y^2)(x + y)$ with x = .4845 and y = .4800, using normalized floating point arithmetic. Compare with value of $(x + y)$. Determine the relative error of the former.

8. Find the value of $(1 + x)^2$ and $(x^2 + 2x) + 1$ *when* $x = \cdot 5099E - 2$. Calculate the relative errors in the two methods of calculating the expression. Which is the preferred method?

9. The function $f(x) = \tan^{-1} x$ can be expanded as

   $$\tan^{-1} x = x - \frac{x^3}{3} + \frac{x^5}{5} - \cdots + (-1)^{n-1} \frac{x^{2n-1}}{2n-1} + \cdots$$

   Find n such that the series determines $\tan^{-1}(1)$ correct the eight significant digits.

10. If $R = 10x^3y^2z^2$ and errors in x, y, z are 0.03, 0.01, 0.02 respectively at x=3., y=1., z=2. Calculate the absolute error, relative error, and percentage error in evaluating R.

11. If $R = \frac{4xy^2}{z^3}$ and errors in *x*, *y*, *z* be 0.001, show that the maximum relative error at $x=y=z$ is 0.006.

12. Given the expansions

$$\sin(t) = t - \frac{t^3}{3!} + \frac{t^5}{5!} + 0(t^7)$$

$$\cos(t) = 1 - \frac{t^2}{2!} + \frac{t^4}{4!} + 0(t^6)$$

Determine the order of approximation for their sum and product.

13. Find the value of $e^x = 1 + x + \frac{x^2}{2!} + \frac{x^3}{3!} + \cdots$ with an absolute error less than 0.005 for x=.2500E0 and x=*.5250E1*, it is given that e=2.7183.

14. Express the following decimal numbers in binary, octal and hexadecimal form

(i) 22.625 (ii) 6.02 (iii) -10.25, (iv) 15.8 (v) -8125.0.2 (vi) 6252.823

15. Express the following numbers in decimal, octal and hexadecimal forms

(i) 1011.1101, (ii) 1110111.00111, (iii) 10011101.0011, (iv) 11011011.101001

CHAPTER 2

ALGEBRAIC AND TRANSCENDENTAL EQUATIONS

2.1 Introduction

Algebraic solutions of quadratic, cubic or biquadratic equations are available by direct methods. But the need often arises to solve higher degree or transcendental equations for which no direct method exists. Such equations can best be solved by approximate methods. In this chapter we shall discuss some numerical methods for the solution of algebraic and transcendental equations which can well be manipulated and managed on digital computers and microprocessors.

Definition 2.1 An expression of the form

$$f(x) = a_0 x^n + a_1 x^{n-1} + a_2 x^{n-2} + \ldots + a_n, \tag{2.1.1}$$

where n is a finite non-negative integer and a_i, $i = 0,1,2,3 \ldots n$ are numbers belonging to same number system, is called a polynomial in x of degree n provided $a_0 \neq 0$. The polynomial $f(x)$ in (2.1.1) is called a real polynomial if $a_i \in R$ for all i, whereas it is called a polynomial with complex coefficient if $a_i \in C$ for $i = 0,1,2,3 \ldots n$.

Definition 2.2 A polynomial equation $f(x) = 0$ is called an algebraic equation of degree n. If $f(x)$ contains some other functions such as trigonometric, logarithmic, exponential etc., then $f(x) = 0$ is called a transcendental equation.

Definition 2.3 Value of the variable (indeterminate) x, which satisfies $f(x) = 0$ is called its root, which geometrically signifies that value of x, where the graph of $y = f(x)$ crosses the x-axis.

Definition 2.4 *A* process of finding the roots of an equation is known as the solution of that equation. This is a problem of great importance in applied mathematics and is of practical significance for engineers and scientists.

2.2 Basic properties of polynomial equations

1. **Factor Theorem:** A number α is a root of the equation $f(x) = 0$ iff $x - \alpha$ is a factor of $f(x)$ i.e. iff $x - \alpha$ exactly divides $f(x)$.
2. **Fundamental Theorem of Algebra:** Every polynomial equation $f(x) = 0$ of degree $n \geq 1$ has at least one root real or imaginary.
3. Every nth degree equation has exactly n roots (real or imaginary). Hence, if a polynomial equation of degree n is satisfied by more than n distinct values of the variable x, then it is an identity i.e. all the coefficients of $f(x)$ must vanish.

4. In an equation with real coefficients, non-real complex roots occur in conjugate pairs i.e. if $\alpha + i\beta$ is a root of a real equation. $f(x) = 0$, then $\alpha - i\beta$ must also be its root.

   (a) Thus every polynomial equation of odd degree with real coefficients has at least one real root, because the number of complex roots for such an equation is always even. The complex roots of an equation $f(x) = 0$, may not occur in conjugate pairs if the equation has non-real (complex) coefficients. For example the equation $x^2 + (3 - 2i)x - 6i = 0$ has $2i$ as its root but $-2i$ is not its root.

   (b) If in an equation $f(x) = 0$, with real coefficients, complex root $\alpha + i\beta$ is repeated m times then its conjugate $\alpha - i\beta$ will also satisfies the equation m times.

5. In an equation with rational coefficients surd roots occurs in conjugate pair's i.e. if $\alpha + \sqrt{\beta}$ is a root of a rational equation $f(x) = 0$, then $\alpha - \sqrt{\beta}$ is also its root.

   (a) Equation of odd degree with rational coefficients may not have any rational root. For example equation $x^3 - 5 = 0$ has the roots

   $$5^{1/3}, 5^{1/3}\omega, 5^{1/3}\omega^2,$$

   none of which is rational.

   (b) If in the equation $f(x) = 0$, with rational coefficients, root $\alpha + \sqrt{\beta}$ (β is not a perfect square) is repeated m-times then its conjugate $\alpha - \sqrt{\beta}$ will also be root of the equation repeated m-times.

   (c) If in an equation $f(x) = 0$, with rational coefficients, one of the root is $\sqrt{\alpha} + \sqrt{\beta}$, α, β are not perfect squares, then $\sqrt{\alpha} - \sqrt{\beta}$, $-\sqrt{\alpha} + \sqrt{\beta}$, $-\sqrt{\alpha} - \sqrt{\beta}$ will also be roots of the equation.

   (d) If in an equation $f(x) = 0$, with rational coefficients one of the root is $\sqrt{\alpha} + i\beta$, α is not a perfect square, then $\sqrt{\alpha} - i\beta$, $-\sqrt{\alpha} + i\beta$ and $-\sqrt{\alpha} - i\beta$ will also be the roots of the equation.

6. Every integral root of a polynomial equation of degree ≥ 1 with integral coefficients and non-zero constant term is a divisor of the constant term.

7. Common roots of two equations $f(x) = 0$ and $g(x) = 0$, are obtained by finding H.C.F. of $f(x)$ and $g(x)$ and equating it to zero and then solving it.

8. A root α is said to be a multiple root of the equation $f(x) = 0$, if α is root of $f(x) = 0$ more than once. If α satisfies $f(x) = 0$, r times repeatedly, then α is said to be root of the equation of multiplicity r and in this case $(x - \alpha)^r$ divides $f(x)$.

   (a) If α is a root of $f(x) = 0$, repeated r times $(r > 1)$, then α is root of $f'(x) = 0$ repeated $r - 1$ times, where $f'(x) = \dfrac{df(x)}{dx}$. α is root of $f''(x) = 0$ repeated $r - 2$ times, that of $f'''(x) = 0$ repeated $r - 3$ times and so on.

9. **Intermediate value Theorem:** If $f(x)$ is continuous in the interval $[a,b]$ and $f(a)$, $f(b)$ are of opposite signs, then the equation $f(x)=0$ has at least one root between $x=a$ and $a=b$.

10. **Rolle's Theorem:** Between two consecutive roots of $f(x)-0$, there lics at lcast onc, and always an odd number of roots of $f'(x)=0$.

    (a) Between two consecutive roots of $f'(x)=0$, lies at most one real root of $f(x)=0$.

11. **Descartes' rule of signs:** In a polynomial equation $f(x)=0$ with real coefficients, the number of positive roots cannot exceed the number of changes of signs of the coefficients in $f(x)$ and number of negative roots cannot exceed the number of changes of sign of the coefficients in $f(-x)$. Thus if an equation of nth degree has at the most p positive roots and at the most q negative roots, then the equation has at least $n-(p+q)$ imaginary roots. For example, consider the equation

$$f(x)=2x^7-x^5+4x^3-5=0 \tag{2.2.1}$$

Then signs of $f(x)$ are $+,\frown,-,\frown,+,\frown,-$ and those of $f(-x)$ are $-,\smile,+,\smile,-,-$. Clearly $f(x)$ has *3* changes of signs and $f(-x)$ has two changes of signs and hence (2.2.1) cannot have more than three positive and two negative roots. Thus the equation (2.2.1) has at least $7-3-2=2$ imaginary (complex) roots.

Example 2.2.1: Solve the equation $x^3+6x+20=0$, one root being $1+3i$.

Solution: The given equation is

$$x^3+6x+20=0, \tag{2.2.2}$$

Since one root is $1+3i$, the other root must be $1-3i$

$$\therefore \qquad x(-1-3i)(x-1+3i)=(x-1)^2+9=x^2-2x+10$$

is divisor of $x^3+6x+20=0$

$$\Rightarrow \qquad x^3+6x+20=(x^2-2x+10)(x+2)$$

Thus the depressed equation is $x+2=0$ and its root is -2.

Hence the roots of the given equation $(2.2.2)$ are $1\pm 3i,\ -2$.

Example 2.2.2: Solve $x^4-2x^3-22x^2+62x-15=0$, given that $2+\sqrt{3}$ is a root.

Solution: The given equation is

$$x^4-2x^3-22x^2+62x-15=0, \tag{2.2.3}$$

Since one root is $2+\sqrt{3}$, the other roots must be $2-\sqrt{3}$

$$\therefore \qquad (x-2-\sqrt{3})(x-2+\sqrt{3})=(x-2)^2-3=(x^2-4x+1)$$

is a divisor of $(2.2.3)$.

$$\Rightarrow \qquad x^4-2x^3-22x^2+62x-15=(x^2-4x+1)(x^2+2x-15)$$

Therefore depressed equation is $x^2+2x-15=0$, whose roots are $x=3,-5$

Thus the roots of (2.2.3) are $3, -5, 2\pm\sqrt{3}$

Example 2.2.3: Find multiple roots of the equation

$$x^4+12x^3+32x^2-24x+4=0 \tag{2.2.4}$$

Solution: Let $f(x)=x^4+12x^3+32x^2-24x+4$

$$\therefore \qquad f'(x)=4x^3+36x^2+64x+24$$

Now the H.C.F. of $f(x)$ and $f'(x)$ is x^2+6x-2

$\therefore$ Common roots of $f(x)$=0 and $f'(x)$=0 are given by

$$x^2+6x-2=0 \quad \Rightarrow x=-3\pm\sqrt{11}$$

Since $-3+\sqrt{11}$ and $-3-\sqrt{11}$ are root of $f'(x)=0$ occurring only once, therefore both of these are roots of $f(x)$=0 repeated twice. As given equation is 4th degree equation, so it has no more roots.

Hence roots of the given equation are $-3+\sqrt{11},-3+\sqrt{11},-3-\sqrt{11},$ and $-3-\sqrt{11}$.

2.3 Relation between roots and coefficients

Let $\alpha_1, \alpha_2, \alpha_3 \ldots\alpha_n$ be the roots of the equation

$$f(x)=a_0x^n+a_1x^{n-1}+a_2x^{n-2}+\ldots+a_{n-1}x+a_n=0,\ a_0\neq 0 \tag{2.3.1}$$

Then

$$\sum\alpha_1=\frac{-a_1}{a_0},\quad \sum\alpha_1\alpha_2=\frac{a_2}{a_0},\quad \sum\alpha_1\alpha_2\alpha_3=-\frac{a_3}{a_0}\ldots$$

$$\alpha_1\alpha_2\alpha_3\ldots\alpha_n=(-1)^n\frac{a_n}{a_0}$$

Proof: Since $\alpha_1,\alpha_2,\alpha_3\ldots\alpha_n$ are the roots of the equation (2.3.1), we have

$$\begin{aligned}&a_0x^n+a_1x^{n-1}+a_2x^{n-2}+\ldots+a_{n-1}x+a_n\\&=a_0(x-\alpha_1)(x-\alpha_2)(x-\alpha_3)\ldots(x-\alpha_n)\\&=a_0\left[x^n-(\alpha_1+\alpha_2+\alpha_3+\ldots+\alpha_n)x^{n-1}+(\alpha_1\alpha_2+\alpha_2\alpha_3+\ldots)x^{n-2}\right.\\&\qquad\left.-(\alpha_1\alpha_2\alpha_3+\alpha_2\alpha_3\alpha_4+\ldots)x^{n-3}+\ldots+(-1)^n\alpha_1\alpha_2\alpha_3\ldots\alpha_n\right]\end{aligned}$$

Dividing both sides by a_0, we get

$$x^n+\left(\frac{a_1}{a_0}\right)x^{n-1}+\left(\frac{a_2}{a_0}\right)x^{n-2}+\ldots+\frac{a_{n-1}}{a_0}x+\frac{a_n}{a_0}$$

$$=x^n-\left(\sum\alpha_1\right)x^{n-1}+\left(\sum\alpha_1\alpha_2\right)x^{n-2}+\ldots+(-1)^n\alpha_1\alpha_2\alpha_3\ldots\alpha_n$$

Equating the coefficients of like powers of x, we obtain the desired result.

Example 2.3.1 Solve $x^4 - 10x^3 + 35x^2 - 50x + 24 = 0$, given that its roots are in A. P.

Solution: Let the roots of given equation

$$x^4 - 10x^3 + 35x^2 - 50x + 24 = 0 \tag{2.3.2}$$

be $\alpha, \beta, \gamma, \delta$. As the roots are in A.P. therefore we have

$$\alpha + \delta = \beta + \gamma$$

But $\alpha + \beta + \gamma + \delta = 10 \Rightarrow \alpha + \delta = \beta + \gamma = 5$

Now $\alpha + \delta = 5$	$\beta + \gamma = 5$
Let $\alpha\beta = p$	Let $\beta\gamma = q$
Then α, δ are the roots of	$\therefore \beta, \gamma$ are the roots of
$x^2 - 5x + p = 0$	$x^2 - 5x + q = 0$

Thus we have

$$x^4 - 10x^3 + 35x^2 - 50x + 24 = (x^2 - 5x + p)(x^2 - 5x + q)$$

$\Rightarrow$ $$p + q + 25 = 35 \text{ and } pq = 24$$

$\Rightarrow$ $$p + q = 10 \text{ and } pq = 24$$

$\therefore p, q$ are the roots of a quadratic equation.

$$t^2 - 10t + 24 = 0 \Rightarrow t = \frac{10 \pm \sqrt{100 - 96}}{2} = \frac{10 \pm 2}{2} = 6,4$$

$\Rightarrow$ $$p = 6,\ q = 4$$

Thus the roots of equation $(2.3.1)$ are same as of the equations

$$x^2 - 5x + 6 = 0 \Rightarrow x = 2,3$$

and $$x^2 - 5x + 4 = 0 \Rightarrow x = 1,4$$

Hence the roots of given equation are $1, 2, 3, 4$.

Example 2.3.2: Solve $x^4 - 6x^3 + 11x^2 - 18x + 9 = 0$, given that the ratio of its two roots is equal to the ratio of the other two.

Solution: The given equation is

$$x^4 - 6x^3 + 11x^2 - 18x + 9 = 0, \tag{2.3.3}$$

Let $\alpha, \beta, \gamma, \delta$ be the roots of $(2.3.3)$, According to the given condition, we have

$$\frac{\alpha}{\beta} = \frac{\gamma}{\delta} \Rightarrow \alpha\delta = \gamma\beta$$

But $\alpha\beta\gamma\delta = 9 \Rightarrow \alpha\delta = \beta\gamma = 3$

Now $\alpha\delta = 3$	$\beta\gamma = 3$
Let $\alpha + \delta = p$	Let $\beta + \gamma = q$
$\therefore \alpha, \delta$ are the roots of	$\therefore \beta, \gamma$ are the roots of
$x^2 - px + 3 = 0$	$x^2 - qx + 3 = 0$

Thus, we have

$$x^4 - 6x^3 + 11x^2 - 18x + 9 = (x^2 - px + 3)(x^2 - qx + 3)$$

$\Rightarrow$ $$p + q + 6 \text{ and } pq = 11 - 6 = 5$$

$\therefore$ p, q are the roots of the quadratic

$$t^2 - 6t + 5 = 0$$
$$\Rightarrow t = 1,\ 5$$

Hence $p = 1,\ q = 5$ and consequently roots of (2.3.3) are given by the equations

$$x^2 - x + 3 = 0 \Rightarrow x = \frac{1 \pm \sqrt{11}i}{2}$$

and

$$x^2 - 5x + 3 = 0 \Rightarrow x = \frac{5 \pm \sqrt{13}}{2}$$

Thus the roots of given equation (2.3.3) are $\dfrac{5 \pm \sqrt{13}}{2}$ and $\dfrac{1 \pm \sqrt{11}i}{2}$.

2.4 Transformation of equations

1. Roots with signs changed : To find an equation whose roots are the negative of the roots of the given equation, we put $y = -x$ or $x = -y$.

Thus

$$f(x) = a_0x^n + a_1x^{n-1} + a_2x^{n-2} + \ldots + a_{n-1}x + a_n = 0\,,\ a \neq 0 \qquad (2.4.1)$$

leads to

$$f(y) = a_0(-y)^n + a_1(-y)^{n-1} + a_2(-y)^{n-2} + \ldots + a_{n-1}(-y) + a_n = 0$$

or $$f(y) = a_0(y)^n - a_1(y)^{n-1} + a_2(y)^{n-2} - a_3(y)^{n-3} + \ldots + (-1)^n a_n = 0 \quad (2.4.2)$$

The roots of $(2.4.1)$ and $(2.4.2)$ will be negative of each other.

2. Roots multiplied by a given number: An equation, whose roots are *m*-times the roots of the given equation, can be obtained by multiplying the successive coefficients of the given equation by $1, m^1,\ m^2,\ m^3 \ldots m^n$ respectively, after writing the given equation in complete, if it is not so, by writing zeros in place of missing powers of x.

3. Reciprocal roots: To find an equation whose roots are reciprocal of the roots of the given equation, we change x to $\dfrac{1}{x}$. For example, consider the equation

$$6x^3 - 11x^2 - 3x + 2 = 0 \qquad (2.3.4)$$

Replacing x with $\dfrac{1}{x}$, we obtain.

$$\frac{6}{x^3}-\frac{11}{x^2}-\frac{3}{x}+2=0$$

$$\Rightarrow \qquad 2x^3-3x^2-11x+6=0 \qquad (2.4.4)$$

If the roots of (2.4.3) are in *H.P.* then the roots of (2.4.4) will be in *A. P.*

4. Reciprocal Equations: If an equation remains unaltered on changing x to $\frac{1}{x}$, it is called a reciprocal equation, such equations are of the following types.

(a) A reciprocal equation of an odd degree having coefficients of terms equidistant from beginning and end equal has a root equal to –1.;

(b) A reciprocal equation of an odd degree having coefficients of terms equidistant from beginning and end equal but opposite in sign has a root equal to 1.

(c) A reciprocal equation of an even degree having coefficients of terms equidistant from the beginning and end equal but opposite in sign, has two roots equal 1 and -1. The substitution $x+\frac{1}{x}=y$ reduces the degree of the equation to half its former degree.

Example 2.4.1 Solve $40x^4-22x^3-21x^2+2x+1=0$, if the roots are in *H.P.*

Solution: The given equation is

$$40x^4-22x^3-21x^2+2x+1=0 \qquad (2.4.5)$$

Changing x to $\frac{1}{x}$ and simplifying, we get

$$x^4+2x^3-21x^2-22x+40=0 \qquad (2.4.6)$$

Since the roots of equation (2.4.5) are in *H.P.*, so the roots of equation (2.4.6) will be in *A. P.*

Let the roots of equation (2.4.6) be $a-3d, a-d, a+d, a+3d$

Then $$4a=\frac{-2}{1}=-2 \Rightarrow a=\frac{-1}{2}$$

Also $$a-3d+a+3d=2a=2(\frac{-1}{2})=-1$$

and $$a-d+a+d=2a=-1$$

Let p be the product of first and fourth and q that of second and third roots, then we have

$$x^4+2x^3-21x^2-22x+40=(x^2+x+p)(x^2+x+q)$$

$$\Rightarrow \qquad p+q=-22 \text{ and } pq=40$$

$\therefore$ p, q are the roots of the quadratic

$$t^2+22t+40=0 \Rightarrow t=-2,-20$$

$\Rightarrow$ $p=-2$ and $q=-20$ and consequently the roots of equation (2.4.6) are given by the equation

$$x^2+x-2=0 \Rightarrow x=-2,1$$

and $$x^2+x-20=0 \Rightarrow x=-5,4$$

Therefore roots of equation (2.4.6) are $1,4,-2,-5$ and hence those of given equation (2.4.5) are $1,\frac{1}{4},\frac{1}{2},\frac{1}{5}$.

Example 2.4.2: Solve $6x^6-25x^5+37x^4-25x+6=0$

Solution: The given equation is reciprocal equation of even degree with opposite signs and hence $x=\pm 1$ is its roots

Dividing the equation by $(x+1)(x-1)=x^2-1$, the reduced equation is

$$6x^4-25x^3+37x^2-25x+6=0$$

or
$$6(x^2+\frac{1}{x^2})-25(x+\frac{1}{x})+37=0$$

(on dividing by x^2 and rearranging the terms)

Put $y=x+\frac{1}{x},$ we get $6y^2-25y+25=0 \Rightarrow y=\frac{5}{2},\frac{5}{3}$

$$\Rightarrow \qquad 2x^2-5x+2=0 \text{ or } 3x^2-5x+3=0$$

$$\Rightarrow \qquad x=2,\frac{1}{2} \text{ Or } x=\frac{5\pm\sqrt{11}i}{6}$$

Hence the required roots are $1.-1,2,\frac{1}{2},\frac{5\pm\sqrt{11}i}{6}$.

2.5 Synthetic division of a polynomial

(a) By linear expression

The division of the polynomial

$$f(x)=a_0x^n+a_1x^{n-1}+a_2x^{n-2}+...+a_{n-1}x+a_n$$

by a polynomial $x-\alpha$ is obtained by synthetic division as follows

	a_0	a_1	a_2	...	a_{n-1}	a_n
α		αb_0	αb_1	...	αb_{n-2}	αb_{n-1}
	a_0 $(=b_0)$	$a_1+\alpha b_0$ $(=b_1)$	$a_2+\alpha b_1$ $(=b_2)$	...	$a_{n-1}+\alpha b_{n-2}$ $(=b_{n-1})$	$a_n+\alpha b_{n-1}$ $(=R)$

Hence the quotient $=b_0x^{n-1}+b_1x^{n-2}+...+b_{n-1}$ and Remainder $=R$

(b) By a quadratic expression

The division of the polynomial $f(x)$ by the quadratic factor $x^2-\alpha x-\beta$ is carried out by the following synthetic scheme.

	a_0	a_1	a_2	a_3	a_{n-1}	a_n
α		αb_0	$\alpha_1 b_1$	αb_2 ...	αb_{n-2}	
β			βb_0	βb_1 ...	βb_{n-3}	βb_{n-2}
	a_0 $(= b_0)$	$a_1 + \alpha b_0$ $(= b)_1$	$a_2 + \alpha_1 b_1 + \beta b_0$ $(= b_2)$	$a_3 + \alpha b_2 + \beta b_1$ $(= b_3)$ ...	$(= b_{n-1})$	$a_n + \beta b_{n-2}$ $(= b_n)$

Hence the quotient $= b_0 x^{n-2} + b_1 x^{n-3} + ... + b_{n-2}$ and the remainder $R = b_{n-1}x + b_n$

(c) Diminishing the roots of an equation by a given number

Let $\alpha_1, \alpha_2, \alpha_3 ... \alpha_n$ be the roots of the equation

$$f(x) = a_0 x^n + a_1 x^{n-1} + a_2 x^{n-2} + ... + a_{n-1}x + a_n = 0 \tag{2.5.1}$$

It is required to find an equation whose roots are

$$\alpha_1 - h, \alpha_2 - h, \alpha_3 - h, ... \alpha_n - h \quad \text{(Say)}$$

Put

$$y = x - h \Rightarrow x = y + h,$$

so that (2.5.1) becomes

$$a_0 (y+h)^n + a_1 (y+h)^{n-1} + a_2 (y+h)^{n-2} + ... + a_n = 0$$

This simplifies to

$$A_0 y^n + A_1 y^{n-1} + A_2 y^{n-2} + ... + A_{n-1} y + A_n = 0 \tag{2.5.2}$$

Example 2.5.1: Use synthetic division to find the quotient and remainder, when $27x^4 - 21x^3 + x^2 - 2x + 2$ is divided by $3x - 1$.

Solution: Writing the divisor $3x - 1$ as $3\left(x - \frac{1}{3}\right)$, we have by synthetic division

	27	− 21	1	− 2	2
$\frac{1}{3}$		9	− 4	− 1	−1
	27	− 12	− 3	− 3	1

On dividing given polynomial by $x - \frac{1}{3}$, quotient is $27x^3 - 12x^2 - 3x - 3$ and remainder is 1.

On dividing given polynomial by $3x - 1$, quotient is

$$\frac{1}{3}\left(27x^3 - 12x^2 - 3x - 3\right) = 9x^2 - 4x^2 - x - 1 \text{ and remainder is 1.}$$

Example 2.5.2: Diminish the roots of

$$x^4 - 5x^3 + 7x^2 - 17x + 11 = 0 \text{ by } 4.$$

Solution: Writing the coefficients of terms of the equation, in order, in a row and supplying the missing terms, if any with zero coefficients, we have

4	1	–5	7	–17	11	
		4	–4	12	–20	
	1	–1	3	–5	–9 = A_4	
		4	12	60		
	1	3	15	55 = A_3		
		4	28			
	1	7	43 = A_2			
		4				
		1 = A_0				

Then the equation whose roots are diminished by 4 is

$$A_0x^4 + A_1x^3 + 43x^2 + A_3x + A_4 = 0$$

i.e.

$$x^4 + 11x^3 + 43x^2 + 55x - 9 = 0$$

Example 2.5.3: Transform $x^4 + 8x^3 + x - 5 = 0$ into one in which the second term is missing.

Solution: The given equation is

$$f(x) = x^4 + 8x^3 + x - 5 = 0 \qquad (2.5.3)$$

Second term is eliminated if we diminish the roots of (2.5.3) by

$$h = -\frac{a_1}{na_0} = -\frac{8}{4.1} = -2$$

Dividing $f(x)$ successively by $x+2$, we has

	1	8	0	1	–5
–2		–2	–12	24	–50
	1	6	–12	25	–55
		–2	–8	40	
	1	4	–20	65	
		–2	–4		
	1	2	–24		
		–2			
	1	0			
	1				

Hence transformed equation is $\quad x^4 + 24x^2 + 65x - 55 = 0$

2.6 Solution of cubic equations–Cardan's method

Consider the equation

$$f(x) = a_0x^3 + 3a_1x^2 + 3a_2x + a_3 = 0 \qquad (2.6.1)$$

We transform equation (2.6.1) into another in which second term is missing and leading coefficient is unity. This is done by diminishing the roots of (2.6.1) by $h = \frac{-3a_1}{3a_0} = \frac{-a_1}{a_0}$ and then dividing the transformed equation by a_0. We obtain

$$y^3 + \frac{3H}{a_0^2} y + \frac{G}{a_0^3} = 0 \tag{2.6.2}$$

where $H = a_0 a_2 - a_1^2$ and $G = a_0^2 a_3 - 3a_0 a_1 a_2 + 2a_1^3$.

If we multiply the roots by a_0, then (2.6.2) leads to

$$z^3 + \frac{3H}{a_0^2} a_0^2 z + \frac{G}{a_0^3} a_0^3 = 0 \quad \text{or} \qquad z^3 + 3Hz + G = 0 \tag{2.6.3}$$

The roots of equation (2.6.1) and (2.6.3) connected by the relation $z = a_0 x + a_1$. Now equation (2.6.3) can be solved by Cardan's method. Let

$$z = u + v \tag{2.6.4}$$

be a root of (2.6.3), so that

$$z = u^3 + v^3 + 3uv(u+v) = u^3 + v^3 + 3uvz$$

$$\text{or} \qquad z^3 - 3uvz - (u^3 + v^3) = 0 \tag{2.6.5}$$

Comparing the coefficients of various powers of z in equation (2.6.3) and (2.6.5), we get

$$uv = -H \Rightarrow u^3 v^3 = -H^3 \quad \text{and} \quad u^3 + v^3 = -G$$

Thus u^3 and v^3 are the roots of quadratic equation

$$t^2 - (u^3 + v^3)t + u^3 v^3 = 0, \text{ or } \qquad t^2 + Gt - H^3 = 0$$

$$\Rightarrow \qquad t = \left[\frac{-G \pm \sqrt{G^2 + 4H^3}}{2}\right]$$

$$\text{Hence} \qquad u^3 = \frac{-G + \sqrt{G^2 + 4H^3}}{2}, v^3 = \frac{-G - \sqrt{G^2 + 4H^3}}{2} \tag{2.6.6}$$

Case 1: If G, H are real and $G^2 + 4H^3$ is positive, from (2.6.6) we conclude that u^3 and v^3 are real. Equations (2.6.6) give three values for each of u and v as $u, u\omega, u\omega^2$ and $v, v\omega, v\omega^2$ respectively, where ω and ω^2 are imaginary cube roots of unity. Corresponding to a value of u obtained above, all the values of v are not admissible to give $u+v$ as a root of (2.6.3). But we are to select that value of v which satisfies the relation $uv = -H$.

Substituting the values of u in the relation $uv = -H$, the corresponding values of v are

$$\frac{-H}{u}, \frac{-H}{\omega u}, \frac{-H}{\omega^2 u}$$

or the corresponding values of v are

$$v, \frac{v}{w}, \frac{v}{w^2} \quad i.e. \quad v, vw^2, vw \qquad (\because v = \frac{-H}{u})$$

The required roots of equation (2.6.3) are

$$u+v, u\omega+v\omega^2, u\omega^2+v\omega$$

Hence the roots of equation (2.6.3) are

$$z_1 = u+v, z_2 = u\omega + v\omega^2 = -\frac{1}{2}(u+v)+\frac{\sqrt{3}}{2}(u-v)i$$

$$z_3 = u\omega^2 + v\omega = -\frac{1}{2}(u+v)-\frac{\sqrt{3}}{2}(u-v)i.$$

In this case one root is real and other two roots are complex.

Case 2: If $G^2+4H^3=0$ and *G, H* are the real numbers.

(a) $G \neq 0, H \neq 0$, We have $\quad u^3 = v^3 = \frac{-G}{2}.$

$\therefore\ u = v$ and hence the roots of (2.6.3) are $2u, -u, -u$.

and thus all three roots are real but two are equal.

(b) If $G = H = 0$ The equation (2.6.3) becomes $z^3 = 0$.

Hence all the three roots are zero in this case.

Case 3: If $G^2+4H^3<0$ or *G* and *H* are complex numbers. In this case, it can be concluded from (2.6.6) that u^3 and v^3 are imaginary quantities and there is no general arithmetical or algebraic method of finding the exact value of the cube roots of imaginary quantities. This case is known as irreducible case of Cardan's method. However, we may find the cube roots with the help of De-Moivre's theorem as under.

Let $$u^3 = \frac{-G+\sqrt{G^2+4H^3}}{2} = a+ib,\ \ a, b \in R$$

Put $a = r\cos\theta,\ b = r\sin\theta$, so that $r^2 = a^2+b^2$ and $\tan\theta = \frac{b}{a}$. Thus we have

$$u^3 = r(\cos\theta + i\sin\theta)$$

or $$u^3 = r[\cos(2\pi n+\theta)+i\sin(2\pi n+\theta)],\ \ n \in I$$

$$\Rightarrow \quad u = r^{\frac{1}{3}}\left[\cos\left(\frac{2\pi n+\theta}{3}\right)+i\sin\left(\frac{2\pi n+\theta}{3}\right)\right], n = 0, 1, 2$$

Hence, the values of u are

$$r^{\frac{1}{3}}\left(\cos\frac{\theta}{3}+i\sin\frac{\theta}{3}\right),\quad r^{\frac{1}{3}}\left[\cos\left(\frac{2\pi+\theta}{3}\right)+i\sin\left(\frac{2\pi+\theta}{3}\right)\right],$$

$$u = r^{\frac{1}{3}}\left[\cos\left(\frac{4\pi+\theta}{3}\right)+i\sin\left(\frac{4\pi+\theta}{3}\right)\right]$$

Having determined *u*, the values of *v* can be obtained from the relation $uv = -H$ which leads us to the required roots. The above process of determining the value of u and v when $G^2+4H^3<0$ is cumbersome. In case *G* and *H* are real quantities the irreducible case of Cardan's method may be evaluated by the use of trigonometry as follows.

We know that

$$4\cos^3\theta - 3\cos\theta = \cos 3\theta, \forall\theta \tag{2.6.7}$$

Put $$z = \cos\theta,\ z^3 - \frac{3}{4}z - \frac{1}{4}\cos 3\theta = 0 \tag{2.6.8}$$

From equation (2.6.3), we obtain

$$z^3 + \frac{3H}{t^2}z + \frac{G}{t^3} = 0 \tag{2.6.9}$$

where we have replaced z with tz in (2.6.3). Comparing (2.6.8) and (2.6.9), we get

$$\frac{3H}{t^2} = -\frac{3}{4} \text{ and } \frac{G}{t^3} = -\frac{1}{4}\cos 3\theta \Rightarrow \qquad t = 2\sqrt{-H}$$

and $$\cos 3\theta = -\frac{4G}{t^3} = -(4G)\frac{1}{(2\sqrt{-H})^3} = \frac{G}{\sqrt{-4H^3}} \tag{2.6.10}$$

Now $$G^2 + 4H^3 < 0 \Rightarrow H < 0 \Rightarrow -4H^3 > 0$$

Again $$G^2 + 4H^3 < 0 \Rightarrow G^2 < -4H^3$$

$\therefore$ $$\frac{G^2}{-4H^3} < 1 \Rightarrow \left|\frac{G}{\sqrt{-4H^3}}\right| < 1$$

This suggests that the value of θ can be determined from (2.6.10) and the roots of (2.6.8) viz. $\cos\theta, \cos\left(\theta + \frac{2\pi}{3}\right)$ and $\cos\left(\theta + \frac{4\pi}{3}\right)$ can be obtained. But roots of $z^3 + 3Hz + G = 0$ are given by $2\sqrt{-H}\cos\theta$, $2\sqrt{-H}\cos\left(\theta + \frac{2\pi}{3}\right)$, $2\sqrt{-H}\cos\left(\theta + \frac{4\pi}{3}\right)$ $\quad (\because z = tz \text{ and } t = 2\sqrt{-H})$

Example 2.6.1 Solve $x^3 - 3x^2 - 10x + 24 = 0$, by Cardan's method.

Solution: The given equation is $$x^3 - 3x^2 - 10x + 24 = 0 \tag{2.6.11}$$

Diminishing the roots of (2.6.11) by $h = -\frac{3}{3} = 1$, so that second term is eliminated

1	1	-3	-10	24
		1	-2	-12
	1	-2	-12	12
		1	-1	
	1	-1	-13	
			1	
	1	0		
	1			

Therefore, the reduced equation is

$$y^3 - 13y + 12 = 0 \tag{2.6.12}$$

Let $y = u + v$ so that

$$y^3 - 3uvy - (u^3 + v^3) = 0 \tag{2.6.13}$$

Comparing coefficients of like powers, we get

$$-3uv = -13 \text{ and } -(u^3 + v^3) = 12$$

or $$u^3v^3 = \frac{169}{27} \text{ and } u^3 + v^3 = -12$$

Thus u^3 and v^3 are roots of the quadratic equation

$$t^2 + 12t + \frac{169}{27} = 0$$

or $$27t^2 + 324t + 169 = 0$$

$\Rightarrow$ $$t = \frac{-324 \pm \sqrt{(324)^2 - 4.27.169}}{54} = -\frac{324}{54} \pm \frac{105\sqrt{3}}{27} i = \frac{1}{27}[-162 \pm 105\sqrt{3}\ i]$$

Let $$u^3 = \frac{-162 + 105\sqrt{3}\ i}{27}$$

and $$v^3 = \frac{-162 - 105\sqrt{3}\ i}{27}$$

$\therefore$ $$u = \frac{1}{3}[-162 + 105\sqrt{3}\ i]^{\frac{1}{3}} \tag{2.6.14}$$

Let $$[-162 + 105\sqrt{3}\ i]^{\frac{1}{3}} = p + \sqrt{3}qi$$

$\Rightarrow$ $$-162 + 105\sqrt{3}\ i = p^3 - 9pq^2 + 3q(p^2 - q^2)$$

$\Rightarrow$ $$p(p^2 - 9q^2) = -162$$

and $$q\ (p^2 - q^2) = 35$$

By inspection we find that $p = -6$ and $q = 1$ and thus

$$u = \frac{1}{3}[-162 + 105\sqrt{3}\ i]^{\frac{1}{3}} = \frac{1}{3}(-6 + \sqrt{3}\ i), \text{ from (2.6.14)}$$

$$= -2 + \frac{1}{\sqrt{3}} i$$

$$\therefore \quad v = -2 - \frac{1}{\sqrt{3}} i$$

One root of equation (2.6.12) is $u + v = -4$

$\Rightarrow$ $x + 4$ is a factor of L.H.S. of (2.6.12). Dividing (2.6.12) by $x + 4$ using synthetic division, we get

–4	1	0	–13	12
		–4	16	–12
	1	4	3	0

Therefore, reduced equation is

$$y^2 - 4y + 3 = 0 \quad \Rightarrow \quad y = 1, 3$$

$\therefore$ Roots of equation $(2.6.12)$ are $1, 3, -4$

Hence the roots of given equation $(2.6.11)$ are $1+1,\ 1+3,\ -4+1$ i.e. $2, 4, -3$.

Example 2.6.2: Solve the cubic $x^3 - 3x + 1 = 0$.

Solution: The given equation is

$$x^3 - 3x + 1 = 0 \tag{2.6.15}$$

Comparing with $x^3 + 3Hx + G = 0$, we get $H = -1, G = -1$

so that $G^2 + 4H^3 = 1 - 4 = -3 < 0$ and hence we solve it by Trigonometric method.

Consider the identity

$$4\cos^3\theta - 3\cos\theta = \cos 3\theta$$

Put $z = \cos\theta$, we get

$$z^3 - \frac{3}{4}z - \frac{1}{4}\cos 3\theta = 0 \tag{2.6.16}$$

We transform $(2.6.15)$ by setting $x = tz$ and obtain

$$t^3z^3 - 3tz + 1 = 0 \Rightarrow z^3 - \frac{3}{t^2}z + \frac{1}{t^3} = 0 \tag{2.6.17}$$

Comparing $(2.6.16)$ and $(2.6.17)$, we get $\frac{-3}{t^2} = \frac{-3}{4}$ and $\frac{1}{t^3} = \frac{-1}{4}\cos 3\theta$

$\Rightarrow t^2 = 4$ or $t = \pm 2$.

Taking $t = 2$, we get $\frac{1}{8} = \frac{-1}{3}\cos 3\theta \Rightarrow \cos 3\theta = -\frac{1}{2} = \cos\left(\frac{2\pi}{9}\right)$

$\therefore$ Roots of (2.6.16) are $\cos\frac{2\pi}{9}$, $\cos\left(\frac{2\pi}{9} + \frac{2\pi}{3}\right)$ and $\cos\left(\frac{2\pi}{9} + \frac{4\pi}{3}\right)$

or $$\cos\frac{2\pi}{9}, \cos\frac{8\pi}{9}, \cos\frac{14\pi}{9}$$

Thus the roots of (2.6.15) are given by

$$2\cos\frac{2\pi}{9},\ 2\cos\frac{8\pi}{9},\ 2\cos\frac{14\pi}{9}, \text{ since } x = tz$$

If we take $t = -2$, then again we arrive at these roots.

2.6.1 Horner's method

This is the best method of finding approximate values of real roots of a numerical polynomial equation. It consists in diminishing the roots of an equation by successive digits occurring in the roots. If the root of an equation lies between a and $a + 1$ the value of the root will be a. bcd… where b, c, d are decimal digits. The method to find these digits proceeds as follows:

Step 1 Diminish the roots of the equation by a so that the root of the new equations is 0.bcd…

Step 2 Multiply the roots of the transformed equation by 10 so that the root of the new equations is b.cd.

Step 3 Diminish the roots now by b and multiply the roots of the resulting equation by 10 so that the root is c.d…

Step 4 Next diminish the root by c and so on.

By continuing this process, the root may be evaluated to any desired degree of accuracy digit by digit.

Example 2.6.2: Find by Horner's method a root of the equation $x^3 - 3x + 1 = 0$, correct to the decimal places lies between 1 and 2.

Solution: The given equation is $f(x) = x^3 - 3x + 1 = 0$. Here $f(1) = 11 - 3 + 1 = -1 < 0$ and $f(2) = 8 - 6 + 1 > 0$ and hence there lies a root between 1 and 2. Diminishing the roots of the given equation by 1 we get,

1	1	0	-3	1
		1	1	-2
	1	1	-2	-1
		1	2	
	1	2	0	
		1		
	1	3		
	1			

The transformed equation is

$$x^3 + 3x^2 - 1 = 0$$

whose root lies between 0 and 1. Multiplying the roots of this equation by 10, we get

$$x^3 + 30x^2 - 1000 = 0.$$

Its root lies between 0 and 10. Clearly

$$f(5) = 125 + 750 - 1000 = 875 - 1000 = -125 < 0$$

and $$f(6) = 216 + 1080 - 1000 = 1296 - 1000 = 296 > 0$$

Diminishing the root by 5, we get

5	1	30	0	−1000
		5	175	− 875
	1	35	175	−125
		1	180	
		36	355	
		1		
	1	37		
	1			

The transformed equation is

$$x^2 + 3x^2 - 1 = 0,$$

which has roots between 0 and 1.
Multiplying its roots by 10, we get

$$x^3 + 30x^2 + 1000 = 0.$$

Its roots lie between 0 and 10. Clearly $f(3) < 0$ and $f(4) > 0$. Diminishing the roots by 3, we get

3	1	30	0	– 1000
		5	175	– 875
	1	35	175	– 125
		1	180	
		36	355	
		1		
	1	37		
	1			

The transformed equation in $x^3 + 375x^2 + 37731x - 15143 = 0$. After multiplying the roots by 10, we get $x^3 + 3750x^2 + 3773100x - 15143000 = 0$. Its roots lie between *0* and *10*, which is nearly equal to $\frac{15143000}{3773100} = 4$. Hence the required root of given equation, is 1.534.

2.7 Iterative numerical methods

2.7.1 Bisection method

This method is based on the repeated application of the Intermediate Value theorem. Let the function $f(x)$ be continuous in [*a*, *b*] and $f(a)$ and $f(b)$ are of opposite signs. To be definite, we assume that $f(a) < 0$ and $f(b) > 0$. Then by Intermediate value theorem $f(x) = 0$ must have a root between *a* and *b*. The first approximation to the root is $x_1 = \frac{1}{2}(a+b)$. If $f(x_1) = 0$, then x_1 is a root of $f(x) = 0$, otherwise the root lies between a and x_1 or x_1 and *b* according as $f(x_1)$ is positive or negative. Suppose $f(x_1) > 0$ so that root lies in *a* and x_1. We again bisect the interval $[a, x_1]$ to find the second approximation to the root as $x_2 = \frac{1}{2}(a + x_1)$. If $f(x_2) < 0$, the root lies between x_1 and x_2. Then the third approximation to the root is obtained as $x_3 = \frac{1}{2}(x_1 + x_2)$ and we continue in a similar way unless approximation to the root of $f(x) = 0$ is obtained to the desired level of accuracy viz the condition $|x_i - x_{i-1}| < \in, \in$ being arbitrarily small number which depends upon our choice, is satisfied. Suppose this happens for $i = N$, then the desired approximate root of $f(x) = 0$ is $x = x_N$.

Convergence of Bisection method (Bisection theorem)

Assume that the function $f(x)$ is continuous on $[a, b]$ and there exists a number $r \in [a, b]$. If $f(a)$ and $f(b)$ are of opposite signs, and $\{x_n\}_{n=0}^{\infty}$ is the sequence of midpoints generated by the bisection process, then $|r - x_n| \le \frac{b-a}{2^{n+1}}$, for $n = 0,1,2,...$ and therefore the sequence $\{x_n\}$ converges to the root $x = r$ i.e. $\lim_{n \to \infty} x_n = r$.

Proof: Since both r and $x_n \in [a_n, b_n]$, the distance between them can not be greater than half the

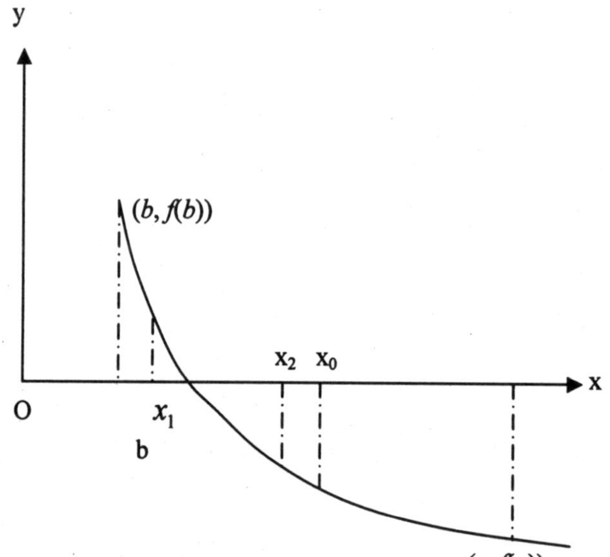

Fig. 2.7.1 The bisection method.

width of this interval. Thus

$$|r - x_n| \le \frac{|b_n - a_n|}{2}, \text{ for all } n.$$

Also the successive interval widths form the pattern.

$$|b_1 - a_1| = \frac{|b_0 - a_0|}{2^1}$$

$$|b_2 - a_2| = \frac{|b_1 - a_1|}{2} = \frac{|b_0 - a_0|}{2^2}$$

$$|b_3 - a_3| = \frac{|b_2 - a_2|}{2} = \frac{|b_0 - a_0|}{2^3}$$

and so on. In general, we have

$$|b_n - a_n| = \frac{|b_0 - a_0|}{2^n}$$

Thus

$$|r - x_n| = \frac{|b_0 - a_0|}{2^{n+1}}, \text{ for all } n.$$

Since $|b_0 - a_0|$ is a fixed number and $\frac{1}{2} < 1$, so $\frac{1}{2^{n+1}} \to 0$, as $n \to \infty$, we have

$$\lim_{n\to\infty} \left| r - x_n \right| \to 0 \Rightarrow \left| r - x_n \right| \to 0 \Rightarrow x_n \to r, \text{ As } n \to \infty.$$

$\therefore \lim_{n\to\infty} x_n = r$, implying that the sequence $\{x_n\}$ converges to the root *r*.

Example 2.7.1 Use bisection method to find a root of the equation $x^3 - 4x - 9 - 0$.

Solution: The given equation is $x^3 - 4x - 9 = 0$

Here $f(2) = 8 - 8 - 9 < 0$ and $f(3) = 27 - 12 - 9 > 0$, hence a root lies between 2 and 3. The first approximation to the root is

$$x_1 = \frac{2+3}{2} = 2.5$$

Now $$f(x_1) = f(2.5) = (2.5)^3 - 4(2.5) - 9 = -3.375 < 0$$

$\therefore$ The root lies between x_1 and 3. Thus the second approximation to the root is

$$x_1 = \frac{1}{2}(x_1 + 3) = \frac{2.5+3}{2} = 2.75$$

Then $$f(x_2) = (2.75)^3 - 4(2.75) - 9 = 0.7969 > 0$$

$\therefore$ Root lies between x_1 and x_2 and hence

$$x_3 = \frac{x_1 + x_2}{2} = \frac{2.5 + 2.75}{2} = 2.625$$

Now $$f(x_3) = -1.4121 < 0$$

The root lies between x_2 and x_3. Thus the fourth approximation to the root is

$$x_4 = \frac{1}{2}(x_2 + x_3) = \frac{2.75 + 2.625}{2} = 2.6875$$

Now $f(x_4) = -0.3391 < 0$ and hence root lies between x_2 and x_4

$$\therefore \quad x_5 = \frac{1}{2}(x_2 + x_4) = \frac{2.75 + 2.6875}{2} = 2.71875$$

Again $f(x_5) = 0.22092 > 0$ and hence root lies between x_4 and x_5

$$\therefore \quad x_6 = 2.70313$$

$f(x_6) = -0.061097 < 0$, root lies between x_5 and x_6

$$x_7 = 2.71094$$

$f(x_7) = -0.92337 < 0$, root lies between x_5 and x_7

$$\therefore \quad x_8 = 2.714850$$

$f(x_8) = 0.150159 > 0$, root lies between x_7 and x_8

$$\therefore \quad x_9 = 2.712895$$

$f(x_9) = 0.114783 > 0$, root lies between x_7 and x_9

$$\therefore \quad x_{10} = 2.7119175$$

Hence the root is 2.7119 approximately.

Algorithm 2.7.1: Bisection method

Given a function $f(x)$ continuous on the interval $[a_0, b_0]$ and such that $f(a_0)f(b_0) \le 0$.

1. for $n = 1,2,3,4\ldots$ until satisfied, do: steps 2 to 5,
2. set $m = (a_n + b_n)/2$

3. if $f(a_n)f(b_m) \le 0$, set $a_{n+1} = a_n, b_{n+1} = m$
4. otherwise, set $a_{n+1} = m, b_{n+1} = b_n$
5. then $f(x)$ has a zero in the interval $[a_{n+1}, b_{n+1}]$

Here the phrase "until satisfied " is used in order to stress that this description of the algorithm is incomplete and this will depend on the part of the specific problem to be solved by the algorithm.

2.7.2 Iteration method

Consider the equation

$$f(x) = 0 \tag{2.7.1}$$

To find the roots of this equation by successive approximations, we rewrite (2.7.1) in the form

$$x = \phi(x) \tag{2.7.2}$$

Let $x = x_0$ be an initial approximation of the desired root α. Then the first approximation x_1 is given by

$$x = \phi(x_0)$$

The second approximation to $x_2 = \phi(x_1)$

The third approximation is $x_3 = \phi(x_2)$ and so on. In general, the *n*th approximation is given by

$$x_n = \phi(x_{n-1}), \text{n} = 1,2,3,\cdots$$

The following theorem gives us the convergence criteria for the iterative sequence of solution.

Theorem 2.7.1: Let α be a root of $f(x) = 0$ which is equivalent to $x = \phi(x), \phi(x)$ is continuously differentiable function in an interval I containing the root $x = \alpha$, if $|\phi'(x)| < 1$, then the sequence of approximations $x_0, x_1, x_2, \ldots x_n$ will converge to the root α provided the initial approximation $x_0 \in I$.

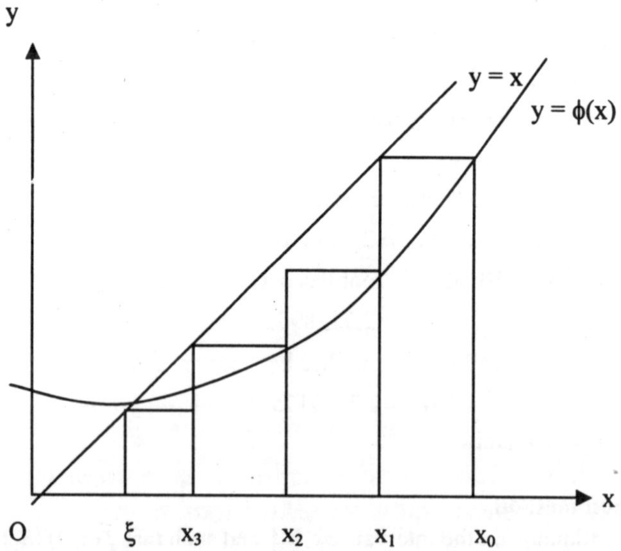

Fig. 2.7.2 Convergence of $x_{x+1} = \phi(x_n)$, for $|\phi'(x)| < 1$.

Proof: Let α be actual root of $x = \phi(x)$, we have

$$\alpha = \phi(\alpha) \qquad (2.7.3)$$

Let $x = x_0$ be the initial approximation to the root, then

$$x_1 = \phi(x_0) \qquad (2.7.4)$$

Subtracting (2.7.3) and (2.7.4), we have

$$\alpha - x_1 = \phi(\alpha) - \phi(x_0)$$

By using Lagrange's mean value theorem, we have

$$\alpha - x_1 = (\alpha - x_0)\phi'(\xi_0), \text{ for } x_0 < \xi_0 < \alpha$$

Similarly

$$\alpha - x_2 = (\alpha - x_1)\phi'(\xi_1), \text{ for } x_1 < \xi_1 < \alpha$$

$$\alpha - x_3 = (\alpha - x_2)\phi'(\xi_2), \text{ for } x_2 < \xi_2 < \alpha.$$

and so on. In general, we have

$$\alpha - x_n = (\alpha - x_{n-1})\phi'(\xi_{n-1}), \text{ for } x_{n-1} < \xi_{n-1} < \alpha$$

Multiplying above equations, we get

$$\alpha - x_n = (\alpha - x_0)\phi'(\xi_0)\phi'(\xi_1)\phi'(\xi_2)...\phi'(\zeta_{n-1})$$

If $|\phi'(x_i)| \le \kappa < 1$, for all *I*, then

$$|x_n - \alpha| \le |x_0 - \alpha||\phi'(\xi_0)|\phi'(\xi_1)...|\phi'(\xi_{n-1})| \le \kappa.\kappa...\kappa|x_0 - \alpha| \le \kappa^n|x_0 - \alpha|$$

As $\kappa < 1$ therefore $k^n \to 0$, as $n \to \infty$, and thus we have $x_n \to \alpha$, provided $x_0 \in I$.
Hence the proof of the theorem.

Error estimate in iteration method:

If we define $\epsilon_n = x_n - \xi$, the error estimate at the nth iteration then $\lim_{n\to\infty}(\epsilon_{n+1}/\epsilon_n) = \phi'(\xi)$ is satisfied. We know that $|x_{n+1} - \xi| = |\phi(x_n) - \phi(\xi)| = |\phi'(\xi_n)||x_n - \xi|$, $\xi_n \in (x_n, \xi)$ by mean value theorem. This implies that

$$\epsilon_{n+1} = \epsilon_n |\phi'(\xi_n)| \Rightarrow \lim_{n\to\infty}(\epsilon_{n+1}/\epsilon_n) = \phi'(\xi).$$

Thus, we obtain the order of convergence as linear. But if $\phi'(\xi) = 0$ *and* $\phi''(\xi) \ne 0$, then the Taylor series expansion of ϕ in a neighborhood of ξ namely

$$\phi(x_n) = \phi(\xi) + (x_n - \xi)\phi'(\xi) + \frac{(x_n - \xi)^2}{\angle 2}\phi''(\xi) + \cdots$$

Implies that

$$\epsilon_{n+1} = \epsilon_n \phi'(\xi) - \frac{\epsilon_n^{\,2}}{2}\phi''(\xi) + \frac{\epsilon_n^{\,3}}{6}\phi'''(\xi) - \cdots$$

on using $x_{n+1} = \phi(x_n)$ and $\epsilon_n = |x_{n+1} - \xi|$.

This leads to $\epsilon_{n+1} = -\frac{\epsilon_n^{\,2}}{2}\phi''(\xi)$, on neglecting the terms containing cubes and higher power of ϵ_n, which means it is a quadratic convergence.

Example 2.7.2: Find a real root of $x^3 - x - 1 = 0$ correct to two decimal places be iterative method.

Solution: The given equation is

$$f(x) = x^3 - x - 1 = 0 \qquad (2.7.5)$$

Here $f(1) = -1 < 0$, $f(2) = 5 > 0$ and hence root of (2.7.5) lies between *1* and *2*. As $f(1) < f(2)$ numerically, we take the initial approximation $x_0 = 1$.

Equation (2.7.5) can be rewritten as

$$x^3 = x + 1 \text{ or } x = (x+1)^{\frac{1}{3}} = \phi(x)$$

$$\therefore \qquad \phi'(x) = \frac{1}{3}(x+1)^{\frac{-2}{3}} \Rightarrow \phi'(x_1) = \phi'(1) = \frac{1}{3}(2)^{\frac{-2}{3}} = 0.21 < 1$$

The successive approximations of the root are given by

$$x_1 = \phi(x_0) = (1+1)^{1/3} = 1.26$$

$$x_2 = \phi(x_1) = (1+1.26)^{1/3} = 1.312$$

$$x_3 = \phi(x_2) = (1+1.312)^{1/3} = 1.322$$

$$x_4 = \phi(x_3) = (1+1.322)^{1/3} = 1.324$$

$$x_5 = \phi(x_4) = (1+1.324)^{1/3} = 1.324$$

Hence the real roots of $f(x) = 0$ is 1.32 correct to two decimal places.

Note: If we write $x = x^3 - 1 = \phi(x)$, then $\phi'(x) = 3x^2, \phi'(x_1) = \phi'(1) = 3 > 1$

$\therefore$ The iteration does not work in this case.

Example 2.7.3: Find by iteration method, a real root of $2x - \log_{10} x = 7$.

Solution: The given equation in

$$f(x) = 2x - \log_{10} x - 7. = 0 \qquad (2.7.6)$$

$$f(3) = 6 - \log_{10} x - 7 = 6 - 0.4771 - 7 = -1.4471 < 0,$$

$$f(4) = 8 \log_{10} 4 - 7 = 8 - 0.602 - 7 = 0.39870$$

$\therefore$ a root of (2.7.6) lie between 3 and 4. Rewriting equation (2.7.6) as

$$x = \frac{1}{2}\left(\log_{10} x + 7\right) = \phi(x)$$

We have $\phi'(x) = \frac{1}{2}\left(\frac{1}{x}\log_{10} e\right) \Rightarrow |\phi'(x)| < 1$, when $3 < x < 4$. $\left(\therefore \log_{10} e = 0.4343\right)$

Since $|f(4)| < |f(3)|$, so we take $x_0 = 3.8$ Thus

$$x_1 = \phi(x_0) = \frac{1}{2}\left(\log_{10} 3.8 + 7\right) = 3.789887, \quad x_2 = \frac{1}{2}\left(\log_{10} 3.781877 + 7\right) = 3.78931$$

$$x_3 = \frac{1}{2}\left(\log_{10} 3.78931 + 7\right) = 3.78928, \qquad x_4 = \frac{1}{2}\left(\log_{10} 3.78928 + 7\right) = 3.78927$$

Thus $x = 3.7892$ is the root of $(2.7.6)$ correct to four decimal places.

2.7.3 Method of false positions (Regula –Falsi method)

Consider the equation $f(x) = 0$.

Let $f(x_0)$ and $f(x_1)$ be of opposite signs *i.e.* the graph of $y = f(x)$ crosses the x-axis between the values x_0 and x_1, and hence a root of $f(x) = 0$ lies between x_0 and x_1 and consequently $f(x_0)f(x_1) < 0$.

The equation of the chord joining the points $(x_{0,}\ f(x_0))$ and $(x_{1,}\ f(x_1))$ is

$$y - f(x_0) = \frac{f(x_1) - f(x_0)}{x_1 - x_0}(x - x_0)$$

Now for x to be an approximation to the root, we must have $y = f(x) = 0$. Hence, we obtain

$$x_2 = x_0 - \frac{x_1 - x_0}{f(x_1) - f(x_0)} f(x_0) \qquad (2.7.7)$$

as a new approximation to the root. Now if $f(x_0)$ and $f(x_2)$ are again of opposite signs then the root lies between x_0 and x_2. (The root may lies between x_1 and x_2 according as $f(x_1)$ and $f(x_2)$ are of opposite signs). So replacing x_1 by x_2 in $(2.7.7)$, we obtain the next approximation $x_3 = x_0 - \dfrac{x_2 - x_0}{f(x_2) - f(x_0)} f(x_0)$ and so on. In general, we have

$$x_{n+1} = x_0 - \frac{x_n - x_0}{f(x_n) - f(x_0)} f(x_0), n = 2,3,4,\ldots \qquad (2.7.8)$$

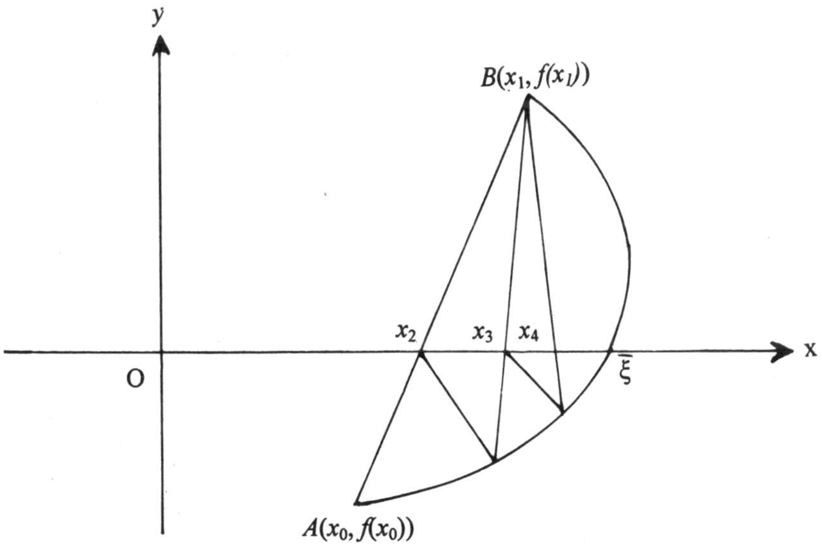

Figure 2.7.3: Method of false positions

This procedure is repeated till the root is found to the desired level of accuracy. The process can not be terminated until and unless the condition $|x_k - x_{k-1}| < \in$, $\in$ being arbitrarily small and depends upon our choice to achieve the desired level of accuracy, is satisfied. Suppose this happens for $k = N$. Then the desired root is $x = x_N$. The iteration process based on above procedure is called method of false positions.

Algorithm 2.7.2: Regula-Falsi method

Given a function $f(x)$ continuous on $[a_0, b_0]$ and such that $f(a_0)f(b_0) < 0$

1. Set $F = f(a_0), G = f(b_0), w_0 = a_0$
2. For $n = 0,1,2,3\ldots$, until satisfied, do: steps 3 to 7
3. Compute $w_{n+1} = (Ga_n - Fb_n)/(G - F)$
4. If $f(w_n)f(w_{n+1}) \le 0$, set $a_{n+1} = a_n, b_{n+1} = w_{n+1}, G = f(w_{n+1})$
5. If also $f(w_n)f(w_{n+1}) > 0$, set $F = F/2$ otherwise, set
   $a_{n+1} = w_{n+1}, F = f(w_{n+1}), b_{n+1} = b_n$
6. If also $f(w_n)f(w_{n+1}) > 0$, set $G = G/2$
7. Then $f(x)$ has a zero in the interval $[a_{n+1}, b_{n+1}]$.

Here the weighted average $w = [f(b_n)a_n - f(a_n)b_n]/[f(b_n) - f(a_n)]$ is the point at which the straight line through the points $(a_n, f(a_n))$ and $(b_n, f(b_n))$ intersects the x-axis. The above algorithm is a modification of simple regular-false algorithm given below:

(i) For $n = 0,1,2,3\ldots$, until satisfied, do: steps (ii) to (iv),

(ii) Calculate $w = [f(b_n)a_n - f(a_n)b_n]/[f(b_n) - f(a_n)]$

(iii) If $f(a_n)f(w_n) \le 0$, set $a_{n+1} = a_n, b_{n+1} = w$

(iv) Otherwise, set $a_{n+1} = w, b_{n+1} = b_n$

Example 2.7.4 Use Regula-Falsi method to find a root of $x e^x = \cos x$, correct to four decimal places.

Solution: The given equation is

$$f(x) = \cos x - x e^x = 0$$

Here $f(0) = 1 > 0, f(1) = \cos 1 - e = -2.17798 < 0$, so that the root lies between 0 and 1.

$\therefore$ We take $x_0 = 0,\ x_1 = 1$, and $f(x_0) = 1$ and $f(x_1) = -2.17798$. The Regula-Falsi method implies that

$$x_2 = x_0 - \frac{x_1 - x_0}{f(x_1) - f(x_0)} f(x_0) = 0 + \frac{1}{3.17798} \times 1 = 0.31467$$

Now $f(0.31467) = 0.51987 > 0$ and hence root lies between 0.31467 and 1.

Again applying the Regula Falsi method by replacing x_0 and x_2, we get

$$x_3 = 0.31467 + \frac{0.68533}{2.69785} \times 0.51987 = 0.44673$$

Repeating this process, the successive approximations are

$$x_4 = 0.44673 + \frac{0.55327}{2.38154} \times 0.20356 = 0.49402$$

$$x_5 = 0.50995,\ x_6 = 0.51520,\ x_7 = 0.51692$$

$$x_8 = 0.51748,\ x_9 = 0.51767,\ x_{10} = 0.51775 \text{ etc.}$$

Hence the root is 0.5177 correct to four decimal places.

Example 2.7.5 Use method of false positions to find a real root of $x^3 - 4x - 9 = 0$. correct to three decimal places

Solution: The given equation is

$$f(x) = x^3 - 4x - 9 = 0.$$

Here $f(2) = 8 - 8 - 9 = -9 < 0$ and $f(3) = 27 - 12 - 9 = 6 > 0$, so the root of the given equation lies between 2 and 3. We take $x_0 = 2,\ x_1 = 3$ so that

$$f(x_0) = -9 \text{ and } f(x_1) = 6.$$

The method of false positions implies that

$$x_2 = x_0 - \frac{x_1 - x_0}{f(x_1) - f(x_0)} f(x_0) = 2 - \frac{3-2}{6+9}(9) = 2 + \frac{9}{15} = 2 + \frac{3}{5} = 2.6$$

Now $f(x_2) = f(2.6) = -1.824$, so root lies between 2.6 and 3.

Therefore, $x_3 = 2.6 - \frac{3 - 2.6}{6 + 1.824}(-1.824) = 2.6 + \frac{0.4}{7.824}(1.824) = 2.69325$

Repeating this process, the successive approximations are

$$x_4 = 2.69325 - \frac{3 - 2.69325}{6 + 0.23725}(-0.23725) = 2.70492$$

$$x_5 = 2.70492 - \frac{3 - 2.70492}{6 + 0.028883}(-0.028883) = 2.7063$$

$$x_6 = 2.7063 - \frac{3 - 2.7063}{6 + 0.004097}(-0.004097) = 2.7065$$

Hence the root is 2.706, correct to three decimal places.

2.7.4 Secant method

A very popular modification of the Regula-Falsi method, called the secant method, retains the use of secants throughout, but may give up the tracking of the root. This method is an improvement over the method of false positions as it does not require the condition

$f(x_0)f(x_1)<0$ of that method. Here the graph of the function $y=f(x)$ is approximated by a secant line but it is not necessary that the interval must contain the root in each of the iteration.

Let x_0 and x_1 be the initial limits of the interval, then the equation of the secant through these points to the curve $y=f(x)$ is

$$y-f(x_1)=\frac{f(x_1)-f(x_0)}{x_1-x_0}(x-x_1)$$

Putting $y=0$, we obtain the new approximation to the root as

$$x_2=x_1-\frac{x_1-x_0}{f(x_1)-f(x_0)}f(x_1)$$

Repeating the process time and again, the general formula for successive approximations is, therefore given by

$$x_{n+1}=x_n-\frac{x_n-x_{n-1}}{f(x_n)-f(x_{n-1})}f(x_n),\ n\geq 1$$

If at any iteration $f(x_n)=f(x_{n-1})$, this method fails and shows that it does not necessarily converge. This is a drawback of the secant method over the method of false position, which always converges. But if the secant method once converges, its rate of convergence is faster than that of the method of false positions.

Definition 2.7.1 Order of convergence: Let $x_0, x_1, x_2 ..., x_n ...$ be a sequence, which converge to a number ξ, and set $e_n=\xi-x_n$. If there exists a number p and a constant $C\neq 0$ so that $\lim\limits_{n\to\infty}\frac{|e_{n+1}|}{|e_n|^p}=C$, then p is called the order of convergence and C is known as the asymptotic error constant of the sequence.

Convergence of secant method

Consider the equation $f(x)=0$. Let x_0 and x_1 be the approximations to the actual root $x=\alpha$ of the equation so that $f(x_1)f(x_2)<0$. Then the general formula for the successive iterations is given by

$$x_{n+1}=x_n-\frac{x_n-x_{n-1}}{f(x_n)-f(x_{n-1})})f(x_n), n=1,2,3,...$$

Let $x_n=\alpha+\epsilon_n$ and $x_{n+1}=\alpha+\epsilon_{n+1}$, where $f(\alpha)=0$ and ϵ_n is the error at the nth iteration. Then, we have

$$\alpha+\epsilon_{n+1}=\frac{(\alpha+\epsilon_{n-1})f(\alpha+\epsilon_n)-(\alpha+\epsilon_n)f(\alpha+\epsilon_{n-1})}{f(\alpha+\epsilon_n)-f(\alpha+\epsilon_{n-1})}$$

which leads to

$$\epsilon_{n+1}=\frac{\epsilon_{n-1}\ f(\alpha+\epsilon_n)-\epsilon_n\ f(\alpha-\epsilon_{n-1})}{f(\alpha+\epsilon_n)-f(\alpha+\epsilon_{n-1})}$$

Applying Taylor's theorem and using the fact that $f(\alpha)\cong 0$, we get

$$\epsilon_{n+1}\cong\frac{\epsilon_{n-1}\epsilon_n}{2!}\frac{f''(\alpha)}{f'(\alpha)}$$

on assuming error to be very small. From the above we note that

$$|\epsilon_{n+1}| = C_n |\epsilon_n \epsilon_{n-1}|, \text{with } \lim_{n\to\infty} C_n = \frac{1}{2}|f''(a)f'(x)| = C$$

For the order of convergence, we find a number p such that $\lim_{n\to\infty} \frac{|\epsilon_{n+1}|}{|\epsilon_n|^p} = C$, for some non zero constant C. Now, we have

$$\frac{|\epsilon_{n+1}|}{|\epsilon_n|^p} = C_n |\epsilon_n|^{1-p} |\epsilon_{n-1}| = C_n \left(\frac{|\epsilon_n|}{|\epsilon_{n-1}|^p} \right)^{\beta} \tag{2.7.9}$$

provided $\beta = 1 - p$ and also $\beta p = -1$, i.e.; provided that $p^2 - p - 1 = 0$

The equation $p^2 - p - 1 = 0$ has the simple positive root $p = (1+\sqrt{5})/2 = 1.618...$

With this choice of p and $\beta = \frac{-1}{p}$, we see that (2.7.9) defines a "fixed point like iteration"

$$y_{n+1} = C_n y^{\frac{-1}{p}}, \text{ here } y_{n+1} = \frac{|\epsilon_{n+1}|}{|\epsilon_n|^p} \text{ and } \lim_{n\to\infty} C_n = C.$$

It follows that y_n converges to the fixed value of the equation $x = Cx^{\frac{-1}{p}}$, whose solution is $C^{\frac{1}{p}}$ since $1 + \frac{1}{p} = p$.

This shows that for the second method $\frac{|\epsilon_{n+1}|}{|\epsilon_n|p} \cong C^{\frac{1}{p}} = \left|\frac{f''(\alpha)}{2f'(\alpha)}\right|^{\frac{1}{p}}$, for large n with p=1.618...

and asymptotic error constant is $\left|\frac{f''(\alpha)}{2f'(\alpha)}\right|^{\frac{1}{p}}$. This means that the secant method converges more rapidly than the usual fixed-point iteration but less rapidly than the Newton Raphson method.

Algorithm 2.7.3: Secant method

Given a continuous function $f(x)$ and two points x_{-1} and x_0.

For $n = 0,1,2,...,$ until satisfied, do:

Compute $x_{n+1} = [f(x_n)x_{n-1} - f(x_{n-1})x_n]/[f(x_n) - f(x_{n-1})]$.

The secant method locates quite rapidly a point at which $|f(x)|$ is 'small' but gives, in general, no feeling for how far from a zero of $f(x)$ this point might be. Also $f(x_n)$ and $f(x_{n-1})$ need not be of opposite sign, so that the expression

$$x_{n+1} = [f(x_n)x_{n-1} - f(x_{n-1})x_n]/[f(x_n) - f(x_{n-1})]$$

is prone to round off error effects. In extreme situation, we might even have $f(x_n) = f(x_{n-1})$, making the evaluation of x_{n+1} impossible. Although this does not cure the trouble and it is better to calculate x_{n+1} from

$$x_{n+1} = x_n - f(x_n)\frac{x_n - x_{n-1}}{f(x_n) - f(x_{n-1})}$$

in which x_{n+1} is obtained from x_n by adding the second term on right hand side as the "correction term".

Example 2.7.6: Use secant method to find a root of $xe^x = \cos x$, correct to four decimal places.

Solution: The given equation is

$$f(x) = xe^x - \cos x = 0$$

Here $f(0) = 1 - 0 = 1 > 0$ and $f(1) = \cos 1 - e = -2.17798 < 0$

So a root lies between 0 and 1. We take $x_0 = 0$, $x_1 = 1$. Then by secant method, we have

$$x_2 = x_1 - \frac{x_1 - x_0}{f(x_1) - f(x_0)} f(x_1)$$

$$= 1 - \frac{1}{3.17798}(-2.17798) = 0.31467$$

Now $f(x_2) = 0.51987 > 0$, so root lies between x_2 and x_1

$$\therefore \qquad x_3 = x_2 - \frac{x_2 - x_1}{f(x_2) - f(x_1)} f(x_2) = 0.44673$$

$f(x_3) = 0.20354 > 0$ Implies that root lies between x_3 and x_1.

$$\therefore \qquad x_4 = x_3 - \frac{x_3 - x_1}{f(x_3) - f(x_1)} f(x_3) = 0.53171$$

Repeating the process, successive approximations are

$$x_5 = 0.51690, \quad x_6 = 0.51775, \quad x_7 = 0.51776$$

Hence the root is 0.5177, correct to four decimal places.

2.7.5 Newton-Raphson method

Consider the equation $f(x) = 0$. Let x_0 be an approximation to the root of $f(x) = 0$. If $x_1 = x_0 + h$ be the exact root then $f(x_1) = 0$.

Therefore expanding $f(x_0 + h) = 0$ by Taylor's theorem, we get

$$f(x_0) + hf'(x_0) + \frac{h^2}{2!} f''(x_0) + \ldots = 0$$

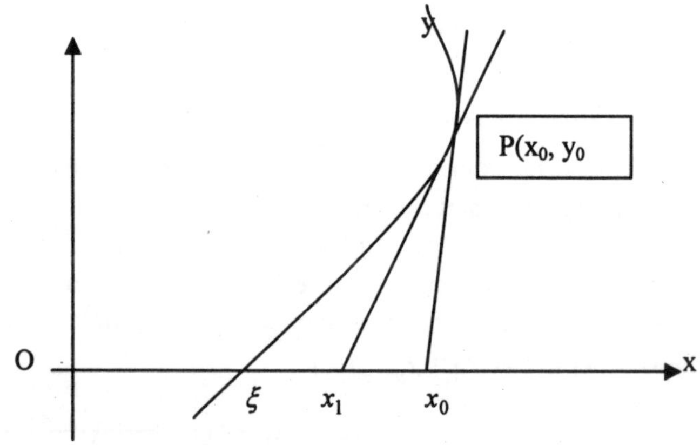

Figure 2.7.4: Newton-Raphson method

Since h is small, neglecting h^2 and higher powers of h, we get

$$f(x_0)+hf'(x_0)=0 \Rightarrow h=-\frac{f(x_0)}{f'(x_0)}$$

∴ A closer approximation to the root is given by

$$x_1 = x_0-\frac{f(x_0)}{f'(x_0)}$$

Similarly, starting with x_1, *a* still better approximation is obtained as

$$x_2 = x_1-\frac{f(x_1)}{f'(x_1)}$$

and so on. In general, we have

$$x_{n+1} = x_n-\frac{f(x_n)}{f'(x_n)}, \quad n=0,1,2,3... \qquad (2.7.10)$$

which is called Newton's iteration formula. In fact, Newton's original ideas on the subject, around 1669, were considerably more complicated. A systematic study and a simplified version of the method are due to Raphson in 1690, so this iteration scheme is also known as the Newton-Raphson method. It has also been described as the tangent method, from its geometric interpretation.

Following observations can be made:

1. Newton's method is useful when the graph of $f(x)$ while crossing the x-axis is nearly vertical i.e when $f'(x)$ is large.
2. Newton's formula converges provided the initial approximation x_0 is chosen sufficiently close to root. Thus a proper choice of initial guess is very important for the success of Newton's method.
3. Newton's method is generally used to improve the result obtained by other methods. It is applicable to the solution of algebraic and transcendental equations and can also be used when the roots are complex.

Conditions and rate of convergence

The general iteration formula (2.7.10) of Newton-Raphson method is

$$x_{n+1}=x_n-\frac{f(x_n)}{f'(x_n)}$$

Comparing (2.7.10) with the relation $x_{n+1}=\phi(x_n)$ of the iteration method, we get

$$\phi(x_n) = x_n-\frac{f(x_n)}{f'(x_n)}$$

Thus, in general

$$\phi(x) = x-\frac{f(x)}{f'(x)}$$

which implies that

$$\phi'(x) = \frac{f(x)f''(x)}{[f'(x_n)]^2}$$

Since the iteration method converges if $|\phi'(x)|<1$, so the Newton-Raphson method converges if

$$|f(x)f''(x)|<|f'(x)|^2$$

Assuming $f(x), f'(x)$ and $f''(x)$ to be continuous and bounded on an interval containing the root $x = \alpha$ of $f(x) = 0.$, if α is a simple root then $f'(\alpha) \neq 0$. Further $f'(x)$ being continuous, implies that $|f'(x)| \leq \in$, for some $\in > 0$ in a suitable neighborhood of α within which we can select an interval I such that $|f(x)f''(x)| < \in^2$ and this is possible because $f(\alpha) = 0$ and $f(x)$ is continuously twice differentiable. Hence in this interval I, we have $|\phi'(x)| < 1$, and thus Newton-Raphson method converges.

To determine the rate of convergence of this method, we note that $f(\alpha) = 0$ so that by Taylor's series expansions around $x = x_n,$ we obtained

$$f(x_n) + (\alpha - x_n) f'(x_n) + \frac{1}{2}(\alpha - x_n)^2 f''(x_n) + \ldots = 0$$

$$\Rightarrow \quad -\frac{f(x_n)}{f'(x_n)} = (\alpha - x_n) + \frac{1}{2}(\alpha - x_n)^2 \frac{f''(x_n)}{f'(x_n)} + \ldots$$

or $$x_{n+1} - x_n = (\alpha - x_n) + \frac{1}{2}(\alpha - x_n)^2 \frac{f''(x_n)}{f'(x_n)} + \ldots \text{ using (2.7.10)}$$

$$\Rightarrow \quad x_{n+1} - \alpha \approx \frac{1}{2}(x_n - \alpha)^2 \frac{f''(x_n)}{f'(x_n)}$$

If we set $\in_n = x_n - \alpha$, then we get

$$\in_{n+1} \approx \frac{1}{2} \in_n^2 \frac{f''(\alpha)}{f'(\alpha)} \qquad (2.7.11)$$

This shows that the subsequent error at each step, is proportional to the square of the previous error and hence the Newton-Raphson method has a quadratic rate of convergence.

Thus for Newton-Raphson method we see that

$$\lim_{n \to \infty} \frac{|\in_{n+1}|}{|\in_n|^2} = \frac{1}{2} \frac{f''(\alpha)}{f'(\alpha)}$$

provided that $f'(\alpha) \neq 0$, so its order of convergence is 2 and the asymptotic error constant is

$$\frac{|f''(\alpha)|}{2|f'(\alpha)|}.$$

Algorithm 2.7.4 Newton-Raphson method

We see that the ratio $[f(x_n) - f(x_{n-1})]/(x_n - x_{n-1})$ is equal to the slop of the tangent to *y= f(x)* at some point between x_{n-1} and x_n if *f(x)* is differentiable. Therefore, it would be reasonable to replace this ratio by the value of *f'(x)* at some point, near x_n and x_{n-1} provided $f(x)$ can be calculated. If $f(x)$ is differentiable, then on replacing the slope of the secant by the slope of the tangent at x_n, we obtain the iteration formula $x_{n+1} = x_n - f(x_n)/f'(x_n)$, of Newton's method.

Now given $f(x)$ continuously differentiable and *a* point x_0

for *n*= 0, 1, 2, 3..., until satisfied, do: compute $x_{n+1} = x_n - f(x_n)/f'(x_n)$.

Example 2.7.7: Use Newton-Raphson method to find the real root of $3x = \cos x + 1$.

Solution: The given equation is

$$f(x) = 3x - \cos x - 1 = 0$$

Here $\quad f(0) = -2 < 0$ and $f(1) = 3 - 05403 - 1 = 1.4597 > 0$

Therefore, a root of equation $f(x) = 0$ lies between 0 and 1. We take $x_0 = 0.6$, as the root is closer to unity. Also, we have

$$f'(x) = 3 + \sin x$$

$\therefore$ Newton-Raphson iteration formula provides us

$$x_{n+1} = x_n - \frac{f(x_n)}{f'(x_n)} = x_n - \frac{3x_n - \cos x_n - 1}{3 + \sin x_n}$$

$$= \frac{x_n \sin x_n + \cos x_n + 1}{3 + \sin x_n}, \; n = 0,1,2,3...$$

Now putting $n = 0$, we get

$$x_1 = \frac{x_0 \sin x_0 + \cos x_0 + 1}{3 + \sin x_0} = \frac{0.6 \sin(0.6) + \cos(0.6) + 1}{3 + \sin(0.6)} = 0.6071$$

Similarly,

$$x_2 = \frac{x_1 \sin x_1 + \cos x_1 + 1}{3 + \sin x_1} = \frac{0.6071 \sin(0.6071) + \cos(0.6071) + 1}{3 + \sin(0.6071)} = 0.6071$$

Since $x_1 = x_{2,}$ so the desired root is 0.6071, correct to four decimal places.

Example 2.7.8: Evaluate $\sqrt{28}$ to four decimal places by Newton's iteration formula.

Solution: Let $x = \sqrt{28}$ so that $x^2 - 28 = 0$

We take $f(x) = x^2 - 28 = 0$, so that $f'(x) = 2x$

The Newton's iteration formula gives

$$x_{n+1} = x_n - \frac{f(x_n)}{f'(x_n)} = x_n - \frac{x_n^2 - 28}{2x_n} = \frac{1}{2}\left(x_n + \frac{28}{x_n}\right) \qquad (2.7.12)$$

Now $f(5) = 25 - 28 = -3 < 0$ and $f(6) = 36 - 28 = 8 > 0$, so a root of $f(x = 0)$ lies between 5 and 6. Taking $x_0 = 5.3$, formula $(2.7.12)$ gives us

$$x_1 = \frac{1}{2}\left(5.3 + \frac{28}{5.3}\right) = 5.2915, \; x_2 = \frac{1}{2}\left(5.2915 + \frac{28}{5.2915}\right) = 5.2915$$

Since $x_1 = x_2$ up to four decimal places, so we have $\sqrt{28} = 5.2915$

Example 2.7.9: Find a root of $x^3 - 5x + 3 = 0$, by Newton-Raphson method.

Solution: The given equation is $f(x) = x^3 - 5x + 3 = 0$

Here $f(1) = 1 - 5 + 3 = -1 < 0$ and $f(2) = 8 - 10 + 3 = 1 > 0$, so a root of $f(x) = 0$ lies between 1 and 2. We take $x_0 = 1.5$ as the root is equidistant from 1 and 2. Also

$$f'(x) = 3x^2 - 5$$

$\therefore$ Newton-Raphson iteration formula leads to

$$x_{n+1} = x_n - \frac{f(x_n)}{f'(x_n)} = x_n - \frac{x_n^3 - 5x_n + 3}{3x_n^2 - 5} = \frac{2x_n^3 - 3}{3x_n^2 - 5}, \; n = 0,1,2,3...$$

Therefore, we have

$$x_1 = \frac{2(1.5)^3 - 3}{3(1.5)^2 - 5} = \frac{3.75}{1.75} = 2.1429\,,\ x_2 = \frac{16.6805}{8.7761} = 1.9007\,,\ x_3 = \frac{10.7332}{5.8380} = 1.8385\,,$$

$$x_4 = \frac{9.4286}{5.14025} = 1.8343\,,\ x_5 = \frac{9.3436}{5.094} = 1.8342$$

$\therefore$ The required root is $x = 1.834$ correct to three decimal places.

Example 2.7.10: Use Newton's method to find a real root of $x^3 - x^2 - 5 = 0$ correct to three decimal places.

Solution: The given equation is $f(x) = x^3 - x^2 - 5 = 0$

By Descartes' rule of signs the equation has only one real root which is positive. Here $f(2) = -1$ and $f(3) = 13 > 0$, therefore root lies between 2 and 3.

Now $f'(x) = 3x^2 - 2x$ *and* $f''(x) = 6x - 2$. clearly $f'(x)$and $f''(x) > 0$ for all values of $x \in [2,3]$

Also $\quad f(2) = -1, f''(2) = 10\}$ and $f(3) = 13, f''(3) = 16\}$

Therefore, $f'(x)$ and $f''(3)$ have the same signs and hence root is closer to *3*. We take $x_0 = 3$.

Then by Newton's method, we get

$$x_1 = 3 - \frac{f(3)}{f'(3)} = 3 - \frac{13}{21} = 3 - 0.619 = 2.381 \approx 2.38$$

Similarly, $x_2 = 2.150, x_3 = 2.1170, x_4 = 2.11635, x_5 = 2.11634$

Therefore the root correct to three decimal places is 2.116.

Example 2.7.11: The equation $f(x) = x - 1.2\sin x - 0.5 = 0$ has exactly one zero between $x_0 = 0.5$ and $x_1 = 1.0$, since $f(0.5)f(1.0) < 0$,while $f'(x)$ does not vanish on [0.5,1.0]. Locate the zero correct to six significant figures using algorithms 2.7.1, 2.7.2, 2.7.3 and 2.7.4.

Solution: The calculation was performed through a FORTRAN-77 program on an Intel Penteium-III 800 *MHz* computer and results are given in the following table:

Table 2.7.1: Comparison of algorithms

n	Algorithm 2.7.1 x_n	Algorithm 2.7.2 x_n	Algorithm 2.7.3 x_n	Algorithm 2.7.4 x_n
-1			1.	
0	0.75	0.75	0.5	0.5
1	0.625	0.80606124	0.61212248	0.61629718
2	0.5625	0.61534080	0.61549349	0.61546820
3	0.59375	0.61701328	0.61546816	0.61546816
4	0.609375	0.61701363	0.615468176	
5	0.6171875	0.61546816		
6	0.61328125	0.61546816		
...	......			
10	0.61547852			
...	...			
19	0.61546850			
20	0.61546820			

In algorithms 2.7.1 and 2.7.2, x_n is the mid point between the lower and the upper levels, a_n and b_n, after n iteration. The rapid and systematic convergence of algorithms 2.7.3 and 2.7.4 is note worthy. The bisection method converges very slowly but steadily, while the Regula Falsi method seems to converge" in jumps", although it does obtain the correct zero rather quickly.

Example 2.7.12 Find a real positive root of the equation $x^3 - x - 1 = 0$, by using algorithms 2.7.1, 2.7.2, 2.7.3 and 2.7.4.

Solution: The given equation is $f(x) = x^3 - x - 1 = 0$

Clearly, its positive root lies between *1* and *2*.

The result obtained from algorithms 2.7.1, 2.7.2, 2.7.3 and 2.7.4 are given in the following table.

Table 2.7.2: Comparison of methods

n	Besection x_n	Regula-Falsi x_n	Secant x_n	Newton-Raphson x_n
-1			1.0	
0	1.5	1.5	2.0	1.0
1	1.25	1.5833333	1.666667	1.5
2	1.375	1.6616541	1.12531120	1.3478261
3	1.3125	1.3249256	1.3372064	1.32522004
4	1.34375	1.3256293	1.3238501	1.3247182
5	1.328125	1.3256305	1.3247079	1.3247180
6	1.3203125	1.3247180	1.3247180	
...	...			
10	1.3247070			
...	...			
20	1.3247180			

2.7.6 Generalized Newton's method for multiple roots

Consider the equation $f(x) = 0$. Let $x = \alpha$ be a root of $f(x) = 0$, with multiplicity m. Then it is also a root of $f'(x) = 0$ with multiplicity $m-1$, of $f''(x) = 0$ with multiplicity $m-2$ and so on. Hence, if initial approximation x_0 is sufficiently close to the root, the expressions

$$x_0 - m\frac{f(x_0)}{f'(x_0)}, \quad x_0 - (m-1)\frac{f'(x_0)}{f''(x_0)}, \quad x_0 - (m-2)\frac{f''(x_0)}{f'''(x_0)}\ldots$$

will have the same values.

For $x = \alpha$ to be a root of $f(x) = 0$ with multiplicity m, the iteration formula is given by

$$x_{n+1} = x_n - m\frac{f(x_n)}{f'(x_n)}$$

which is called the generalized Newton's formula. It reduces to Newton's-Raphson formula for $m = 1$.

Example 2.7.13: Find the double root of the equation $x^3 - x^2 - x + 1 = 0$

Solution: The given equation is

$$f(x) = x^3 - x^2 - x + 1 = 0$$

So that $$f'(x) = 3x^2 - 2x - 1, \quad f''(x) = 6x - 2$$

We take $x_0 = 0.9$, so that

$$x_0 - 2\frac{f(x_0)}{f'(x_0)} = 0.9 - \frac{2(0.019)}{-0.37} = 1.003$$

and $$x_0 - (2-1)\frac{f(x_0)}{f'(x_0)} = 0.9 - \frac{(-0.37)}{3.4} = 1.009$$

The closeness of these values implies that there is a double root near $x = 1$,

$\therefore$ Choosing $x_1 = 1.01$ for the next approximation, we get

$$x_1 - 2\frac{f(x_1)}{f'(x_1)} = 1.01 - \frac{2(0.002)}{0.0403} = 1.0001$$

$$x_1 - (2-1)\frac{f'(x_1)}{f''(x_1)} = 1.01 - \frac{0.0403}{4.06} = 1.0001$$

This shows that there is a double root at $x = 1.0001$ which is quite near to the actual root $x = 1$.

2.7.7 Lin-Bairstow's method

This method is often used for finding quadratic factors of polynomials. Let the polynomial be given by

$$f(x) = a_3x^3 + a_2x^2 + a_1x + a_0 = 0$$

Let $x^2 + Rx + S$ be a quadratic factor of *f(x)* and also let an approximate factor be $x^2 + rx + s$. Usually, first approximations to *r* and *s* can be obtained from the last three terms of the given polynomial. Thus $r = \frac{a_1}{a_2}$ and $s = \frac{a_0}{a_2}$

Let $f(x) = (x^2 + rx + s)(B_2x + B_1) + Cx + D$

$$= B_2x^3 + (B_2r + B_1)x^2 + (C + B_1r + sB_2)x + (B_1s + 1) \qquad (2.7.13)$$

where the constants B_1, B_2, C and D are to be determined. Equating coefficients of the like powers of *x* in the given equation and (2.7.13), we get

$$B_2 = a_3 \;, B_2 = a_2 - rB_2, \quad C = a_1 - rB_1 - sB2 \text{ and } D = a_0 - sB_1$$

Thus, it is clear that the coefficients B_1, B_2, C and D are functions of *r* and *s*. Since $x^2 + Rx + S$ is factor of the given polynomial, therefore

$$C(R,S) = 0 \quad \text{and} \quad D(R,S) = 0$$

Letting $R = r + \Delta r$ and $S = s + \Delta s$ and using Taylor series expansions, we get

$$\left.\begin{aligned} C(R,S) &= C(r,s) + \Delta r\frac{\partial D}{\partial r} + \Delta s\frac{\partial C}{\partial s} = 0 \\ D(R,S) &= D(r,s) + \Delta r\frac{\partial D}{\partial r} + \Delta s - \frac{\partial D}{\partial s} = 0 \end{aligned}\right\} \qquad (2.7.14)$$

where the derivatives are to be computed at *r* and *s*. The above equations (2.7.14) can then be solved for Δr and Δs to obtain the next approximation to *R* and *S*, respectively. The process can be repeated until successive values of R and S show no significant change.

Example 2.7.14 Find the quadratic factor of the polynomial $f(x) = x^3 - 2x^2 + x - 2$.

Solution: The given polynomial is $f(x) = x^3 - 2x^2 + x - 2$

Here $a_3 = 1$ $a_2 = -2$, $a_1 = 1$ and $a_0 = -2$

Clearly, we get $r = -\frac{1}{2}, s = 1$

Therefore, we obtain

$$B_2 = 1 \ , \ B_1 = -2 - r$$

$$C = -1 - r(-2 - r) - s = 1 + 2r + r^2 - s$$

and

$$D = -2 - s(-2 - r) = -2 + 2s + rs$$

Also

$$[C(r,s)]_{\left(-\frac{1}{2},1\right)} = 1 - 1 + \frac{1}{4} - 1 = \frac{-3}{4}$$

$$[D(r,s)]_{\left(-\frac{1}{2},1\right)} = -2 + 2 - \frac{1}{2} = \frac{-1}{2}$$

$$\left(\frac{\partial C}{\partial r}\right)_{\left(-\frac{1}{2},1\right)} = 2 + 2r = 1, \quad \left(\frac{\partial C}{\partial s}\right)_{\left(\frac{-1}{2},1\right)} = -1$$

$$\left(\frac{\partial D}{\partial r}\right)_{\left(-\frac{1}{2},1\right)} = s = 1 \quad and \quad \left(\frac{\partial D}{\partial s}\right)_{\left(--\frac{1}{2},1\right)} = 2 + r = \frac{3}{2}$$

Equation (2.7.14) gives us

$$\Delta r - \Delta s = \frac{3}{4}$$

Hence

$$\Delta r = \frac{13}{20} \qquad \Delta s = -\frac{1}{10}$$

Thus

$$R = -\frac{1}{2} + \frac{13}{20} = \frac{3}{20} = 0.15 \quad \text{and} \quad S = 1 - \frac{1}{10} == 0.9$$

Therefore the quadratic factor is $x^2 + 0.15x + 0.9$. For the second approximation, the approximate quadratic factor is $x^2 + 0.15x + 0.9$ so that $r = 0.15$ and $s = 0.9$.

Now

$$C = 1 + 2.15(0.15) - 0.9 = 0.4225,$$

$$D = -2 + 2.15(0.9) = 0.065$$

$$\frac{\partial C}{\partial r} = 2 + 2(0.15) = 2.30$$

$$\frac{\partial C}{\partial s} = -1 \ , \ \frac{\partial D}{\partial r} = 0.9 \ and \ \frac{\partial D}{\partial s} = 2 + r = 2.15$$

Hence, equation (2.7.14) gives us

$$2.3\Delta r - \Delta s = -0.4225$$

and

$$0.9\Delta r + 2.15 \ \Delta s = -0.065$$

Solving these equations we get

$$\Delta r = -0.1665312 \quad and \quad \Delta s = 0.0394782$$

Therefore

$$R = 0.15 - 0.1665312 = -0.0165312$$

$$S = 0.15 - 0.0394783 = 0.9394783$$

Thus, the second approximation to the factor of *f(x)* is $x^2 - 0.0165312x + 0.9394783$ and it can be seen that R and *S* are approaching their actual values 0 and 1, respectively.

2.7.8 Mü ller method

In this method $y = f(x)$ is approximated by a second-degree curve passing through three points $(x_{i-2}, y_{i-2}), (x_{i-1}, y_{i-1})$ and (x_i, y_i) in the vicinity of the root. Then a root of this quadratic is taken as the next approximation to x_{i+1}, the root of $f(x) = 0$.

Consider the equation $$f(x) = 0 \tag{2.7.15}$$

Let x_{i-2}, x_{i-1}, x_i be three approximations to the root of the equation (2.7.15) and y_{i-2}, y_{i-1} y_i be the corresponding values of $f(x)$.

We assume that the parabola

$$y = ax^2 + bx + c \tag{2.7.16}$$

passes through the points (x_{i-2}, y_{i-2}), (x_{i-1}, y_{i-1}) and (x_i, y_i). Then we have

$$\left.\begin{aligned} y_{i-2} &= ax_{i-2}^2 + bx_{i-2} + c \\ y_{i-1} &= ax_{i-1}^2 + bx_{i-1} + c \\ y_i &= ax_i^2 + bx_i + c \end{aligned}\right\} \tag{2.7.17}$$

Eliminating a ,b ,c from equations (2.7.16) and (2.7.17),we obtain

$$\begin{vmatrix} y & x^2 & x & 1 \\ y_{i-2} & x_{i-2}^2 & x_{i-2} & 1 \\ y_{i-1} & x_{i-1}^2 & x_{i-1} & 1 \\ y_i & x_i^2 & x_i & 1 \end{vmatrix} = 0$$

which can be written as

$$y = \frac{(x - x_{i-1})(x - x_i\}}{(x_{i-2} - x_{i-1})(x_{i-2} - x_i)} y_{i-2} + \frac{(x - x_{i-2})(x - x_i)}{(x_{i-2} - x_{i-1})(x_{i-1} - x_i)} y_{i-1} + \frac{(x - x_{i-2})(x - x_{i-1})}{(x_i - x_{i-2})(x_{i-} x_{i-1})} y_i \tag{2.7.18}$$

We now define

$$\lambda = \frac{x - x_i}{x_i - x_{i-1}}, \quad \lambda_i = \frac{x_i - x_{i-1}}{x_{i-1} - x_{i-2}}, \quad \delta_i = \frac{x_i - x_{i-2}}{x_{i-1} - x_{i-2}} \tag{2.7.19}$$

Then (2.7.18) simplifies to

$$y = \frac{(y_{i-2}\lambda_i + y_{i-1}\delta_i + y_i)}{\delta_i} \lambda_i \lambda^2 + \frac{y_{i-2}\lambda_i^2 - y_{i-1}\delta_i^2 + y_i(\lambda_i + \delta_i)}{\delta_i} \tag{2.7.20}$$

First equation of (2.7.19) provides us that

$$x = x_i + \lambda(x_i - x_{i-1}) \tag{2.7.21}$$

To find λ we get $y = 0$ in (2.7.20) giving

$$\left(y_{i-2}\lambda_i - y_{i-1}\delta_i + y_i\right)\lambda_i\lambda^2 + \mu_i\lambda + \delta_i y_i = 0$$

where
$$\mu_i = y_{i-2}\lambda_i^2 - y_{i-1}\delta_i^2 + y_i(\lambda_i + \delta_i)$$

Solving for $\frac{1}{\lambda}$, we get

$$\frac{1}{\lambda} = \left\{-\mu_i \pm \left[\mu_i^2 - 4y_i\delta_i\lambda_i(y_{i-2}\lambda_i - y_{i-1}\delta_i + y_i)^{1/2}\right]\right\}/2y_i\delta_i$$

Because x is close to x_i, λ should be small in magnitude and hence the sign should be so chosen that the number becomes largest in magnitude. Then the equation (2.7.21) gives a better approximation to the root. This method is iterative and converges for almost all initial approximations quadratic ally. In case no better approximations are known, we take $x_{i-2} = -1,\ x_{i-1} = 0$ and $x_i = 1$.

Algorithm 2.7.5 Muller's method

1. Let x_0, x_1, x_2 be three approximations to a zero α of $f(x)$ compute $f(x_0), f(x_1), f(x_2)$.
2. Compute $h_2 = x_2 - x_1, h_1 = x_1 - x_0$

   $f(x_1, x_2) = [f(x_2) - f(x_1)]/h_2$

   $f(x_1, x_0) = [f(x_1) - f(x_0)]/h_1$
3. Set $i = 2$
4. Compute $f(x_i, x_{i-1}, x_{i-2}) = [f(x_i, x_{i-1}) - f(x_{i-1}, x_{i-2})]/(h_i + h_{i-1})$

   $$c_i := f(x_i, x_{i-1}) + h_i f(x_i, x_{i-1}, x_{i-2})$$
5. Compute $h_{i+1} := -2f(x_i)/c_i \pm \sqrt{c_i^2 - 4f(x_i)f(x_i, x_{i-1}, x_{i-2})}$.

   Choosing the sign so that the denominator is largest in magnitude
6. Set $x_{i+1} = xi + h_{i+1}$
7. Compute $f(x_{i+1})$ and $f(x_{i+1}, x_i) = f(x_{i+1} - f_{xi})/h_{i+1}$
8. Set $i = i + 1$ and repeat steps 4 to 7 until either of the following holds for given $\in_1, \in_2$: (a)

   $|x_i - x_{i-1}| < \in_1 |x_i|$, (b) $|f(x_i)| < \in_2$, or until the maximum number of iteration is

exceeded.

Example 2.7.15: Apply Muller's Method to find the root of the equation $\cos x = xe^x$, which lies between 0 and *1*.

Solution: The given equation is

$$y = f(x) = \cos x - xe^x = 0$$

We take the initial approximation as

$$x_{i-2} = -1, x_{i-1} = 0, x_i = 1$$

so that

$$y_{i-2} = \cos 1 - e^{-1}, y_{i-1} = 1, y_i = \cos 1 - e$$

$$\lambda = x - 1, \lambda_i = 1, \delta_i = 2$$

and

$$\mu_i = \cos 1 + e^{-1} - 4 + 3(\cos 1 - e) = 4\cos 1 + e^{-1} + 3e - 4$$

$$\lambda^{-1} = \left\{ -\mu_i \pm \left[\mu_i^2 - 4y_i\delta_i\lambda_i (y_{i-2}\lambda_i - y_{i-1}\delta_i + y_i) \right]^{\frac{1}{2}} \right\} \bigg/ 2y_i\delta_i$$

$$= \left\{ -\mu_i \pm \left[\mu_i^2 - 4.1.2(\cos 1 - e)(\cos 1 - e^{-1} - 2 + \cos 1 - e) \right]^{\frac{1}{2}} \right\} \bigg/ 2.2(\cos 1 - e)$$

$$= \left\{ -\mu \pm \left[\mu_i^2 - 8(\cos 1 - e)(2\cos 1 - e - e^{-1} - 2) \right]^{\frac{1}{2}} \right\} \bigg/ 4(\cos 1 - e)$$

We choose the $-ve$ sign and so that the numerator in λ^{-1} is largest in magnitude and hence $\lambda = -0.5585$. The next approximation is

$$x_{i+1} = x_i + \lambda(x_i - x_{i-1}) = 1 - 0.5585 = 0.4415$$

Repeating the above process, we get

$$x_{i+2} = 0.5125, x_{i+3} = 0.5177, x_{i+4} = 0.5177$$

Hence the root is 0.5177 correct to four decimal places.

2.7.9 Graeffe's root squaring method

This method requires no prior information about the root of an equation and hence has an advantage over the other methods. But it is only applicable to polynomial equations and is capable of giving all the roots. Consider the polynomial equation,

$$f(x) = x^n + a_1x^{n-1} + a_2x^{n-2} + ... + a_{n-1}x + a_n = 0 \tag{2.7.22}$$

Separating the even and odd power of x and squaring, we get

$$\left(x^n + a_2x^{n-2} + a_4x^{n-4} + ...\right)^2 = \left(a_1x^{n-1} + a_3x^{n-3}a_5x^{n-5} + ...\right)^2$$

On simplification this equation leads to

$$y^n + b_1y^{n-1} + \cdots + b_{n-1}y + b_n = 0 \tag{2.7.23}$$

where we have put $x^2 = y$ and

$$\left.\begin{aligned} b_1 &= -a_1^2 + 2a_2 \\ b_2 &= a_2^2 + 2a_1a_3 + 2a_4 \\ \vdots \quad & \quad \vdots \quad \vdots \quad \vdots \\ b_n &= (-1)^n a_n^2 \end{aligned}\right\} \tag{2.7.24}$$

Let $\alpha_1, \alpha_2, \alpha_3 ... \alpha_n$ be the roots of (2.7.22) so that the roots of equation (2.723) are $\alpha_1^2, \alpha_2^2, \alpha_3^2 ... \alpha_n^2$. After m squaring, let the new transformed equation be

$$z^n + c_1 z^{n-1} + c_2 z^{n-2} + \dots c_{n-1} z + c_n = 0 \qquad (2.7.25)$$

Whose roots $\beta_1, \beta_2, \beta_3 \dots \beta_n$ be such that $\beta_i = \alpha_i^{2m}, i = 1,2,3,\dots n.$

Assuming that $|\alpha_1| > |\alpha_2| > \dots > |\alpha_n|$ then

$$|\beta_1| >> |\beta_2| >> \dots > |\beta_n|. \qquad (2.7.26)$$

Thus $\frac{|\beta_2|}{|\beta_1|} = \frac{\beta_2}{\beta_1}, \dots \frac{|\beta_n|}{|\beta_{n-1}|} = \frac{\beta_n}{\beta_{n-1}}$ are negligible as compared to unity. Also β_i being an even power of α_i is always positive. From equation (2.7.25), we have

$$\sum \beta_1 == -c_1 \Rightarrow c_1 = -\beta_1 (1 + \frac{\beta_2}{\beta_1} + \frac{\beta_3}{\beta_1} + \dots)$$

$$\sum \beta_1 \beta_2 = c_2 \Rightarrow c_2 = \beta_1 \beta_2 (1 + \frac{\beta_3}{\beta_1} + \dots)$$

$$\sum \beta_1 \beta_2 \beta_3 = -c_3 \Rightarrow c_3 = -\beta_1 \beta_2 \beta_3 (1 + \frac{\beta_4}{\beta_1} + \dots)$$

$$\vdots \qquad \vdots \qquad \vdots \qquad \vdots$$

$$\beta_1 \beta_2 \beta_3 \dots \beta_n = (-1)^n c_n \Rightarrow c_n = -\beta_1 \beta_2 \beta_3 \dots \beta_n .$$

Hence by (2.7.26), we get

$$c_1 = -\beta_1, c_2 = -\beta_1 \beta_2, c_3 = -\beta_1 \beta_2 \beta_3 \dots, \; c_n = (-1)^n \beta_1 \beta_2 \beta_3 \dots \beta_n$$

$$\Rightarrow \beta_1 = -c_1, \beta_2 = -\frac{c_2}{c_1}, \beta_3 = -\frac{c_3}{c_2}, \dots, \beta_n = -\frac{c_n}{c_{n-1}}$$

Therefore, since $\beta_i = \alpha_i^{2m} \Rightarrow \alpha_i = (\beta_i)^{\frac{1}{2m}} = |c_i / c_{i-1}|$, we can determine the roots

$$\alpha_1, \alpha_2, \alpha_3 \dots \alpha_n$$

of the given equation (2.7.22).

Double roots

If the magnitude of c_i is half the square of magnitude of the corresponding coefficient in the previous equation after a few squaring, then it shows that α_i is a double root of (2.7.22) this double root can be determined as follows.

$$\beta_k \approx \frac{c_k}{c_{k+1}} \quad \text{and} \quad \beta_{k+1} \approx -\frac{c_{k+1}}{c_k}$$

$$\therefore \qquad \beta_k \beta_{k+1} \approx \beta_k^{2m} = \left| \frac{c_{k+1}}{c_{k-1}} \right| \Rightarrow \alpha_k^{2m} \beta_k^2 = \left| \frac{c_{k+1}}{c_{k-1}} \right|$$

This gives us the magnitude of the double root and its substitution in (2.7.22) provides us its sign.

Complex roots

If α_r and α_{r+1} are a complex pair $\rho_r \exp(\pm i\phi_r)$, then the coefficients of x^{n-r} in successive squaring would fluctuate both in magnitude and sense by an amount

$$\rho_r^{2(2m)} = \left|\frac{c_{r+1}}{c_{r-1}}\right|, \; 2\rho_r \cos(m\phi_r) = \frac{c_{r+1}}{c_{r-1}}.$$

If (2.7.22) has one pair of complex roots say $\rho_r e^{\pm i\phi_r} = \lambda + i\eta$, then we can first find all the real roots. There after λ is given by $\alpha_1 + \alpha_2 + ... + \alpha_{r-1} + 2\lambda + \alpha_{r+2} + ... + \alpha_n = -\alpha_1$ and η is given by $\rho_r^2 = \lambda^2 + \eta^2$ or $\eta = \sqrt{\rho_r^2 - \lambda^2}$. Hence the method.

$$x^4 - 3x + 1 = 0$$

Solution: The given equation is

$$f(x) = x^4 - 3x + 1 = 0 \tag{2.7.27}$$

Clearly, it has two changes in sign, therefore by Descartes, rule, it has two positive roots and no negative roots since $f(-x)$ has no change in sign. Thus the remaining two roots are complex. Now rewriting $f(x) = 0$ as

$$x^4 + 4 = 3x$$

and squaring, we get

$$(y^2 + 1)^2 = 9y, \text{ where } y = x^2$$

Squaring again and putting $y^2 = z$, we get

$$(z+1)^4 = 81z \Rightarrow z^4 + 6z^2 + 1 = -z(4z^2 - 77) \tag{2.7.28}$$

Squaring once again and setting $z^2 = u$, we obtain

$$(u^2 + 6u + 1)^2 = u(4u - 77)^2$$

or

$$u^4 - 4u^3 + 65u^2 - 5917u + 1 = 0 \tag{2.7.29}$$

If $\alpha_1, \alpha_2, \alpha_3, \alpha_4$ be the roots of given equation $f(x) = 0$, then roots of last resulting equation are $\alpha_1^8, \alpha_2^8, \alpha_3^8$ and α_4^8. Thus we have

$$\alpha_1^8 = 4, \; \alpha_2^8 = \frac{654}{4} = 163.5, \; \alpha_3^8 = \frac{5917}{654} = 9.0474, \; \alpha_4^8 = \frac{1}{5917} = 0.00017$$

Let $\alpha_1 = 1.189, \; \alpha_2 = 1.891, \; \alpha_3 = 1.3169, \; \alpha_4 = 0.3379$

From (2.7.28) and (2.7.29) we can see that the magnitude of c_1 and c_4 have become constant which indicates that α_1, α_4 are the real roots and α_2, α_3 are a pair of complex roots. Thus the real roots are $\alpha_1 = 1.1892$ and $\alpha_4 = 0.3379$. Let the complex roots be $\rho_2 e^{\pm\phi_2} = \lambda + i\eta$. From (2.7.29), the magnitude is given by

$$\rho_2^{(2^3)} \approx \frac{c_{2+1}}{c_{2-1}} \Rightarrow \rho_2^{16} = \frac{5917}{4} = 1479.25 \Rightarrow \rho_2 = 1.5781$$

Also from (2.7.27), we have $\alpha_1 + 2\lambda + \alpha_4 = 0$

$$\Rightarrow \lambda = \frac{-1}{2}(\alpha_1 + \alpha_4) = -0.7636 \quad \text{and} \quad \eta = \sqrt{\rho_2^2 - \lambda^2} = 1.381$$

Hence the complex roots are $-0.7636 \pm 1.381i$ Thus the roots of given equation (2.7.27) are

$$1.1892, \quad 0.3379, \quad -0.7636 \pm 1.381i$$

2.8 Solution of non-linear simultaneous equations

Consider the equations

$$f(x, y) = 0, \quad g(x, y) = 0 \tag{2.8.1}$$

Let (x_0, y_0) be an initial approximation to the solution of (2.8.1). Suppose $x_1 = x_0 + h$, $y_1 = y_0 + k$ be a better approximation to the solution of (2.8.1) then

$$f(x_0 + h, y_0 + k) = 0 \;,\; g(x_0 + h, y_0 + h) = 0$$

Expanding these functions in Taylor's series and retaining only first degree terms, we get

$$\left.\begin{aligned} f_0 + h\frac{\partial f}{\partial x_0} + k\frac{\partial f}{\partial y_0} = 0 \\ g_0 + f\frac{\partial g}{\partial x_0} + k\frac{\partial g}{\partial y_0} = 0 \end{aligned}\right\} \tag{.2.8.2}$$

where $f_0 = f(x_0, y_0), \; \dfrac{\partial f}{\partial x_0} = \left(\dfrac{\partial f}{\partial x}\right)_{(x_0, y_0)}$ etc.

If the Jacobian

$$J(f, g) = \begin{vmatrix} f_x & f_y \\ g_x & g_y \end{vmatrix} \neq 0 \tag{2.8.3}$$

Then the linear equations (2.7.14) possess a unique solution given by

$$h = \begin{vmatrix} -f & f_y \\ -g & g_y \end{vmatrix} \Big/ J(f, g)$$

and

$$R = \begin{vmatrix} f_x & -f \\ g_x & -g \end{vmatrix} \Big/ J(f, g) \tag{2.8.4}$$

New approximation to the root is then given by $x_1 = x_0 + h$, $y_1 = y_0 + k$. This process is repeated till we get the solution to the desired level of accuracy. This method will not converge unless the initial solutions roots are chosen quite close to the actual roots. The method can be extended to three variable equations, but the process of finding a meaningful solution is very cumbersome unless the entire information about the equation and their physical context is available. It is possible, one of the variables may be eliminated from the given equations giving a single polynomial equation in other variable and this variable is determined to desire degree of accuracy by Newton-Raphson method. Sometimes the root of this equation can be found by trial if so, we reduce this equation to the next lower degree equation and find its other root by Newton-Raphson method. Having found this variable to required degree of accuracy, the other variable can at once, be determined from one of the given equation.

Example 2.8.1: Solve the equations $x^2 + y = 11, \quad y^2 + x = 7$

Solution: An initial approximation to the solution is taken as

$$x_0 = 3.5 \text{ and } y_0 = -1.8$$

We take

$$f = x^2 + y - 11 \text{ and } g = y^2 + x - 7$$

so that

$$\frac{\partial f}{\partial x} = 2x, \quad \frac{\partial f}{\partial y} = 1$$

$$\frac{\partial g}{\partial x} = 1, \quad \frac{\partial g}{\partial y} = 2y$$

By Newton-Raphson method, we get

$$7h + k = 0.55 \quad , \quad h - 3.6k = 0.26$$

Solving these equations, we obtained

$$h = 0.0855 \quad , \quad k = -0.0485$$

$\therefore$ The better approximation to the root is

$$x_1 = x_0 + h = 3.5855 \quad , \quad y_1 = y_0 + k = -1.8485$$

Repeating the above process with (x_1, y_1)instead of (x_0, y_0), we get

$$x_2 = 3.5844 \quad , \quad y_2 = -1.8482.$$

$$x_3 = 3.58423 \quad , y_3 = -1.84815$$

Thus the required solution is

$$x = 3.584 \quad , y = -1.848,$$

correct up-to three decimal places.

2.8.1 Method of iteration

Let the given equations be

$$\begin{aligned} f(x, y) &= 0 \\ g(x, y) &= 0 \end{aligned} \qquad (2.8.6)$$

whose real roots are required within a specified accuracy. We assume that the equations (2.8.6) may be written in the form

$$\begin{aligned} x &= F(x, y) \\ y &= G(x, y) \end{aligned} \qquad (2.8.7)$$

where the functions F and G satisfy the conditions

$$\left|\frac{\partial F}{\partial x}\right| + \left|\frac{\partial F}{\partial y}\right| < 1$$

and

$$\left|\frac{\partial G}{\partial x}\right| + \left|\frac{\partial G}{\partial y}\right| < 1 \qquad (2.8.8)$$

in the neighborhood of the root.

Let (x_0, y_0)be the initial approximation to a root (ξ, η) of the system $(2.8.6)$. We construct the successive approximations according to the formulae

$$\begin{aligned} x_1 &= F(x_0, y_0), & y_1 &= G(x_0, y_0) \\ x_2 &= F(x_1, y_1), & y_2 &= G(x_1, y_1) \\ x_3 &= F(x_2, y_2), & y_3 &= G(x_2, y_2) \\ &\cdots\cdots & &\cdots\cdots \\ x_{n+1} &= F(x_n, y_n), & y_{n+1} &= G(x_n, y_n) \end{aligned} \tag{2.8.9}$$

For faster convergence, most recent values of x_i may be used in the evaluation of y_i in (2.8.9). If the iteration process (2.8.9) converges, then we obtain

$$\xi = F(\xi, \eta) \text{ and } \eta = G(\xi, \eta) \tag{2.8.10}$$

in the limit. Thus ξ and η are the roots of the system (2.8.7) and hence of the system (2.8.6). Condition (2.8.8) is sufficient for the convergence of the iteration (2.8.9).

Example: 2.8.2 Find a real root of the equations

$$x = 0.2x^2 + 0.8, y = 0.3xy^2 + 0.7$$

Solution: Here

$$F(x, y) = 0.2x^2 + 0.8$$

$$G(x, y) = 0.3xy^2 + 0.7$$

Therefore, we have $F_x = 0.4x, F_y = 0, G_x = 0.3y^2$, and $G_y = 0.6xy.$

Clearly $x = 1$ and $y = 1$, as seen from the given equations we take $x_0 = \frac{1}{2} = y_0$ and find that

$$|F_x|_{(x_0, y_0)} + |F_y|_{(x_0, y_0)} = 0.2 < 1$$

and

$$|G_x|_{(x_0, y_0)} + |G_y|_{(x_0, y_0)} = 0.225 < 1$$

Thus the conditions (2.8.8) are satisfied. Hence

$$x_1 = F(x_0, y_0) = \frac{0.2}{4} + 0.8 = 0.85$$

$$y_1 = G(x_0, y_0) = \frac{0.3}{8} + 0.7 = 0.74$$

The second approximation is obtained as

$$x_2 = F(x_1, y_1) = 0.2(0.85)^2 + 0.8 = 0.9445$$

$$y_2 = G(x_1, y_1) = 0.3(0.85)(0.74)^2 + 0.7 = 0.81$$

Thus the convergence to the root (1,1) is obvious.

2.8.2 Complex roots by Newton-Raphson method

Let us consider the problem of finding a root z^* of the equation $f(z) = 0$, where $z \in C$. Define x, y, u and v such that $z = x + iy$ and $f(z) = u(x, y) + iv(x, y), i = \sqrt{-1}.$

By the Newton-Raphson method, we have

$$z^{(n+1)} = z^{(n)} - f\left(z^{(n)}\right) / f'\left(z^{(n)}\right) \tag{2.8.11}$$

But we have

$$f'(z)=\frac{df}{dz}=\frac{\partial u}{\partial x}+i\frac{\partial v}{\partial x}$$

and by the Cauchy-Riemann equations

$$\frac{\partial u}{\partial x}=\frac{\partial v}{\partial y},\ \frac{\partial u}{\partial y}=-\frac{\partial v}{\partial x}$$

this implies that

$$\frac{df}{dz}=\frac{\partial u}{\partial x}-i\frac{\partial u}{\partial y}$$

Therefore

$$\frac{f(z)}{f'(z)}=\frac{u+iv}{u_x-iu_y}=\frac{u\,u_x-vu_y}{u_x^2+u_y^2}+i\frac{v\,u_x+u\,u_y}{u_x^2+u_y^2} \tag{2.8.12}$$

Now equations (2.8.11) and (2.8.12) after simplifications and on equating real and imaginary parts provide us

$$x^{n+1}=x^n-\frac{u\,u_x-v\,u_y}{u_x^2+u_y^2},\quad y^{n+1}=y^n-\frac{v\,u_x+u\,u_y}{u_x^2+u_y^2} \tag{2.8.13}$$

Thus it is noticed that the iteration process (2.8.11) defined in C can be replaced by two simultaneous iterations in R.

Exercises

1. Obtain a root, correct to three decimal places, of the following equations using the bisection method

   (i) $x^3+x^2+x+7=0$ (v) $x^3-18=0$ (ix) $x\log_{10} x=1.2$

   (ii) $x^3-9x+1=0$ (vi) $x^3-5x+3=0$ (x) $x^3+x-1=0$

   (iii) $x^3-4x-9=0$ (vii) $x^3-3x-5=0$

   (iv) $x^3-x-11=0$ (viii) $x^4-x-10=0$

2. Use the method of false positions to find a root, correct to three decimal places, of the following.

   (i) $x^3-4x-9=0$ (v) $x^3-x-1=0$ (ix) $x^3-2x^2+3x-5=0$

   (ii) $x^6-x^4-x^3-1=0$ (vi) $xe^x=2$ (x) $\tan x=4x$

   (iii) $x^3-18=0$ (vii) $\cos x=3x-1$

   (iv) $x^3-x^2-1=0$ (viii) $x\log_{10} x=1.2$

3. Find a root of the following equations correct to three decimal places by the secant method.

   (i) $x^3+x^2+x+7=0$ (iii) $x^6-x^4-x^3-1=0$ (v) $\cos x=3x-1$

   (ii) $x^3-18=0$ (iv) $x-e^{-x}=0$ (vi) $x^3-4x-9=0$

4. Use the iteration method to find, correct to four significant figures, a real root of the following.

   (i) $x^3+x^2-1=0$ (v) $e^{-x}=10x$ (ix) $5x^3-20x+3=0$

   (ii) $\cos x=3x-1$ (vi) $x\sin x=1$ (x) $\tan x=x$

   (iii) $x=(5-x)^{\frac{1}{3}}$ (vii) $1+x^2=x^3$ (xi) $x^3+x^2-100=0$

   (iv) $\sin x=10(x-1)$ (viii) $e^x\tan x=1$ (xii) $e^x=x^2$

5. Use Newton-Raphson method to obtain a root, correct to three decimal places, of the following equations.

   (i) $x^3-3x-5=0$ (vii) $x^3+3x^2-3=0$ (xii) $\sin x=\frac{1}{2}x$

   (ii) $\sin x=1-x$ (viii) $x^4+x^2-80=0$ (xiii) $x^3-2x^2+3x-5=0$

   (iii) $x^3-5x+3=0$ (ix) $x^5-5x^2+3=0$ (xiv) $x+\log x=2$

   (iv) $x^3-3x+1=0$ (x) $x\sin x+\cos x=0$, near $x=\pi$

   (v) $x^4-x-13=0$ (xi) $x-\cos x=0$ (xv) $\cos x=xe^x$

   (vi) $x^4+x^3-7x^2-x+5=0$ (xvi) $e^x\sin x=1$

6. Establish the formula $x_{i+1}=\frac{1}{2}\left(x_i+\frac{N}{x_i}\right)$ and hence or otherwise compute.

   (i) $\sqrt{2}$ (ii) $\sqrt{41}$ (iii) $\sqrt[3]{10}$

   (iv) $\sqrt[4]{32}$

7. Find a double root of the equation $x^3 + 15.87x + 24.34 = 0$, correct to three decimal places.
8. Find the multiplicity and the multiple root of: $x^4 - 11x^3 + 36x^2 - 16x - 64 = 0$, near $x = 3.9$.
9. Use Muller's method to find a root of the equations

   (i) $x^3 - x - 1 = 0$ (ii) $x^3 - x^2 - x - 1 = 0$ (iii) $x^3 - 3x - 5 = 0$

   (iv) $\log x = x - 3$ taking $x_0 = 0.25$, $x_1 = 0.5$ and $x_2 = 1$
10. Apply Graeffe's root squaring method to find the approximate roots of the equations:

    (i) $x^3 - 3x^2 - 6x + 8 = 0$ (iii) $x^2 - 7x^2 + 10x - 2 = 0$ (v) $x^3 - 8x^2 + 17x - 10 = 0$

    (ii) $x^3 - 6x^2 + 11x - 6 = 0$ (iv) $x^3 - 4x^2 + 5x - 2 = 0$ (vi) $x^4 - 3x + 1 = 0$
11. Solve the following systems of nonlinear equations by Newton's method:

    (i) $x^2 - y^2 = 4,\ y^2 + x^2 = 16$

    (ii) $x^3 = y + 100,\ y^3 = x + 150$

    (iii) $x^2 = 3xy - 7,\ \ y = 2(x + 1)$

CHAPTER 3

SYSTEM OF SIMULTANEOUS ALGEBRAIC EQUATIONS

3.1 Introduction

Simultaneous algebraic linear equations occur quite often in engineering and science applications. The analysis of electronic circuits consisting of invariant elements, analysis of a network under sinusoidal steady-state conditions, determination of the output of a chemical plant, finding the cost of chemical reactions, study of civil engineering structures, and problems of layered media in elasticity and seismology, are some of the problems which involve the solutions of simultaneous linear algebraic equations. Also there are two numerical problems, which arise in connection with the matrices, one the problem of finding the inverse of a matrix and other the problem of finding the eigenvalues and the corresponding eigenvectors of a matrix. In fact, the computation of the eigenvalues is required in many engineering and scientific problems. For example, the frequencies of the vibrations of beams, system of masses and springs, annular membrane and oscillations of a triple pendulum, torsional oscillations of a uniform cantilever and multi-cylinder eigenvalues etc., are the eigenvalues of a matrix. The solutions of such equations can be obtained by Direct or Iterative methods. We discuss some of such methods of solution in this chapter after briefly outlining some basic concepts and properties of matrices and determinants.

3.2 Matrices

Definition 3.2.1: A rectangular arrangement of numbers (real or complex) in m rows and n columns is called an m×n matrix. It is denoted by

$$A = \begin{bmatrix} a_{11} & a_{12} & a_{13} & \cdots & a_{1n} \\ a_{21} & a_{22} & a_{23} & \cdots & a_{2n} \\ a_{31} & a_{32} & a_{33} & \cdots & a_{3n} \\ \vdots & \vdots & \vdots & & \vdots \\ a_{m1} & a_{m2} & a_{m3} & \cdots & a_{mn} \end{bmatrix} = (a_{ij})_{mxn}$$

Here a_{ij}'s are called elements of the matrix A and the number m×n is called order of the matrix A, for example a_{ij} is called *jth* element of *ith* row or *ith* element of the *jth* column.

3.2.1 Types of matrices

1. **Row and column matrices:** A matrix having a single row (column) is called row (column) matrix or a row (column) vector.

2. **Square matrix:** A matrix in which the number of rows equals the number of columns is called a square matrix.

3. **Null matrix:** A matrix $(a_{ij})_{m\times n}$ in which all the elements are zero is called a null or zero or void matrix. Mathematically, in such matrices $a_{ij} = 0$ for all i and j.

4. **Diagonal matrix:** A matrix in which all the elements except the main diagonal entries are zero, is called a diagonal matrix e.g.

$$A = \begin{bmatrix} a & 0 & 0 \\ 0 & b & 0 \\ 0 & 0 & c \end{bmatrix}$$

5. **Scalar matrix:** A diagonal matrix, in which all the diagonal entries are equal, is called a scalar matrix e.g.

$$A = \begin{bmatrix} \lambda & 0 & 0 \\ 0 & \lambda & 0 \\ 0 & 0 & \lambda \end{bmatrix}$$

6. **Unit matrix:** A diagonal matrix, in which all the diagonal elements are unity, is called an unit or identity matrix.

7. **Triangular matrix:** A square matrix (a_{ij}) in which all the elements below the principal diagonal are zero i.e. $a_{ij} = 0, \ \forall \text{ i} > \text{j},$ is called an upper triangular matrix. A square matrix (a_{ij}) in which all the elements above the principal diagonal are zero i.e., $a_{ij} = 0, \ \forall \text{ i} < \text{j},$ is called a lower triangular matrix. A square matrix is called a triangular matrix if it is either upper triangular or lower triangular.

8. **Equal matrices:** Two matrices $A = (a_{ij})_{m\times n}$ and $B = (b_{ij})_{p\times q}$ are said to be equal if

   (i) they are of the same order i.e. $m = p, n = q$ and

   (ii) their corresponding entries are same i.e. $a_{ij} = b_{ij}, \forall \text{ i, j}.$

9. **Equivalent matrices:** Two matrices A and B are said to be equivalent if one can be obtained from the other by a sequence of elementary row or column transformations.

10. **Symmetric and skew symmetric matrices:** A matrix obtained from a given matrix A by interchanging its rows and columns, is called the transpose of A and it is denoted by A'. Thus here $(a_{ij})' = (a_{ji})$. If a matrix A equals its transpose i.e. if $A = A'$, then it is called a symmetric matrix. If $A = (a_{ij})$ is symmetric then $a_{ij} = a_{ji}, \ \forall \ i$ and j If a matrix A equals

negative of its transpose i.e. $A' = -A$, then it is called a skew symmetric matrix. If $A = (a_{ji})$ is skew symmetric, then $a_{ij} = -a_{ji}, \forall \; i$ and j.

It can also be noted here that the transpose of product of two matrices is the product of their transpose taken in the reverse order *i.e*, $(AB)' = B'A'$

Also every square matrix can be expressed as the sum of a symmetric and a skew symmetric matrix.

3.2.2 Operations on matrices

1. **Addition and subtraction of matrices**: Two matrices A and B are said to be confirmable for addition (subtraction) if they are of the same order and their sum (difference) is the matrix of the sum (subtraction) of their corresponding elements. Mathematically, if $A = (a_{ij})_{m\times n}$ $B = (b_{ij})_{p\times q}$, then $p = m, q = n$ and $A + B = (c_{ij})_{m\times n}$, where $c_{ij} = a_{ij} + b_{ij}$. Also $A - B = (d_{ij})_{m\times n}$, $d_{ij} = a_{ij} - b_{ij}$.

2. **Multiplication of a matrix by a scalar**: The product of a matrix $A = (a_{ij})$ by a scalar k is the matrix B, obtained by multiplying each and every element of A by k, i.e., $B = kA = (ka_{ij})$.

3. **Multiplication of matrices**: Two matrices $A = (a_{ij})_{m\times n}$ and $B = (b_{ij})_{p\times q}$ are said to be conformable for multiplication if the number of columns in the premultiplier is equal to the number of rows in the post multiplier. Thus AB is defined if $n = p$ and the product will be a matrix of order $m\times q$ i.e., $AB = C = c_{ij\,m\times q}$, where $c_{ij} = \sum_{k=1}^{m} b_{ik} a_{kj}$. In general, $AB \neq BA$.

3.3 Determinants

Definition 3.3.1 For every square matrix ($A = (a_{ij})$, we associate a number called determinant of the matrix A denoted by det A or $|A|$. The expression $\begin{vmatrix} a_1 & b_1 \\ a_2 & b_2 \end{vmatrix}$ is called a determinant of the second order and equals to $a_1 b_2 - a_2 b_1$. Similarly, $\begin{vmatrix} a_1 & b_1 & c_1 \\ a_2 & b_2 & c_2 \\ a_3 & b_3 & c_3 \end{vmatrix}$ is called a determinant of the third order. In general, a determinant of nth order matrix is of the form

$$|A| = \begin{vmatrix} a_{11} & a_{12} & a_{13} & \cdots & a_{1n} \\ a_{21} & a_{22} & a_{23} & \cdots & a_{2n} \\ a_{31} & a_{32} & a_{33} & \cdots & a_{3n} \\ \vdots & \vdots & \vdots & & \vdots \\ a_{m1} & a_{m2} & a_{m3} & \cdots & a_{mn} \end{vmatrix}$$

The matrices, which are not square, do not have determinants.

3.3.1 Basic properties of determinants

1. A determinant remains unaltered by changing its rows into columns and columns into rows.
2. If any two rows or columns of a determinant are interchanged, the determinant retains its numerical value but changes its sign.
3. If any two rows or columns of a determinant are identical, then its value is zero.
4. If each element of a row or column of a determinant is multiplied by a constant k, then its value gets multiplied by k. Thus, if A is a square matrix of order n, then $|kA| = k^n|A|$.
5. If two rows (or columns) of a determinant are proportional, then its value is zero.
6. If some or all the elements of a row (or column) of a determinant are expressed as sum of two or more terms, then the determinant can be expressed as sum of two or more determinants.
7. If to any row or column of a determinant a multiple of another row or column is added, the value of the determinant remains unaltered.
8. The product of two determinants of the same order is itself a determinant of that order i.e.

$$\begin{vmatrix} a_1 & b_1 & c_1 \\ a_2 & b_2 & c_2 \\ a_3 & b_3 & c_3 \end{vmatrix} \times \begin{vmatrix} l_1 & m_1 & n_1 \\ l_2 & m_2 & n_2 \\ l_3 & m_3 & n_3 \end{vmatrix}$$

$$= \begin{vmatrix} a_1l_1 + b_1m_1 + c_1n_1 & a_2l_1 + b_2m_1 + c_2n_1 & a_3l_1 + b_3m_1 + c_3n_1 \\ a_1l_2 + b_1m_2 + c_1n_2 & a_2l_2 + b_2m_2 + c_2n_2 & a_3l_2 + b_3m_2 + c_3n_2 \\ a_1l_3 + b_1m_3 + c_1n_3 & a_2l_3 + b_2m_3 + c_2n_3 & a_3l_3 + b_3m_3 + c_3n_3 \end{vmatrix}$$

Definition 3.3.2: Minors and cofactors: Let $A = (a_{ij})$ be a square matrix. If we suppress the ith row and jth column of the determinant $|A|$, we get a determinant M_{ij} of order (n-1) called minor of the element a_{ij}. The cofactor c_{ij} of a_{ij} is defined to be $(-1)^{i+j} M_{ij}$.

Definition 3.3.3: Adjoint of a matrix: The adjoint of a square matrix A is the transpose of the matrix of cofactors of A.

Definition 3.3.4: Inverse of a matrix: If A be a non singular matrix of order n, then a square matrix B of the same order such that $AB = BA = 1$, is called the inverse of A, I being a unit matrix.

Definition 3.3.5 A matrix is said to be of rank r if

(i) it has at-least one non-zero minor of order r, and

(ii) Every minor of order higher than r, is zero.

3.4 Elementary transformations of a matrix

Following rows and columns operations are called elementary transformations of a matrix

1. The interchange of any two rows (columns) denoted by $R_{ij}\,(C_{ij})$
2. The multiplication of any row (column) by a non-zero number denoted by $kR_i(kC_i)$
3. The addition of constant multiple of the elements of any row (column) to the corresponding elements of any other row (column). It is denoted by $R_i + pR_j(C_i + pC_j)$

The elementary transformations do not change either the order or the rank of a matrix. While the value of the minors may get change by the transformations 1 and 2, their zero or non-zero character remains unaffected.

Definition 3.4.1: Equivalent matrices: Two matrices A and B are said to be equivalent if one can be obtained from the other by a sequence of elementary transformations. Such matrices have same order and same rank and the symbol ~ is used for equivalence.

Definition 3.4.2: Elementary matrix: In elementary matrix is that, which is obtained from a unit matrix, by subjecting it to any of the elementary transformations.

Definition 3.4.3 Normal form of a matrix: Every non-zero matrix A of rank r can be reduced by a sequence of elementary transformations to the form $\begin{bmatrix} I_r & 0 \\ 0 & 0 \end{bmatrix}$, which is called the normal form of A.

3.5 System of simultaneous linear equations

3.5.1 Criterion for consistency

Non-homogeneous equations: Consider the system of equations

$$\begin{aligned} a_{11}x_1 + a_{12}x_2 + a_{13}x_3 + \ldots + a_{1n}x_n &= b_1 \\ a_{21}x_1 + a_{22}x_2 + a_{23}x_3 + \ldots + a_{2n}x_n &= b_2 \\ a_{31}x_l + a_{32}x_2 + a_{33}x_3 + \ldots + a_{3n}x_n &= b_3 \\ \vdots \qquad \vdots \qquad \vdots \qquad \vdots \qquad & \vdots \\ {}_{n1}x_1 + a_{n2}x_2 + a_{n3}x_3 + \ldots + a_{nn}x_n &= b_n \end{aligned} \tag{3.5.1}$$

This can be rewritten as

$$AX = B \tag{3.5.2}$$

where
$$A = \begin{bmatrix} a_{11} & a_{12} & a_{13} & \cdots & a_{1n} \\ a_{21} & a_{22} & a_{23} & \cdots & a_{2n} \\ \vdots & \vdots & \vdots & \vdots & \vdots \\ a_{n1} & a_{n2} & a_{n3} & \cdots & a_{nn} \end{bmatrix}, \quad B = \begin{bmatrix} b1 \\ b_2 \\ \vdots \\ b_n \end{bmatrix}, \quad X = \begin{bmatrix} x_1 \\ x_2 \\ \vdots \\ x_n \end{bmatrix} \tag{3.5.3}$$

We define the matrix
$$[A/B] = \left[\begin{array}{ccccc|c} a_{11} & a_{12} & a_{13} & \ldots & a_{1n} & b_1 \\ a_{21} & a_{22} & a_{23} & \ldots & a_{2n} & b_2 \\ \vdots & \vdots & \vdots & \vdots & \vdots & \vdots \\ a_{n1} & a_{n2} & a_{n3} & \ldots & a_{nn} & b_n \end{array}\right] \tag{3.5.4}$$

called augmented matrix.

1. If rank (A) $\neq$ rank $[A|B]$, then the system of equations (3.5.1) is inconsistent and hence have no solution.
2. If rank (A) = rank $[A|B]$ = n, the order of A, the equations (3.5.1) are consistent and have a unique solution.
3. If rank (A) = rank $[A|B]$ >n, the order of A, the equations (3.5.1) are consistent and have an infinite number of solutions.

Homogeneous equations

If the matrix B on right hand side of equation (3.5.2) is a null matrix i.e. $B = [0 \; 0 \; 0 \; \ldots \; 0]'$, then the system of equations (3.5.2) and hence (3.5.1) is called a homogeneous system. This type of system of equations is represented in matrix from by

$$AX = 0 \qquad (3.5.5)$$

We find the rank ρ of the coefficient matrix A by reducing it to the triangular form by elementary row operations.

1. If $\rho = n$ the equations (3.5.5) have only a trivial solution $x_1 = x_2 \ldots = x_n = 0$.
2. If $\rho < n$, the equation (3.5.5) have (n–ρ) independent solutions.
3. When $m < n$ i.e. number of equations is less than the number of variables, the solution is always other than $x_1 = x_2 = x_3 \ldots = x_n = 0$ *viz* non-trivial.
4. When $m = n$, the number of equations equals to number of variables, the necessary and sufficient condition for non-trivial solution is that $|A| = 0$ i.e. A must be a singular matrix.

Example 3.5.1: Test the consistency and solve the equations

$$5x + 3y + 7z = 4, \quad 3x + 26y + 2z = 9, \quad 7x + 2y + 10z = 5.$$

Solution: The system of equations can be rewritten as

$$\begin{bmatrix} 5 & 3 & 7 \\ 3 & 26 & 2 \\ 7 & 2 & 7 \end{bmatrix} \begin{bmatrix} x \\ y \\ z \end{bmatrix} = \begin{bmatrix} 4 \\ 9 \\ 5 \end{bmatrix}$$

i.e.,

$$AX = B.$$

The augmented matrix is

$$[A|B] = \begin{bmatrix} 5 & 3 & 7 \\ 3 & 26 & 2 \\ 7 & 2 & 10 \end{bmatrix} \begin{bmatrix} 4 \\ 9 \\ 5 \end{bmatrix} \overset{5R_2 - 3R_1}{\cong} \begin{bmatrix} 15 & 9 & 21 \\ 0 & 121 & -11 \\ 7 & 2 & 10 \end{bmatrix} \begin{bmatrix} 12 \\ 33 \\ 5 \end{bmatrix}$$

$$\underset{\frac{1}{11}R_1}{\overset{5R_3-\frac{7}{8}R_1}{\cong}}\begin{bmatrix}35 & 21 & 49\\ 0 & 11 & -1\\ 0 & 0 & 0\end{bmatrix}\begin{bmatrix}4\\3\\0\end{bmatrix}\overset{\frac{1}{7}R_1}{\cong}\begin{bmatrix}5 & 3 & 7\\ 0 & 11 & -1\\ 0 & 0 & 0\end{bmatrix}\begin{bmatrix}4\\3\\0\end{bmatrix}.$$

Thus rank of matrix $A, \rho(A) = 2 = \rho(A/B)$, rank of $(A|B)$ and hence the equations are consistent. Also the equivalent system of equations is

$$5x + 3y + 7z = 4,\ \ 11y - z = 3$$

$$\Rightarrow\ y = \frac{3}{11} + \frac{k}{11} \text{ and } x = \frac{7}{11} - \frac{16}{11}k, z = k, \text{ k is being a parameter.}$$

Here $x = \frac{7}{11}$, $y = \frac{3}{11}$, $z = 0$ is a particular solution.

Example 3.5.2: Investigate the values of λ and μ for which the equations

$$2x + 3y + 5z = 9,\ \ 7x + 3y - 2z = 8,\ \ 2x + 3y + \lambda z = \mu,$$

have (*i*) no solution (*ii*) a unique solution and (*iii*) infinitely many solutions.

Solution: The given system of equations is

$$\begin{bmatrix}2 & 3 & 5\\ 7 & 3 & -2\\ 2 & 3 & \lambda\end{bmatrix}\begin{bmatrix}x\\y\\z\end{bmatrix} = \begin{bmatrix}9\\8\\\mu\end{bmatrix}$$

or $$AX = B.$$

Now

$$[A|B] = \left[\begin{array}{ccc|c}2 & 3 & 5 & 9\\ 7 & 3 & -2 & 8\\ 2 & 3 & \lambda & \mu\end{array}\right]\underset{R_3-R_1}{\overset{R_2-\frac{7}{2}R1}{\cong}}\left[\begin{array}{ccc|c}2 & 2 & 5 & 9\\ 0 & \frac{-15}{2} & \frac{-39}{2} & \frac{-47}{2}\\ 0 & 0 & \lambda-5 & \mu-9\end{array}\right]$$

$$\overset{-2R_2}{\cong}\left[\begin{array}{ccc|c}2 & 3 & 5 & 9\\ 0 & 15 & 39 & 47\\ 0 & 0 & \lambda\text{-}5 & \mu-9\end{array}\right]$$

Following cases arise:

(i) If $\lambda \neq 5$ and μ may have any value, then $\rho(A) = 3 = \rho(A|B)$ = number of variables. Therefore the system of equations is consistent and has unique solution in this case.

(ii) If $\lambda = 5$, and $\mu \neq 9$, then $\rho(A) = 2 \neq 3 = \rho(A|B)$, the system of the equations is inconsistent and has no solution.

(iii) If $\lambda = 5$, and $\mu = 9$, then $\rho(A) = 2 = \rho(A|B) < 3$ = number of variables. The system of equations is consistent and has an infinite number of solutions in this case.

3.6 Direct methods of solutions

3.6.1 Cramer's rule

This method is firstly developed and used by a Swiss mathematician Gabriel Cramer (1704-1752) and hence it is known as Cramer's rule after his name. In order to explain the method we consider a system of three equations as

$$\begin{aligned} a_1x+b_1y+c_1z &= d_1 \\ a_2x+b_2y+c_2z &= d_2 \\ a_3x+b_3y+c_3z &= d_3 \end{aligned} \tag{3.6.1}$$

We calculate the determinants Δ, Δ_1, Δ_2 and Δ_3, respectively of coefficient matrix and the coefficient matrices in which first, second and third columns are replaced with the column vector $\begin{bmatrix} d_1 \\ d_2 \\ d_3 \end{bmatrix}$. Thus, we have

$$\Delta = \begin{vmatrix} a_1 & b_1 & c_1 \\ a_2 & b_2 & c_2 \\ a_3 & b_3 & c_3 \end{vmatrix}, \quad \Delta_1 = \begin{vmatrix} d_1 & b_1 & c_1 \\ d_2 & b_2 & c_2 \\ d_3 & b_3 & c_3 \end{vmatrix},$$

$$\Delta_2 = \begin{vmatrix} a_1 & d_1 & c_1 \\ a_2 & d_2 & c_2 \\ a_3 & d_3 & c_3 \end{vmatrix}, \quad \Delta_3 = \begin{vmatrix} a_1 & b_1 & d_1 \\ a_2 & b_2 & d_2 \\ a_3 & b_3 & d_3 \end{vmatrix}$$

Then solution of equations (3.6.1) is given by Cramer's rule as

$$x = \frac{\Delta_1}{\Delta}, \quad y = \frac{\Delta_2}{\Delta}, \quad z = \frac{\Delta_3}{\Delta}$$

provided $\Delta \neq 0$. If $\Delta = 0$, then Cramer's rule fails. This method is quite general but involves a lot of labour when the number of equations exceeds four. For a 10×10 system, it requires, about 7×10^7 multiplications and as such, the Cramer's rule is not at all suitable for large system of equations.

Example 3.6.1 Apply Cramer's rule to solve the equations

$$3x+y+2z=3,\ 2x-3y-=-3,\ x-2y+z=4$$

$$\Delta = \begin{vmatrix} 3 & 1 & 2 \\ 2 & -3 & -1 \\ 1 & -2 & 1 \end{vmatrix} = 8,$$

$$\Delta_1 = \begin{vmatrix} 3 & 1 & 2 \\ -3 & -3 & -1 \\ 4 & -2 & 1 \end{vmatrix} = 8,$$

$$\Delta_2 = \begin{vmatrix} 3 & 1 & 2 \\ 2 & -3 & -1 \\ 4 & 4 & 1 \end{vmatrix} = 6,$$

$$\Delta_3 = \begin{vmatrix} 3 & 1 & 3 \\ 2 & -3 & -3 \\ 1 & -2 & 4 \end{vmatrix} = -8,$$

$\therefore$ By Cramer's rule, we have

$$x = \frac{\Delta_1}{\Delta} = \frac{8}{8} = 1, \quad y = \frac{\Delta_2}{\Delta} = \frac{16}{8} = 2, \quad z = \frac{\Delta_3}{\Delta} = \frac{-8}{8} = -1,$$

Hence the required solution is $x = 1, y = 2, z = -1$.

3.6.2 Matrix inversion method

Consider the equations

$$\begin{aligned} a_{21}x_1 + a_{22}x_2 + a_{23}x_3 + \ldots + a_{2n}x_n &= b_2 \\ a_{21}x_1 + a_{22}x_2 + a_{23}x_3 + \ldots + a_{2n}x_n &= b_2 \\ a_{31}x_1 + a_{32}x_2 + a_{33}x_3 + \ldots + a_{3n}x_n &= b_3 \\ \vdots \qquad \vdots \qquad \vdots \qquad & \vdots \quad \vdots \\ a_{n1}x_1 + a_{n2}x_2 + a_{n3}x_3 + \ldots + a_{nn}x_n &= b_n \end{aligned} \qquad (3.6.2)$$

i.e. $$AX = B$$

$\Rightarrow$ $$A^{-1}\, AX = A^{-1}B$$

or $$IX = A^{-1}B$$

or $$X = A^{-1}B$$

i.e. $$\begin{bmatrix} x_1 \\ x_2 \\ x_3 \\ \\ x_n \end{bmatrix} = \frac{1}{|A|} \begin{bmatrix} c_{11} & c_{21} & \cdots & c_{n1} \\ c_{12} & c_{22} & \cdots & c_{n2} \\ c_{13} & c_{23} & \cdots & c_{n3} \\ \cdots & \cdots & \cdots & \cdots \\ c_{1n} & c_{2n} & \cdots & c_{nn} \end{bmatrix} \begin{bmatrix} b_1 \\ b_2 \\ b_3 \\ . \\ b_n \end{bmatrix} \qquad (3.6.3)$$

where c_{ij} are the cofactors of a_{ij} in A and $|A|$ is the determinant of A. Equation (3.6.3) on equating the corresponding elements in the product on the right side leads to the solution of equations (3.6.2). This method fails when A is a singular matrix i.e. $|A| = 0$. Although this method is quite general but it is not suitable for large systems because evaluation of A^{-1} by cofactors becomes very cumbersome. We shall now discuss some methods, which can be applied to any number of equations.

3.6.3 Gauss elimination method

Complete pivoting. It is more complicated and in this method the unknowns are eliminated successively and the system of equations is reduced to an upper triangular system from which the variables are determined by back substitution. This method is well suited for computer operations and is quite frequent in use.

Definition 3.6.1 Two linear systems $AX = B$ and $\widetilde{A}X = \widetilde{B}$ of equations are said to be equivalent if any solution of one is a solution of other.

Theorem 3.6.1 Let $AX = B$ be a linear non-homogenous system of n-equations Suppose we subject this system to the system of following operations.

(i) Multiplication of one equation by a non-zero constant.

(ii) Addition of a multiple of one equation to another equation.

(iii) Interchange of two equations.

If this sequence of operations produces the new system $\widetilde{A}X = \widetilde{B}$, the both the systems $AX = B$ and $\widetilde{A}X = \widetilde{B}$ are equivalent. In particular, then A is invertible iff $\widetilde{A}$ is invertible. We adopt this theorem without proof but the elimination process is based on this. We explain it by considering a system of three equations as

$$\begin{aligned} a_{11}x_1 + a_{12}x_2 + a_{13}x_3 &= a_{14} \\ a_{21}x_1 + a_{22}x_2 + a_{23}x_3 &= a_{24} \\ a_{31}x_1 + a_{32}x_2 + a_{33}x_3 &= b_{34} \end{aligned} \tag{3.6.4}$$

or

$$AX = B$$

where

$$A = \begin{vmatrix} a_{11} & b_{12} & c_{13} \\ a_{21} & b_{22} & c_{23} \\ a_{31} & b_{32} & c_{33} \end{vmatrix}, \quad B = \begin{bmatrix} a_{14} \\ a_{24} \\ a_{34} \end{bmatrix}, \quad X = \begin{bmatrix} x_1 \\ x_2 \\ x_3 \end{bmatrix} \tag{3.6.5}$$

Elimination process

Here, we have

$$[A|B] = \begin{bmatrix} a_{11} & a_{12} & a_{13} \\ a_{21} & a_{22} & a_{23} \\ a_{31} & a_{32} & a_{33} \end{bmatrix} \begin{bmatrix} a_{14} \\ a_{24} \\ a_{34} \end{bmatrix} \underset{R_3 - \frac{a_{31}}{a_{11}}R_1}{\overset{R_1 - \frac{a_{21}}{a_{11}}R_1}{\cong}} \left[\begin{array}{ccc|c} a'_{11} & a'_{12} & a'_{13} & a'_{14} \\ 0 & a'_{22} & a'_{23} & a'_{24} \\ 0 & a_{32} & a'_{33} & a'_{34} \end{array}\right] \tag{3.6.6}$$

where $a'_{ij} = a_{ij} - \dfrac{a_{ik}}{a_{kk}} a_{kj}$, k = 1 , j = 2,3,4, i = 2,3 and $a'_{ij} = a_{ij}$, j = 1,2,3,4

$$R_3 - \frac{a'_{32}}{a'_{22}} R_2 \cong \begin{bmatrix} a''_{11} & a''_{12} & a''_{13} & a''_{14} \\ 0 & a''_{22} & a''_{23} & a''_{24} \\ 0 & 0 & a''_{33} & a''_{34} \end{bmatrix} \tag{3.6.7}$$

where $a''_{ij} = a'_{ij} - \frac{a'_{ik}}{a'_{kk}} a'_{kj}$, k = 2, i = 3, j = 3,4 $a''_{ij} = a'_{ij}$, j = 1,2,3.

Thus the reduced system of equation is

$$\begin{aligned} a''_{11}x_1 + a''_{12}x_2 + a''_{13}x_3 &= a''_{14} \\ a''_{22}x_2 + a''_{23}x_3 &= a''_{24} \\ a''_{33}x_3 &= a''_{34} \end{aligned} \tag{3.6.8}$$

Back substitution process

From third equation, we get

$$x_3 = a''_{34} / a''_{33}$$

Using this value of x_3 in the second equation of (3.6.8), we get

$$x_2 = \left[a''_{24} - a''_{23} \frac{a''_{34}}{a''_{33}} \right] / a''_{22}$$

Upon using x_2 and x_3 in the first equation, we get

$$x_3 = \left[a''_{14} - \frac{a''_{12}}{a''_{22}} \left(a''_{24} - \frac{a''_{23}a''_{34}}{a''_{33}} \right) - \frac{a''_{13}a''_{34}}{a''_{33}} \right] \Big/ a''_{11}$$

Here the elements a'_{11}, a''_{22} and a''_{33} are called pivot elements and the corresponding equations are called pivotal equations. The Gauss elimination process derives from a given linear system $AX = B$ of order n, a sequence of equivalent system $A^{(k)}X = B^{(k)}$, $k = 0,1,2,3,\ldots,(n-1)$. Here $A^{(0)}X = B^{(0)}$ is just the original (given) system. The ith system has the form

$$\begin{aligned} a_{11}^{(i)}x_1 + a_{12}^{(i)}x_2 + \ldots + a_{1i}^{(i)}x_i + a_{1(i+1)}^{(i)}x_{i+1} + \ldots + a_{1n}^{(i)}x_n &= b_1^{(i)} \\ a_{22}^{(i)}x_2 + \ldots + a_{2i}^{(i)}x_i + a_{2(i+1)}^{(i)}x_{i+1} + \ldots + a_{2n}^{(i)}x_n &= b_2^{(i)} \\ \vdots \quad \vdots \quad \vdots \quad \vdots \quad \vdots \quad \vdots \quad \vdots & \\ a_{ii}^{(i)}x_i + a_{i,i+1}^{(i)}x_{i+1} + \ldots + a_{in}^{(i)}x_n &= b_i^{(i)} \\ a_{i+1,i+1}^{(i)}x_{i+1} + \ldots + a_{i+1n}^{(i)}x_n &= b_{i+1}^{(i)} \\ \vdots \quad \vdots \quad \vdots \quad \vdots \quad \vdots & \\ a_{n,i+1}^{(i)}x_{i+1} + \ldots + a_{nn}^{(i)}x_n &= b_n^{(i)} \end{aligned}$$

This shows that the first $(i+1)$ equations are already in upper triangular form, while the last $n-(i+1)$ equations involves only the unknowns $x_{i+1}, x_{i+2}, \ldots x_n$. from this, the $(i+1)$ system $A^{(i+1)}X = B^{(i+1)}$ is derived during $(i+1)$ step of Gauss elimination as follows:

The first $(i+1)$ equations are left unchanged and if the coefficient $a_{i+1,i+1}$ of x_{i+1} in $(i+1)$ equation is non-zero, then $m_{k(i+1)} = a_{k,i+1}/a_{i+1,i+1}$ times equations $(i+1)$, is equation k, $k = i+2, i+3 \cdots n$. the resulting system $A^{(i+1)}X = B^{(i+1)}$ is clearly equivalent to $A^{(i)}X = B^{(i)}$, hence by induction to the original system, further the $(i+1)$ system has its first $(i+2)$ equations in upper triangular form. After $(n-1)$ steps of this procedure, one arrives at the system $A^{(n-1)}X = B^{(n-1)}$, whose coefficients matrix is upper triangular, so that this system can now be solved quickly by back substitution as discussed below.

In general, for a particular system of n-equations in n-unknown of the type

$$AX = B \tag{3.6.9}$$

where A is a square matrix of order n, the solution vector $X = (x_1, x_2, x_3, \ldots, x_n)$ can be obtained without difficulty in case the matrix A is reduced to a upper triangular with all diagonal entries non-zero. Then the system (3.6.9) has the form

$$\begin{aligned}
a_{11}x_1 + a_{12}x_2 + a_{13}x_3 + \ldots + a_{1n}x_n &= b_1 \\
a_{22}x_2 + a_{23}x_3 + \ldots + a_{2n}x_n &= b_2 \\
a_{33}x_3 + \ldots + a_{3n}x_n &= b_3 \\
\vdots \quad \vdots \quad \vdots \quad \vdots \quad & \vdots \\
a_{n-1,n-1}x_{n-1} + \ldots + a_{(n-1)n}x_n &= b_{n-1} \\
a_{nn}x_n &= b_n
\end{aligned} \tag{3.6.10}$$

In particular the last equation involves only x_n, we have $x_n = b_n/a_{nn}$, since $a_{nn} \neq 0$. Since we know x_n, the second last equation gives us

$x_{n-1} = (b_{n-1} - a_{n-1,n}x_n)/a_{n-1,n-1}$, $a_{n-1,n-1} \neq 0$. With x_n and x_{n-1}, now determined, the third last equation leads to $x_{n-2} = (b_{n-2} - a_{n-2,n-1}x_{n-1} - a_{n-2,n}x_n)/a_{n-2,n-2}$, $a_{n-2,n-2} \neq 0$ In general $x_{k+1}, x_{k+2}, \ldots x_n$ already computed, the kth equation can be uniquely solved for x_k, since $a_{kk} \neq 0$ to give $x_k = \left(b_k - \sum_{j=k+1}^{n} a_{kj}x_j\right)\Big/a_{kk}$ This process of determine the solution of system (3.6.9) is called back substitution.

This method fails if any one of the pivots $a_{kk}^{(i)}, k = 1,2,3,\ldots n$ becomes zero. In such cases, we rewrite the equations in a different order so that the pivots are non-zero. In the first step, the numerically largest coefficient of x_1 is chosen from all the equations and brought as the first pivot by interchanging the first equation with the one having the largest coefficient of x_1. In the second step, the numerically largest coefficient of x_2 is chosen from the remaining equations except the first equation and brought as the second pivot by interchanging the second equation with the equation having the largest coefficient of x_2 and so on. This process is continued till we arrive at the equation with the single variable. This modified procedure is called *pivoting*. If we are not interested about the elimination of x_1, x_2, x_3 in a specified order, then we can choose at each stage the numerically largest coefficient of the entire matrix of coefficients. This required the interchange of both of the equation as well as the position of the variables. This method of elimination is called complete pivoting. It is more complicated and does not appreciably improve the accuracy.

Example: 3.6.2 Apply Gauss elimination method to solve the equations

$$x+4y-z=-5,\ x+y-6z=-12,\ 3x-y-z=4.$$

Solution: The given system of equations is rewritten a

$$\begin{bmatrix}1 & 4 & -1\\ 1 & 1 & -6\\ 3 & -1 & -1\end{bmatrix}\begin{bmatrix}x\\ y\\ z\end{bmatrix}=\begin{bmatrix}-5\\ -12\\ 4\end{bmatrix},\ \text{i. e.}\quad AX = B.$$

$$\text{Now } [A|B]=\left[\begin{array}{ccc|c}1 & 4 & -1 & -5\\ 1 & 1 & -6 & -12\\ 3 & -1 & -1 & 4\end{array}\right]\begin{array}{c}R_2-R_1\\ \simeq\\ R_3-3R_1\end{array}\left[\begin{array}{ccc|c}1 & 4 & -1 & -5\\ 0 & -3 & -5 & -7\\ 0 & -13 & 2 & 19\end{array}\right]\begin{array}{c}R_3-\frac{13}{3}R_2\\ \simeq\end{array}\left[\begin{array}{ccc|c}1 & 4 & -1 & -5\\ 0 & -3 & -5 & -7\\ 0 & 0 & \frac{71}{3} & \frac{148}{3}\end{array}\right]$$

The reduced system of the equations is

$$x+4y-z=-5$$
$$3y+5z=7$$
$$\frac{71}{3}z=\frac{148}{3}$$

By back substitution, we get

$$z=\frac{148}{71},\quad y=\frac{7}{3}-\frac{5}{3}\left(\frac{148}{71}\right)=-\frac{81}{71},\quad x=-5-4\left(\frac{-81}{71}\right)+\frac{148}{71}=\frac{117}{71}$$

Hence the required solution is

$$x=\frac{117}{71},\quad y=-\frac{81}{71},\quad z=\frac{148}{71.}.$$

3.6.4 Gauss-Jordan method

This method is a modification of the Gauss elimination method. In this method the elimination of variables is performed by reducing the coefficient's matrix of the system of equations into normal/diagonal form by using elementary row operations from the resulting system of equations in which each equation involves only one unknown, the values of the variables can be obtained readily. Thus in this method, the labour of back substitution for finding the unknowns is saved at the cost of additional calculations. Here, in order to explain this method; we again consider the system of equations (3.6.4) so that we have

$$[A|B]=\left[\begin{array}{ccc|c} a_{11} & a_{12} & a_{13} & a_{14} \\ a_{21} & a_{22} & a_{23} & a_{24} \\ a_{31} & a_{32} & a_{33} & a_{34} \end{array}\right] \simeq \left[\begin{array}{ccc|c} a'_{11} & a'_{12} & a'_{13} & a'_{14} \\ 0 & a'_{22} & a'_{23} & a'_{24} \\ 0 & 0 & a'_{33} & a'_{34} \end{array}\right], \text{ on operating } \begin{pmatrix} R_2 - \dfrac{a_{21}}{a_{11}} R_1 \\ R_3 - \dfrac{a_{31}}{a_{11}} R_1 \end{pmatrix}$$

where $a'_{ij} = a_{ij} - \dfrac{a_{ik}}{a_{kk}} a_{kj}, k = 1, j = 2,3,4$ and $a'_{ij} = a_{ij}, j = 1,2,3$

$$\simeq \left[\begin{array}{ccc|c} a''_{11} & 0 & a''_{13} & a''_{14} \\ 0 & a''_{22} & a''_{23} & a''_{24} \\ 0 & 0 & a''_{33} & a''_{34} \end{array}\right], \text{ on operating } \begin{pmatrix} R_1 - \dfrac{a'_{12}}{a'_{22}} R_2 \\ R_3 - \dfrac{a'_{32}}{a'_{22}} R_2 \end{pmatrix}$$

where $\quad a''_{ij} = a'_{ij} - \dfrac{a'_{ik}}{a'_{kk}} a'_{kj}, k = 2, i = 1,3, j = 1,3,4$

$$\simeq \left[\begin{array}{ccc|c} a'''_{11} & 0 & 0 & a'''_{14} \\ 0 & a'''_{22} & 0 & a'''_{24} \\ 0 & 0 & a'''_{33} & a'''_{34} \end{array}\right], \text{ on operating } \left(R_1 - \dfrac{a''_{13}}{a''_{33}} R_3, R_2 - \dfrac{a''_{23}}{a''_{33}} R_3 \right)$$

where $a'''_{ij} = a''_{ij} - \dfrac{a''_{ik}}{a''_{kk}} a''_{kj} \quad , k = 3; i = 1,2; j = 1,2,4$

this implies that $x_1 = a'''_{14} / a'''_{11}, \;\; x_2 = a'''_{24} / a'''_{22}, \; x_3 = a'''_{34} / a'''_{33}$ which is the required solution of the system (3.6.4). The method can also be explained with the help of following example.

Example: 3.6.3 Apply Gauss – Jordan method to solve the equations

$$x + y + z = 9, \;\; 2x - 3y + 4z = 13, \;\; 3x + 4y + 5z = 40.$$

Solution: The given system of the equations is given by

$$\begin{bmatrix} 1 & 1 & 1 \\ 2 & -3 & 4 \\ 3 & 4 & 5 \end{bmatrix} \begin{bmatrix} x \\ y \\ z \end{bmatrix} = \begin{bmatrix} 9 \\ 13 \\ 40 \end{bmatrix},$$

i.e., $AX = B$

Now $$[A|B]=\left[\begin{array}{ccc|c} 1 & 1 & 1 & 9 \\ 2 & -3 & 4 & 13 \\ 3 & 4 & 5 & 40 \end{array}\right] \begin{array}{c} R_2 - 2R_1 \\ \simeq \\ R_3 - 3R_1 \end{array} \left[\begin{array}{ccc|c} 1 & 1 & 1 & 9 \\ 0 & -5 & 2 & -5 \\ 0 & 1 & 2 & 13 \end{array}\right]$$

$$\begin{matrix} R_1+\frac{1}{5}R_2 \\ \simeq \\ R_3+\frac{1}{5}R_2 \end{matrix}\left[\begin{array}{ccc|c} 1 & 0 & \frac{7}{5} & 8 \\ 0 & -5 & 2 & -5 \\ 0 & 0 & \frac{12}{5} & 12 \end{array}\right] \quad \begin{matrix} \frac{5}{12}R_3 \\ \simeq \\ -\frac{1}{5}R_2 \end{matrix} \left[\begin{array}{ccc|c} 1 & 0 & \frac{7}{5} & 8 \\ 0 & 1 & -\frac{2}{5} & -5 \\ 0 & 0 & 1 & 12 \end{array}\right]$$

$$\begin{matrix} R_1-\frac{7}{5}R_3 \\ \simeq \\ R_2+\frac{2}{5}R_3 \end{matrix}\left[\begin{array}{ccc|c} 1 & 0 & 0 & 1 \\ 0 & 1 & 0 & 3 \\ 0 & 0 & 1 & 5 \end{array}\right]$$

This resulting system of equations implies that $x=1,\ \ y=3,\ \ z=5$. This is the required solution.

For a system of 10 equations, the number of multiplication's required for Gauss-Jordan method is about 500 where as for Gauss elimination method we need only about 3 multiplication. This shows that though Gauss-Jordan method appears to be easier but requires approximately 50% more operations than the Gauss elimination method. As such Gauss elimination method is preferred for large system of equations.

Algorithm: 6.3.1 Gauss elimination method

Elimination: Given the $n \times (n+1)$ matrix $W=[A|B]$ corresponding to the system of equations (3.6.9) *viz* AX= B.

1. Initialize the n-vector $\vec{p}$ to have $p_i = i,\ i=1,2,3,...n.$
2. For $k=1,2,3,...,n-1,$ do: in steps 3 to 9
3. Find (the smallest) $i \geq k$ such that $w_{ik} \neq 0$.
4. If no such i exit, signal that A is inevitable and stop.
5. Otherwise, exchange the contents of p_k and p_i and of row k and i of W and continue
6. For $i = k+1, k+2,..., n,$ do: in steps 7 to 9
7. $m = w_{ik} - w_{ik}/w_{kk}$
8. for $j = k+1, k+2,..., n+1,$ do: in steps 8 to 9
9. $w_{ij} = w_{ij} - m\ \ w_{kj}$
10. If $w=0$, signal that A is not inevitable and stop.
11. Otherwise, the original system $AX = B$ is now known to be equivalent to the system $UX = Y$, where U and Y are given in terms of the final entries of W by

$$u_{ij} = \begin{Bmatrix} w_{ij} & i \leq j \\ 0 & i > j \end{Bmatrix}, \quad y_i = w_{i,\ n+1}, i = 1,2,3...n.$$

In particular, U is an upper-triangular matrix with all diagonal entries non-zero, and algorithm for back substitution can be used to calculate the solution X.

Back substitution:

Given the upper triangular matrix A of order n with none of the diagonal entries equal to zero and the n-vector B, the entries $x_n, x_{n-1}, x_{n-2} \ldots x_2, x_1$ of the solution vector X of $AX = B$ can then be computed (in that order) by

for $k = n, n-1, \ldots 2, 1$ do

$$x_k = \left(b_k - \sum_{j=k+1}^{n} a_{kj} x_j \right) / a_{kk}$$

where for $x = n$, the summation is interpreted as the sum over no terms and gives by convention, the value 0.

Tridiagonal system of equations

It is often possible to reduce the computational work for solving $AX = B$ by taking into account special features of the coefficient matrix A such as symmetry and sparseness. We say that then matrix $A = (a_{ij})$ of the order n is tridiagonal if $a_{ij} = 0$, wherever $|i - j| > 1$. In the other words, A is tridiagonal if the only non zero entries of A lie on the diagonal of $A, i = 1,2,3,\ldots n$ or the sub-diagonal of A, $a_{i,i+1}, i = 2,3,\ldots n$ or the super-diagonal of A, $a_{i,i+1}, i = 1,2,\ldots n-1$. for example the matrices

$$\begin{bmatrix} 3 & 1 & 0 & 0 \\ 1 & 3 & 1 & 0 \\ 0 & 2 & 2 & 1 \\ 0 & 0 & 1 & 6 \end{bmatrix}, \begin{bmatrix} 1 & 0 & 0 & 0 \\ 0 & 1 & 1 & 0 \\ 0 & 0 & 1 & 0 \\ 0 & 0 & 0 & 0 \end{bmatrix}, \begin{bmatrix} 4 & 7 & 0 & 0 \\ 0 & 8 & 1 & 0 \\ 0 & 1 & 1 & 0 \\ 0 & 0 & 0 & 1 \end{bmatrix} \text{ are tridiagonal.}$$

More specifically, we now consider the following tri-diagonal system of order n as.

$$\begin{aligned} d_1 x_1 + c_1 x_2 &= b_1 \\ a_2 x_1 + d_2 x_2 + c_2 x_3 &= b_2 \\ a_3 x_2 + d_3 x_3 + c_3 x_4 &= b_3 \\ \vdots \qquad \vdots \qquad \vdots & \\ a_{n-1} x_{n-2} + d_{-1} x_{n-1} + c_{n-1} x_n &= b_{n-1} \\ a_n x_{n-1} + d_n x_n &= b_n \end{aligned}$$

Assuming $d_1 \neq 0$, we eliminate x_1 from second equation, to get the new equation

$$d_2' x_2 + c_2 x_3 = b_2'$$

where $d_2' = d_2 - \dfrac{a_2}{d_1} c_1$, $b_2' = b_2 - \dfrac{a_2}{d_1} b_1$.

Next assuming $d_2' \neq 0$, we use this equation to eliminate x_2 from the third equation, get the new equation $d_3' x_3 + c_3 x_4 = b_3'$ with $d_3' = d_3 - \dfrac{a_3}{d_2^1} c_2$, $b_3' = b_3 - \dfrac{a_3}{d_2^1} b_2'$.

Continuing in this way, we eliminate, during step k, x_k from the $(k-1)$st equations (assuming $d'_k \neq 0$) and get

$$d'_{k+1}\, x_{k+1} + c_{k+1}\, x_{k+2} = b'_{k+1}$$

with $d'_{k+1} = d_{x+1} - \dfrac{a_{X+1}}{d^1_k} c_k, b'_{k+1} = b_{k+1} - \dfrac{a_{x+1}}{d^1_k} b'_k,$ for $k = 1,2,3,\ldots,n-1$

During back substitution, we first get, assuming $d'_n \neq 0,$ $x_n = b'_n / d'_n$ and then for $k = n-1, n-2, \ldots, 2, 1$, we get

$$x_k = \frac{b'_x - c_k x_{k+1}}{d^1_k}.$$

Algorithm 3.6.2: Given the coefficients a_i, d_i, c_i and then right hand side b_i of the tri-diagonal system $a_i x_{i-1} + d_i x_i + c_i x_{i+1} = b_i, i = 1,2,\ldots n$ with $a_i = c_n = 0$

1. For $k = 2,3,\ldots.n,$ do: steps 2 to 5
2. If $d_{k-1} = 0,$ signal failure and stop
3. Otherwise, $m = \dfrac{a_k}{d_{k-1}}$ and continue.
4. $d_k = d_k - m * c_{k-1}$
5. $b_k = b_k - m * b_{k-1}$
6. If $d_n = 0,$ signal failure and stop.
7. Otherwise $x_n = \dfrac{b_n}{d_n}$ and continue.
8. For $k = n-1, n-2, \ldots 2,1$ do: step 9
9. $x_k = \dfrac{b_k - c_k * x_{k+1}}{d_k}$

3.6.5 Crout's (L U) method

This method was discovered by *P.D.* Crout of USA in 1941 and is based on the fact that every square matrix *A* can be expressed as the product of a lower triangular matrix and an upper triangular matrix, provided all the principal minors of *A* are non-singular. Also such a factorization, if exists, is unique. This method was also used by the French engineer *A. L.* Cholesky's and for that reason it is also knows as ***Cholesky's*** method.

Consider the system of equations

$$a_{11}x_1 + a_{12}x_2 + a_{13}x_3 = a_{14}$$
$$a_{21}x_1 + a_{22}x_2 + a_{23}x_3 = a_{24}$$
$$a_{31}x_1 + a_{32}x_2 + a_{33}x_3 = a_{34}$$

This can be rewritten as $AX = B$

where

$$A = \begin{bmatrix} a_{11} & a_{12} & a_{13} \\ a_{21} & a_{22} & a_{23} \\ a_{31} & a_{32} & a_{33} \end{bmatrix}, \quad B = \begin{bmatrix} a_{14} \\ a_{24} \\ a_{34} \end{bmatrix}, \quad X = \begin{bmatrix} x_1 \\ x_2 \\ x_3 \end{bmatrix}.$$

Let us suppose that

$$A = LU = \begin{bmatrix} 1 & 0 & 0 \\ l_{21} & 1 & 0 \\ l_{31} & l_{32} & 1 \end{bmatrix} \begin{bmatrix} u_{11} & u_{12} & u_{13} \\ 0 & u_{22} & u_{23} \\ 0 & 0 & u_{33} \end{bmatrix}$$

$$\Rightarrow \quad \begin{bmatrix} a_{11} & a_{12} & a_{13} \\ a_{21} & a_{22} & a_{23} \\ a_{31} & a_{32} & a_{33} \end{bmatrix} = \begin{bmatrix} u_{11} & u_{12} & u_{13} \\ l_{21}u_{11} & l_{21}u_{12} + u_{22} & l_{21}u_{13} + u_{23} \\ l_{31}u_{11} & l_{31}u_{12} + l_{32}u_{22} & l_{31}u_{13} + l_{32}u_{23} + u_{33} \end{bmatrix}$$

This provides us

$$u_{11} = a_{11} \qquad u_{12} = a_{12} \qquad u_{13} = a_{13}$$

$$l_{21} = \frac{a_{21}}{a_{11}} \qquad u_{22} = a_{22} - \frac{a_{21}}{a_{11}} a_{12} \qquad u_{23} = a_{23} - \frac{a_{21}}{a_{11}} a_{13}$$

$$l_{31} = \frac{a_{31}}{a_{11}} \qquad l_{32} = \left(a_{32} - \frac{a_{31}}{a_{11}} a_{12} \right) \Big/ \left(a_{22} - \frac{a_{21}}{a_{11}} a_{12} \right)$$

$$u_{33} = a_{33} - \frac{a_{31}}{a_{11}} a_{13} - l_{32} u_{23}.$$

Thus the matrices L and U become known. Now $AX = B$ becomes

$$LUX = B \Rightarrow \quad \text{LY} = \text{B, where Y} = \text{UX}.$$

From $LY = B$, we compute the column vector Y which is then used in $UX = Y$ to obtain the solution X. Also it is quite easy to compute A^{-1} from $A = LU$ which implies that $A^{-1} = U^{-1}L^{-1}$.

This method is often used to find the solution of system of linear equations and for finding inverse of a matrix, being a superior to Gauss elimination method. The number of operations involved in terms of multiplication for a system of 10 equations by this method is about one-third to that of Gauss elimination method. This method is also preferred as the software for computers.

Example: 3.6.4 Apply Crout's method to solve the equations

$$3x + 2y + 7z = 4 \ , \quad 2x + 3y + z = 5 \ , \quad 3x + 4y + z = 7$$

Solution : The given system of equations is

$$\begin{bmatrix} 3 & 2 & 7 \\ 2 & 3 & 1 \\ 3 & 4 & 1 \end{bmatrix} \begin{bmatrix} x \\ y \\ z \end{bmatrix} = \begin{bmatrix} 4 \\ 5 \\ 7 \end{bmatrix}$$

i.e., $$AX = B.$$

Let

$$A = LU = \begin{bmatrix} 1 & 0 & 0 \\ l_{21} & 1 & 0 \\ l_{31} & l_{32} & 1 \end{bmatrix} \begin{bmatrix} u_{11} & u_{12} & u_{13} \\ 0 & u_{22} & u_{23} \\ 0 & 0 & u_{33} \end{bmatrix}$$

$$\Rightarrow \quad \begin{bmatrix} 3 & 2 & 7 \\ 2 & 3 & 1 \\ 3 & 4 & 1 \end{bmatrix} = \begin{bmatrix} u_{11} & u_{12} & u_{13} \\ l_{21}u_{11} & l_{21}u_{12} + u_{22} & l_{21}u_{13} + u_{23} \\ l_{31}u_{11} & l_{31}u_{12} + l_{32}u_{22} & l_{31}u_{13} + l_{32}u_{23} + u_{33} \end{bmatrix}$$

$$\Rightarrow \quad u_{11} = 3,\ u_{12} = 2,\ u_{13} = 7$$

$$l_{21} = \frac{2}{3},\ l_{31} = \frac{3}{3} = 1$$

$$u_{22} = 3 - \frac{2}{3}(2) = 3 - \frac{4}{3} = \frac{5}{3},$$

$$u_{23} = 1 - \frac{2}{7}(7) = \frac{3-4}{3} = -\frac{11}{3}$$

$$u_{32} = 3(4 - 1(2))/5 = \frac{3(4-2)}{5} = \frac{6}{5},$$

$$u_{33} = 1 - \frac{6}{5}\frac{(-11)}{3} - 1(7) = 1 + \frac{22}{5} - 7 = -\frac{8}{5}$$

Thus $$A = \begin{bmatrix} 1 & 0 & 0 \\ 2/3 & 1 & 0 \\ 1 & 6/5 & 1 \end{bmatrix} \begin{bmatrix} 3 & 2 & 7 \\ 0 & 5/3 & -11/3 \\ 0 & 0 & -8/5 \end{bmatrix}$$

We write $UX = Y$ which gives us $LY = B$

$$\Rightarrow \quad \begin{bmatrix} 1 & 0 & 0 \\ 2/3 & 1 & 0 \\ 1 & 6/5 & 1 \end{bmatrix} \begin{bmatrix} y_1 \\ y_2 \\ y_3 \end{bmatrix} = \begin{bmatrix} 4 \\ 5 \\ 7 \end{bmatrix}$$

$$\Rightarrow \quad y_1 = 4,\ y_2 = \frac{7}{3},\ y_3 = \frac{1}{5}$$

Hence the original system reduces to

$$\begin{bmatrix} 3 & 2 & 7 \\ 0 & 5/3 & -11/3 \\ 0 & 0 & -8/5 \end{bmatrix} \begin{bmatrix} x \\ y \\ z \end{bmatrix} = \begin{bmatrix} 4 \\ 7/3 \\ 1/5 \end{bmatrix}$$

$$\Rightarrow \quad 3x + 2y + 7z = 4, \quad \frac{5}{3}y - \frac{11}{3}z = \frac{7}{3}, \quad -\frac{8}{5}z = \frac{1}{5}$$

$$\Rightarrow \qquad z = -\frac{1}{8},\ y = \frac{9}{8},\ x = \frac{7}{8}$$

which is the required solution.

3.7 Iterative methods of solution

An iterative method is that in which we start from an initial approximation to the actual solution and obtain better and better approximations from a computation cycle repeated time and again for achieving a desired accuracy. Thus in an iterative method, the amount of computation depends on the degree of accuracy required. For large systems, iterative methods may be faster than the direct methods and even the round off errors in iterative methods are smaller. In fact, iteration is a self-correcting process and any error made at any stage of computation gets automatically corrected in the subsequent steps. Simple iterative method can be devised for systems in which the coefficients of the leading diagonal are large as compared to others. In this section we describe three such methods namely, Jacobi, Gauss- Seidal and Relaxation methods.

3.7.1 Jacobi's method

Consider the system of equations

$$\begin{aligned} a_{11}x_1 + a_{12}x_2 + a_{13}x_3 + \ldots + a_{1n}x_n &= b_1 \\ a_{21}x_1 + a_{22}x_2 + a_{23}x_3 + \ldots + a_{2n}x_n &= b_2 \\ a_{31}x_1 + a_{32}x_2 + a_{33}x_3 + \ldots + a_{3n}x_n &= b_3 \\ \ldots\ldots\ldots\ldots\ldots\ldots\ldots\ldots\ldots \\ \ldots\ldots\ldots\ldots\ldots\ldots\ldots\ldots\ldots \\ a_{n1}x_1 + a_{n2}x_2 + a_{n3}x_3 + \ldots + a_{nn}x_n &= b_n \end{aligned} \tag{3.7.1}$$

The equations (3.7.1) can be rewritten as

$$\begin{aligned} x_1 &= [b_1 - (a_{12}x_2 + a_{13}x_3 + \ldots + a_{1n}x_n)]/a_{11} \\ x_2 &= [b_2 - (a_{21}x_1 + a_{23}x_3 + \ldots + a_{2n}x_n)]/a_{22} \\ x_3 &= [b_3 - (a_{31}x_1 + a_{32}x_2 + \ldots + a_{3n}x_n)]/a_{33} \\ \ldots\ldots\ldots\ldots\ldots\ldots\ldots\ldots\ldots \\ \ldots\ldots\ldots\ldots\ldots\ldots\ldots\ldots\ldots \\ x_n &= [b_n - (a_{n2}x_2 + a_{n3}x_3 + \ldots + a_{nn}x_n)]/a_{nn} \end{aligned} \tag{3.7.2}$$

Let us suppose that $\left(x_1^{(0)}, x_2^{(0)}, x_3^{(0)}, \ldots x_n^{(0)}\right)$ be the set of initial solution. Using this initial solution simultaneously in equations (3.7.2) we obtain a current approximation to the solution as $\left(x_1^{(1)}, x_2^{(1)}, x_3^{(1)}, \ldots\ x_n^{(1)}\right)$ called first approximation. Again using this solution set of approximations in (3.7.2) we obtain second approximation to the solution as $\left(x_1^{(2)}, x_2^{(2)}, x_3^{(2)}, \ldots x_n^{(2)}\right)$ and so on. We continue to repeat the process time and again till the solution become stable to the desired level of accuracy. Thus the procedure will be stopped as and when the condition $\left|x_i^{(k)} - x_i^{(k-1)}\right| < \in$

is satisfied. Here $\in$ is arbitrarily small number to be selected at random depending upon the desired level of accuracy. Suppose this condition is satisfied for $k = m$, then the final required solution of system of equations (3.7.1) will be $x_1 = x_1^{(m)}, x_2 = x_2^{(m)}, x_3 = x_3^{(m)}, \ldots x_n = x_n^{m}$. In the absence of better estimates of $\left(x_1^{(0)}, x_2^{(0)}, x_3^{(0)}, \ldots x_n^{(0)}\right)$, we may take each of these as zero. This method is also known as method of simultaneous displacements and firstly developed and used by **Jacobi** and hence is known as **Jacobi's** method.

Example 3.7.1 Use Jacobi's method to solve the equations

$$\begin{aligned} 20x + y - 2z &= 17 \\ 3x + 20y - z &= -18 \\ 2x - 3y + 20z &= 25 \end{aligned} \tag{3.7.1}$$

These equations can be rewritten as

$$\begin{aligned} x &= (17 - y + 2z)/20 \\ y &= (-18 - 3x + z)/20 \\ z &= (25 - 2x + 3y)/20 \end{aligned} \tag{3.7.2}$$

We take $x^{(0)} = y^{(0)} = z^{(0)} = 0.$

Using these approximations simultaneously in each equation of (3.7.2), we obtain

$$\begin{aligned} x^{(1)} &= 17/20 = 0.85 \\ y^{(1)} &= -18/20 = -0.9 \\ z^{(1)} &= 25/20 = 1.25 \end{aligned}$$

Putting these values in right hand side of (3.7.2), we get

$$\begin{aligned} x^{(2)} &= 1.02 \\ y^{(2)} &= -0.965 \\ z^{(2)} &= 1.1515 \end{aligned}$$

Again using these values in (3.7.2), we obtain

$$\begin{aligned} x^{(3)} &= 1.0134 \\ y^{(3)} &= -0.9954 \\ z^{(3)} &= 1.0032 \end{aligned}$$

Use of these on right hand side of (3.7.2) leads to

$$x^{(4)} = 1.0009, y^{(4)} = -1.0018, z^{(4)} = 0.9993$$

and so on. We obtain

$$x^{(5)} = 1.0000, y^{(5)} = -1.0002, z^{(5)} = 0.9996$$

$$x^{(6)} = 1.0000, y^{(6)} = -1.0000, z^{(6)} = 1.0000$$

$$x^{(7)} = 1.0000, y^{(7)} = -1.0000, z^{(7)} = 1.0000$$

Because the 6th and 7th iterations are same up to four decimal of places, so we can stop the process. Hence the required solution of (3.7.3) is

$$x = 1 \ , \ y = -1 \ , z = 1.$$

3.7.2 Gauss-Seidal method

Consider the system of equations

$$\begin{aligned}
a_{11}x_1 + a_{12}x_2 + a_{13}x_3 + \ldots + a_{1n}x_n &= b_1 \\
a_{21}x_1 + a_{22}x_2 + a_{23}x_3 + \ldots + a_{2n}x_n &= b_2 \\
a_{31}x_1 + a_{32}x_2 + a_{33}x_3 + \ldots + a_{3n}x_n &= b_3 \\
\ldots \quad \ldots \quad \ldots \quad \ldots & \quad \ldots \\
\ldots \quad \ldots \quad \ldots \quad \ldots & \quad \ldots \\
a_{n1}x_1 + a_{n2}x_2 + a_{n3}x_3 + \ldots + a_{nn}x_n &= b_n
\end{aligned}$$

This system of equations can be rewritten as

$$\begin{aligned}
x_1 &= \left(b_1 - \sum_{j=2}^{n} a_{1j}x_j \right) / a_{11} \\
x_2 &= \left(b_2 - \sum_{\substack{j=1 \\ j\neq 2}}^{n} a_{2j}x_j \right) / a_{22} \\
x_3 &= \left(b_3 - \sum_{\substack{j=2 \\ j\neq 3}}^{n} a_{3j}x_j \right) / a_{33} \qquad (3.7.4) \\
\ldots \quad \ldots & \quad \ldots \quad \ldots \\
\ldots \quad \ldots & \quad \ldots \quad \ldots \\
x_n &= \left(b_n - \sum_{j=2}^{n-1} a_{ij}x_j \right) / a_{nn}
\end{aligned}$$

Let us suppose that $\left(x_1^{(0)}, x_2^{(0)}, x_3^{(0)}, \ldots x_n^{(0)}\right)$ be the set of initial approximations to the solution of (3.7.4) and hence of the given equations. Upon using this on the *r.h .s* of first equation of (3.7.4), we get the solution set as $\left(x_1^{(1)}, x_2^{(0)}, x_3^{(0)}, \cdots x_n^{(0)}\right)$. Then using this current set of solution in the second equation of (3.7.4), it modifies to $\left(x_1^{(1)}, x_2^{(1)}, x_3^{(0)}, \cdots x_n^{(0)}\right)$. Upon using this recent set in the third equation we obtain the new set as $\left(x_1^{(1)}, x_2^{(1)}, x_3^{(1)}, x_4^{(0)} \ldots x_n^{(0)}\right)$ and so on. Repeating this procedure with each and every equation successively up to last equation, we obtain the first approximation to the solution as $\left(x_1^{(1)}, x_2^{(1)}, x_3^{(1)}, \ldots x_n^{(1)}\right)$. This completes first-iteration. We repeat the above procedure with the recent modified solution to obtain $\left(x_1^{(2)},x_2^{(2)},x_3^{(2)},\ldots.x_n^{(2)}\right)$ at the end of second iteration. Similarly after the third iteration we get the approximation solution set of (3.7.4) as $\left(x_1^{(3)},x_2^{(3)},x_3^{(3)},\ldots.x_n^{(3)}\right)$ and so on. We continue to repeat this process time and again till the solution become stable to the desired level of accuracy. The process will be stopped when the condition $\left|x_i^{(k)}-x_i^{(k-1)}\right|<\in$ is satisfied Here $\in$ is arbitrarily small number and can be selected at random depending upon the desired level of accuracy. Suppose this requirement is met for $k=N$. Then the final solution set of (3.7.4) and hence of the given system of the equations is $\{x_j=x_j^{(N)}, j=1,2,3,\ldots.n\}$.

The method is also known as method of successive approximations. This method is firstly developed and used by Gauss and Seidal and hence it is called Gauss-Seidal method. Since the most recent approximations of the variables are used while proceeding to the next step, the convergence of the Gauss-Seidal method is twice as fast as in Jacobi's method. The Gauss-Seidal and Jacobi methods converge for any choice of the initial approximations if in each equation of the system, the absolute value of the largest coefficient is greater than the sum of the absolute values of all the remaining coefficients Mathematically,

$$\sum_{\substack{j=1\\ j\neq 1}}^{n}\left|\frac{a_{ij}}{a_{ii}}\right|\le 1, \quad i=1,2,3,\ldots.n.$$

where inequality (<) holds in case of at-least one equation.

Example 3.7.2 Use Gauss-Seidal method to solve the equations

$$\left.\begin{aligned} 10x_1-2x_2-x_3-x_4&=3\\ -2x_1+10x_2-x_3-x_4&=15\\ -x_1-x_2+10x_3-2x_4&=27\\ -x_1-x_2-2x_3+10x_4&=-9 \end{aligned}\right\}\qquad(3.7.4)$$

The equations (3.7.4) can be rewritten as

$$\left.\begin{aligned} x_1&=.3+0.2x_2+0.1x_3+0.1x_4\\ x_2&=1.5+0.2x_1+0.1x_3+0.1x_4\\ x_3&=2.7+0.1x_1+0.1x_3+0.2x_4\\ x_4&=-0.9+0.1x_1+0.1x_3+0.2x_3 \end{aligned}\right\}\qquad(3.7.5)$$

We assume the initial solution as $(0,0,0,0)$. Using this in the first equation of (3.7.5), we get $x_1^{(1)}=0.3$ and solution set becomes $(0.3,0,0,0)$ which on substitution in second equation of (3.7.5) leads to the solution set $(0.3,1.56,0,0)$. This when used in third equation provides us $(0.3,1.56,2886,0)$. Putting this in the last equation of (3.7.5) at the end of the first iteration we have $(0.3,\ 1.56,\ 2.886,\ -0.1368)$. Similarly repeating the procedure with this solution we obtain the $(0.3, 1.56, 2.886, -0.1368)$ new approximation to the solution of (3.7.5) as $(0.8869, 1.9523, 2.9566, -0.0248)$ at the end of second iteration. Repeating this procedure, we obtain

Third iteration: $(0.9836, 1.9899, 2.9924, -0.0042)$

Fourth iteration: $(0.9968, 1.9982, 2.9987, -0.0008)$

Fifth iteration: $(0.9994, 1.9999, 2.9999, -0.0001)$

Sixth iteration: $(0.9999, 1.9999, 2.9999, -0.0001)$

Seventh iteration: $(0.9999, 1.9999, 2.9999, -0.0001)$

Hence the required solution of the equations (3.7.5) and hence of the given system of equations is

$$x_1 = 0.999\ ,\ x_2 = 1.9999,\ x_3 = 2.9999,\ x_4 = -0.0001$$

correct to four decimal places.

3.7.3 Relaxation method

Consider the equations

$$\begin{aligned}
a_{11}x_1 + a_{12}x_2 + a_{13}x_3 + \ldots + a_{1n}x_n &= b_1 \\
a_{21}x_1 + a_{22}x_2 + a_{23}x_3 + \ldots + a_{2n}x_n &= b_2 \\
a_{31}x_1 + a_{32}x_2 + a_{33}x_3 + \ldots + a_{3n}x_n &= b_3 \\
\ldots \qquad \ldots \qquad & \ldots \\
\ldots \qquad \ldots \qquad & \ldots \\
a_{n1}x_1 + a_{n2}x_2 + a_{n3}x_3 + \ldots + a_{nn}x_n &= b_n
\end{aligned} \tag{3.7.6}$$

We define the residuals $R(x_1), R(x_2), R(x_3) \ldots R(x_n)$ as

$$R(x_1) = b_1 - \sum_{j=1}^{n} a_{1j}x_j\ ,$$

$$R(x_2) = b_2 - \sum_{j=1}^{n} a_{2j}x_j\ ,$$

$$R(x_3) = b_3 - \sum_{j=1}^{n} a_{3j}x_j\ ,$$

$$\vdots \qquad \vdots \qquad \vdots$$

$$R(x_n) = b_n - \sum_{j=1}^{n} a_{nj}x_j \tag{3.7.7}$$

To start with we assume that $x_1 = x_2 = x_3 = x_4 = \ldots x_n = 0$ and calculate the initial residuals. Thereafter the residuals are reduced step by step, by giving increments to the variables and we construct the following table of operations for the purpose:

	δRx_1	δRx_2	$\delta Rx_3 \cdots \delta Rx_n$
$\delta x_1 = 1$	$-a_{11}$	$-a_{21}$	$-a_{31} \cdots -a_{n1}$
$\delta x_2 = 1$	$-a_{12}$	$-a_{22}$	$-a_{32} \cdots -a_{n2}$
...	...	...	... ...
...	...	...	... ...
$\delta x_n = 1$	$-a_{1n}$	$-a_{n2}$	$-a_{n3} \cdots -a_{nn}$

It can be noted from equation (3.7.7) that if x_1 is increased by unity keeping $x_j, j = 1,2,3,\ldots,n$ constant, then $R(x_1), R(x_2), R(x_3)\ldots R(x_n)$ decrease by $a_{11}, a_{21}, a_{31} \ldots a_{n1}$ respectively. This is shown in the above table along with the effect on the residuals when $x_2, x_3, x_4 \ldots x_n$ are given unit increments. At each step, the numerically largest residual is reduced to almost zero. To reduce a particular residual, the value of the corresponding variable is changed, e. g. to reduce $R(x_1)$ by p, x_1 should be increased by p/a_{11}. When all the residuals almost reduced to zero, the increments in $x_j, j = 1,2,3,\ldots n$ are added separately to give the desired solution. Relaxation method can be applied successfully only if the diagonal elements of the coefficient matrix dominate the other coefficients in the corresponding row, i.e. if in the equation (3.7.6)

$$|a_{11}| > \sum_{j=2}^{n} |a_{1j}|,\ |a_{22}| > \sum_{\substack{j=2 \\ j\neq 2}}^{n} |a_{2j}|,\ \ldots\ \ldots\ \ldots\ \ldots,\ |a_{nn}| > \sum_{j=1}^{n-1} |a_{nj}|$$

where > sign should be valid for atleast one row.

Example: Apply Relaxation method to solve the equations

$$9x - 2y + z = 50,\ x + 5y - 3z = 18,\ -2x + 2y + 7z = 19.$$

Solution: We define the residuals as

$$R_x = 50 - 9x + 2y - z,\ R_y = 18 - x - 5y + 3z,\ R_z = 19 + 2x - 2y - 7z$$

The operation table is

	δR_x	δR_y	δR_z
$\delta x = 1$	-9	-1	2
$\delta y = 1$	2	-5	-2
$\delta z = 1$	-1	3	-7

The relaxation table is

	R_x	R_y	R_z	
$x = y = z = 0$	50	18	19	
$\delta x = 5$	5	13	29	(i)
$\delta z = 4$	1	25	1	(ii)
$\delta y = 5$	11	0	-9	(iii)
$\delta x = 1$	2	-1	-7	
$\delta z = -1$	3	-4	0	(iv)
$\delta y = -0.8$	1.4	0	1.6	
$\delta z = 0.23$	1.17	0.69	-0.9	
$\delta x = 0.13$	0	0.56	0.17	(v)
$\delta y = 0.112$	0.224	0	-0.054	

Here $\sum \delta x = 6.13, \quad \sum \delta y = 4.31, \quad \sum \delta y = 2.23.$

Thus, $x = 6.13, \; y = 4.31, \; z = 2.23$

This is the required solution.

Note: Initially, to reduce the largest residual 50 we increase x by $\delta x = 5$ (50/9=5.555..~5) and the corresponding residuals are obtained by adding δx multiple of first row in operation table to the corresponding elements of first row of the relaxation table and are shown in (i). Of these $R_z = 29$ is the largest so we increase z by $\delta z = 4$ to get the residuals in the second row (ii). Then $R_y = 25$ is largest and we increase y by $\delta y = 5$ to get third row (iii) and so on tables (i the v) and (v) are obtained. This procedure is repeated time and again till all the residuals almost reduce to zero.

3.7.4 Ill-conditioned system of equations

A linear system of equations is said to be ill-conditioned if small changes in the coefficients of the equations results in large changes in the value of the unknowns. Thus a system is well-conditioned, if small changes in the coefficients of the system also produce small changes in the solution of the system. Ill-conditioning of a system is usually expected when the determinant of the coefficient matrix is small. The coefficient matrix of a ill-conditioned system is called an ill-conditioned matrix. During the process of solving simultaneous equations, we also come across two forms of instabilities:

(i) **Inherent:** Inherent instability of a system is the property of given problem and occurs due to the problem being ill-conditioned; it can be avoided by reformulation of the problem suitably.

(ii) **Induced:** Induced instability occurs because of the incorrect choice of method of solution. The accuracy of an ill-conditioned system can be improved with the help of following iterative procedure. Consider the system of the equations

$$a_{11}x_1 + a_{12}x_2 + \cdots + a_{1n}x_n = b_1$$

$$a_{21}x_1 + a_{22}x_2 + \cdots + a_{2n}x_n = b_2$$

$$\cdots \quad \cdots \quad \cdots \quad \cdots$$

$$\cdots \quad \cdots \quad \cdots \quad \cdots$$

$$a_{n1}x_1 + a_{n2}x_2 + \cdots + a_{nn}x_n = b_n \tag{3.7.8}$$

Let x'_j , $j = 1,2,3,4\cdots,n$ be an approximation to the solution. Let $b'_j, j = 1,2,3,\cdots,n$ be the new values of $b'_j s$ obtained after substituting x'_j on the left hand sides of (3.7.8). Subtracting (3.7.8) and the resulting system on replacing x_j by $x'_j, j = 1,2,3,\cdots n$ in it, we get

$$a_{11}x_1^e + a_{12}x_2^e + \cdots + a_{1n}x_n^e = b_1^e$$

$$a_{21}x_1^e + a_{22}x_2^e + \cdots + a_{2n}x_n^e = b_2^e$$

$$\cdots \quad \cdots \quad \cdots \quad \cdots$$

$$\cdots \quad \cdots \quad \cdots \quad \cdots \tag{3.7.9}$$

$$a_{n1}x_1^e + a_{n2}x_2^e + \cdots + a_{nn}x_n^e = b_n^e$$

where $x_j^e = x_j - x'_j$, $b_j^e = b_j - b'_j$ $j = 1,2,3,\cdots n$. We solve the system (3.7.9) to find x_j^e so that $x_j = x'_j + x_j^e, \forall j$, will be a better approximation for $x_j, j = 1,2,3,\cdots n$. The procedure can be repeated for improving the accuracy.

3.8 Matrix inversion

The inverse of a square matrix is written as A^{-1} so that $AA^{-1} = A^{-1}A = I$. The inverse of a matrix exists if it is a non-singular matrix. Also inverse of a matrix, if it exists, is unique. There are many methods of finding the inverse of a matrix, of which, the method of obtaining inverse with the help of an adjoint requires a lot of calculations. Therefore we shall now discuss some other methods which require less of computational labour and can be easily extended to matrices of higher order.

3.8.1 Gauss elimination method

Consider a 3*3 matrix
$$A = \begin{bmatrix} a_{11} & a_{12} & a_{13} \\ a_{21} & a_{22} & a_{23} \\ a_{31} & a_{32} & a_{33} \end{bmatrix}$$

Suppose A^{-1} exists and $$A^{-1} = \begin{bmatrix} x_1 & x_2 & x_3 \\ y_1 & y_2 & y_3 \\ z_1 & z_2 & z_3 \end{bmatrix}$$

Then, we have $AA^{-1} = I$

$$\Rightarrow \begin{bmatrix} a_{11} & a_{12} & a_{13} \\ a_{21} & a_{22} & a_{23} \\ a_{31} & a_{32} & a_{33} \end{bmatrix} \begin{bmatrix} x_1 & x_2 & x_3 \\ y_1 & y_2 & y_3 \\ z_1 & z_2 & z_3 \end{bmatrix} = \begin{bmatrix} 1 & 0 & 0 \\ 0 & 1 & 0 \\ 0 & 0 & 1 \end{bmatrix}$$

This is equivalent to systems of equations

$$AX_1 = I_1, \quad X_1 = \begin{bmatrix} x_1 \\ y_1 \\ z_1 \end{bmatrix}, \quad I_1 = \begin{bmatrix} 1 \\ 0 \\ 0 \end{bmatrix}$$

$$AX_2 = I_2, \quad X_2 = \begin{bmatrix} x_2 \\ y_2 \\ z_2 \end{bmatrix}, \quad I_2 = \begin{bmatrix} 0 \\ 1 \\ 0 \end{bmatrix}$$

$$AX_3 = I_3, \quad X_3 = \begin{bmatrix} x_3 \\ y_3 \\ z_3 \end{bmatrix}, \quad I_3 = \begin{bmatrix} 0 \\ 0 \\ 1 \end{bmatrix}$$

We solve this system of equations by using Gauss elimination method by adopting the procedure outline in section 3.6.3 to find X_1, X_2 and X_3 in order to determine A^{-1}. The elimination procedure can be applied on the composite augmented matrix $[A/I]$ in one stretch to reduce the matrix A to upper triangular matrix and applying the same operations on the unit matrix I. Ultimately when A is transformed into a unit matrix, the adjoint matrix, emerged out from unit matrix I, gives us the inverse of A, to increase the accuracy, the largest element of A is taken as the pivot element.

3.8.2 Gauss-Jordan method

This is similar to the Gauss elimination method except that the matrix A is directly reduced to unit matrix instead of upper triangular form. In practice, the two matrices A and I are written side by side and the same transformations are performed on both. As soon as A reduces to I, the unit matrix I reduces to represent A^{-1}.

Example 3.8.1: Use Gauss-Jordan method to find the inverse of the matrix $A = \begin{bmatrix} 1 & 1 & 3 \\ 1 & 3 & -3 \\ -2 & -4 & -4 \end{bmatrix}$.

Solution: We have

$$[A/I]=\left[\begin{array}{ccc|ccc}1 & 1 & 3 & 1 & 0 & 0\\ 1 & 3 & -3 & 0 & 1 & 0\\ -2 & -4 & -4 & 0 & 0 & 1\end{array}\right]\begin{array}{c}R_2-R_1\\ \cong\\ R_3+2R_1\end{array}\left[\begin{array}{ccc|ccc}1 & 1 & 3 & 1 & 0 & 0\\ 0 & 2 & -6 & -1 & 1 & 0\\ 0 & -2 & 2 & 2 & 0 & 1\end{array}\right]$$

$$\begin{array}{c}1/2R_2\\ \cong\\ 1/2R_3\end{array}\left[\begin{array}{ccc|ccc}1 & 1 & 3 & 1 & 0 & 0\\ 0 & 2 & -3 & -1/2 & 1/2 & 0\\ 0 & -1 & 1 & 1 & 0 & 1/2\end{array}\right]\begin{array}{c}R_1-R_2\\ \cong\\ R_3+R_2\end{array}\left[\begin{array}{ccc|ccc}1 & 0 & 6 & 3/2 & -1/2 & 0\\ 0 & 1 & -3 & -1/2 & 1/2 & 0\\ 0 & 0 & -2 & 1/2 & 1/2 & 1/2\end{array}\right]$$

$$\begin{array}{c}\\ -1/2R_2\\ \\ \end{array}\begin{array}{c}R_1-3R_3\\ \cong\\ R_2+3/2R_3\end{array}\left[\begin{array}{ccc|ccc}1 & 0 & 0 & 3 & 1 & 3/2\\ 0 & 1 & 0 & -5/4 & -1/4 & -3/4\\ 0 & 0 & 1 & -1/4 & -1/4 & -1/4\end{array}\right]$$

Therefore

$$A^{-1}=\begin{bmatrix}3 & 1 & 3/2\\ -5/4 & -1/4 & -3/4\\ -1/4 & -1/4 & -1/4\end{bmatrix}$$

3.8.3 Crout's method

This method is also called triangularisation or factorization method. Hence, we factorize the given matrix as A = LU, where L is a lower triangular matrix with unit diagonal elements and U is an upper triangular matrix as defined in section 3.8.5. Then $A^{-1}=(LU)^{-1}=U^{-1}L^{-1}$.

Example 3.8.2: Use LU-method to find out the inverse of the matrix

$$A=\begin{bmatrix}1 & 2 & 3\\ 3 & 2 & 1\\ 2 & 1 & 3\end{bmatrix}$$

Solution: Let $A=LU=\begin{bmatrix}1 & 0 & 0\\ l_{21} & 1 & 0\\ l_{31} & l_{32} & 1\end{bmatrix}\begin{bmatrix}u_{11} & u_{12} & u_{13}\\ 0 & u_{22} & u_{23}\\ 0 & 0 & u_{33}\end{bmatrix}$

$$\Rightarrow\quad\begin{bmatrix}1 & 2 & 3\\ 3 & 2 & 1\\ 2 & 1 & 3\end{bmatrix}=\begin{bmatrix}u_{11} & u_{12} & u_{13}\\ l_{21}u_{11} & l_{21}u_{12}+u_{22} & l_{21}u_{13}+u_{23}\\ l_{31}u_{11} & l_{31}u_{12}+l_{32}u_{22} & l_{31}u_{13}+l_{32}u_{23}+u_{33}\end{bmatrix}$$

$$\Rightarrow\quad\begin{array}{lll}u_{11}=1 & u_{12}=2 & u_{13}=3\\ l_{21}=3 & u_{22}=-4 & u_{23}=-8\\ l_{31}=2 & l_{32}=3/4 & u_{33}=3\end{array}$$

$$\therefore \qquad L = \begin{bmatrix} 1 & 0 & 0 \\ 3 & 1 & 0 \\ 2 & 3/4 & 1 \end{bmatrix}, \qquad U = \begin{bmatrix} 1 & 2 & 3 \\ 0 & -4 & -8 \\ 0 & 0 & 3 \end{bmatrix}$$

Now suppose $L^{-1} = X \Rightarrow LX = I$ which implies that

$$x_{11} = 1,\ x_{22} = 2, x_{21} = -3\ \ x_{31} = 1/4,\ x_{32} = -3/4,\ x_{33} = 1$$

$$\therefore \qquad L^{-1} = X = \begin{bmatrix} 1 & 0 & 0 \\ -3 & 1 & 0 \\ 1/4 & -3/4 & 1 \end{bmatrix}$$

Similarly, let $U^{-1} = Y \Rightarrow YU = \mathrm{I}$ which implies that

$$\begin{bmatrix} y_{11} & y_{12} & y_{13} \\ 0 & y_{22} & y_{23} \\ 0 & 0 & y_{33} \end{bmatrix} \begin{bmatrix} 1 & 2 & 3 \\ 0 & -4 & -8 \\ 0 & 0 & 3 \end{bmatrix} = \begin{bmatrix} 1 & 0 & 0 \\ 0 & 1 & 0 \\ 0 & 0 & 1 \end{bmatrix}$$

$$\Rightarrow \qquad y_{11} = 1,\ y_{12} = 1/2, y_{13} = 1/3, y_{22} = -1/4,\ \ y_{23} = -2/3,\ y_{33} = 1/3$$

Therefore, $U^{-1} = \begin{bmatrix} 1 & 1/2 & 1/3 \\ 0 & -1/4 & 1/3 \\ 0 & 0 & 1/3 \end{bmatrix}$

$$A^{-1} = U^{-1}L^{-1} = \begin{bmatrix} 1 & 1/2 & 1/3 \\ 0 & -1/4 & -2/3 \\ 0 & 0 & 1/3 \end{bmatrix} \begin{bmatrix} 1 & 0 & 0 \\ -3 & 1 & 0 \\ 1/4 & -3/4 & 1 \end{bmatrix} = \begin{bmatrix} \frac{-5}{12} & \frac{1}{4} & \frac{1}{3} \\ \frac{7}{12} & \frac{1}{4} & \frac{-2}{3} \\ \frac{1}{12} & -\frac{1}{4} & \frac{1}{3} \end{bmatrix}$$

This is the required inverse of given matrix A.

3.8.4 Partition method

This method is applied to find the inverse of a matrix of order n+1 when the inverse of matrix of order *n* is known. This can be done by adding *(n+1)th* row and *(n+1)th* column to A.

Suppose $A = \begin{bmatrix} A_1 & A_2 \\ A_3' & \alpha \end{bmatrix}$ and $A^{-1} = \begin{bmatrix} X_1 & X_2 \\ X_3' & x \end{bmatrix}$ where A_2, X_2 are column vectors and A_3', X_3' are row vectors and α, x are numbers.

We assume that A_1^{-1} is known. Then $AA^{-1} = I_{n+1}$ gives us

$$A_1X_1 + A_1X_3' = I_n$$
$$A_1X_2 + A_2x = 0 \Rightarrow X_2 = -A_1^{-1}A_2x$$
$$A_3'X_1 + \alpha xX_3' = 0$$
$$A_3'X_2 + \alpha x = 1 \Rightarrow \left(\alpha - A_3'A_1^{-1}A_2\right)x = 1$$

Hence x and then X_2 can be determined. Also first equation provides us $X_1 = A_1^{-1}(I_n - A_2X_3^1)$. Upon using this in third equation we get $(\alpha - A_3'A_1^{-1}A_2)X_3' = -A_3'A_1^{-1}$.

Therefore X_1 and X_3' can be determined. Thus, having found X_1, X_2, X_3' and x, A^{-1} is completely known.

This method is also known as **Escalator method**.

Example 3.8.3: Apply partition method to find out the inverse of the matrix

$$A = \begin{bmatrix} 1 & 3 & 3 & 2 \\ 1 & 4 & 3 & 4 \\ 1 & 3 & 4 & 5 \\ 2 & 5 & 3 & 2 \end{bmatrix}.$$

Solution: We have

$$A = \begin{bmatrix} 1 & 3 & 3 & 2 \\ 1 & 4 & 3 & 4 \\ 1 & 3 & 4 & 5 \\ 2 & 5 & 3 & 2 \end{bmatrix} = \begin{bmatrix} A_1 & A_2 \\ A_3' & \alpha \end{bmatrix}.$$

where $A_1 = \begin{bmatrix} 1 & 3 & 3 \\ 1 & 4 & 3 \\ 1 & 3 & 4 \end{bmatrix}$, $A_2 = \begin{bmatrix} 2 \\ 4 \\ 5 \end{bmatrix}$, $\alpha = 2, A_3' = [2,5,3]$

We find that $A_1^{-1} = \begin{bmatrix} 7 & -3 & -3 \\ -1 & 1 & 0 \\ -1 & 0 & 1 \end{bmatrix}$

Let $A^{-1} = \begin{bmatrix} X_1 & X_2 \\ X_3' & x \end{bmatrix}$. Then $AA^{-1} = I$ and hence

$$A_3'A_1^{-1}A_2 = [2\ 5\ 3]\begin{bmatrix} 7 & -3 & -3 \\ -1 & 1 & 0 \\ -1 & 0 & 1 \end{bmatrix}\begin{bmatrix} 2 \\ 4 \\ 5 \end{bmatrix} = [2\ 5\ 3]\begin{bmatrix} 14-12-15 \\ -2+4+0 \\ -2+0+5 \end{bmatrix}$$

$$= [2\ 5\ 3]\begin{bmatrix} -13 \\ 2 \\ 3 \end{bmatrix} = -26 + 10 + 9 = -7$$

$$\therefore \qquad (\alpha - A_3' A_1^{-1} A\ 2)x = 1 \Rightarrow (2-(-7))x = 1 \Rightarrow x = 1/9$$

Also $X_2 = -A_1^{-1} A_2 x = -\begin{bmatrix} 7 & -3 & -3 \\ -1 & 1 & 0 \\ -1 & 0 & 1 \end{bmatrix}\begin{bmatrix} 2 \\ 4 \\ 5 \end{bmatrix}(1/9) = -(1/9)\begin{bmatrix} 14-12-15 \\ -2+4+0 \\ -2+0+5 \end{bmatrix} = \begin{bmatrix} 13/9 \\ -2/9 \\ -3/9 \end{bmatrix}$

Then $\qquad (\alpha - A_3' A_1^{-1} A_2) X_3' = -A_3' A_1^{-1}$

$$\Rightarrow \qquad (2+7)X_3' = -[2\ 5\ 3]\begin{bmatrix} 7 & -3 & -3 \\ -1 & 1 & 0 \\ -1 & 0 & 1 \end{bmatrix} = [-6\ 1\ 3]$$

$$\Rightarrow \qquad X_3' = [-2/3 \quad 1/9 \quad 1/3]$$

Finally, $\qquad X_1 = A_1^{-1}(I - A_2 X_3') = \begin{bmatrix} 7 & -3 & -3 \\ -1 & 1 & 0 \\ -1 & 0 & 1 \end{bmatrix}\left(\begin{bmatrix} 1 & 0 & 0 \\ 0 & 1 & 0 \\ 0 & 0 & 1 \end{bmatrix} - \begin{bmatrix} 2 \\ 4 \\ 5 \end{bmatrix}[-2/3 \quad 1/9 \quad 1/3]\right)$

$$= \begin{bmatrix} 7 & -3 & -3 \\ -1 & 1 & 0 \\ -1 & 0 & 1 \end{bmatrix}\begin{bmatrix} 7/3 & -2/9 & -2/3 \\ 8/3 & 5/9 & -4/3 \\ 10/3 & -5/9 & -2/3 \end{bmatrix} = (1/9)\begin{bmatrix} 7 & -3 & -3 \\ -1 & 1 & 0 \\ -1 & 0 & 1 \end{bmatrix}\begin{bmatrix} 21 & -2 & -6 \\ 24 & 5 & -12 \\ 30 & -5 & -6 \end{bmatrix}$$

$$= (1/9)\begin{bmatrix} -15 & -14 & 12 \\ 3 & 7 & -6 \\ 9 & -3 & 0 \end{bmatrix}$$

$$\therefore \quad A^{-1} = \begin{bmatrix} X_1 & X_2 \\ X_3' & x \end{bmatrix} = \begin{bmatrix} -5/3 & -14/9 & 4/3 & 13/9 \\ 1/3 & 7/9 & -2/3 & -2/9 \\ 1 & -1/3 & 0 & -3/9 \\ -2/3 & 1/9 & 1/3 & 1/9 \end{bmatrix} = (1/9)\begin{bmatrix} -15 & -14 & 12 & 13 \\ 3 & 7 & -6 & -2 \\ 9 & -3 & 0 & -3 \\ -6 & 1 & 3 & 1 \end{bmatrix}$$

which is required inverse of the given matrix A.

3.9. Eigenvalues and eigenvectors

Let $A = (a_{ij})$ be a square matrix of order *n*. If we can find a column vector X and a scalar λ such that

$$AX = \lambda X$$

$$\text{or } [A - \lambda I]X = 0 \qquad (3.9.1)$$

then the determinant equation

$$|A - \lambda I| = 0 \tag{3.9.2}$$

is nth degree equation in λ, called characteristic equation of the matrix A. The roots $\lambda_i, i = 1,2,3,\cdots n$, of equation (3.9.2) are called eigenvalues or latent roots and corresponding to each eigenvalue, the equation (3.9.1) will have a non-trivial solution $X = [x_1, x_2, \cdots x_n]'$, which is known as the eigenvector.

3.9.1 Cayley-Hamilton theorem

Every square matrix satisfies its own characteristic equation i.e. if the characteristic equation for the nth order square matrix A is

$$|A - \lambda I| = (-1)^n + a_1\lambda^{n-1} + a_2\lambda^{n-2} \cdots a_n = 0 \tag{3.9.3}$$

then

$$(-1)^n A^n + a_1 A^{n-1} + a_2 A^{n-2} + \cdots a_n I_n = 0 \tag{3.9.4}$$

3.9.2 Properties of eigenvalues

Some of the important properties of eigenvalues for ready reference are stated below.

1. The sum of the eigenvalue of a matrix A, is the sum of the elements of its principal diagonal.
2. If λ is an eigenvalue of a matrix, then $1/\lambda$ is the eigenvalue of A^{-1}.
3. If λ is an eigenvalue of an orthogonal matrix, then $1/\lambda$ is also its eigenvalue.
4. If $\lambda_1, \lambda_2 \cdots \lambda_n$ are the eigenvalues of a matrix A, then $\lambda_1^m, \lambda_2^m, \cdots \lambda_n^m$ are the eigenvalues of the matrix A^m, m being a positive integer.
5. The eigenvalues of symmetric and Hermitian matrices are real.
6. The eigenvalues of skew symmetric and skew Hermitian matrices are purely imaginary.
7. For every square matrix A having n linearly independent eigenvectors, a matrix P can be found such that $P^{-1}AP$ is a diagonal matrix whose diagonal elements are the eigenvalues of A. The transformation of matrix A by a non-singular matrix P to $P^{-1}AP$ is called a similarity transformation.
8. Every similarity transformation applied to a matrix leaves its eigenvalues unchanged.
9. If λ be an eigenvalue of matrix of A, then there exists some $k(1 \le k \le n)$ such that

$$|\lambda - a_{kk}| \le |a_{k1}| + |a_{k2}| + \cdots + |a_{kn}| = \rho_k \text{ (say)} \tag{3.9.5}$$

i.e. all the eigenvalue of matrix A lie in the union of the n circles with centers a_{kk} and radii ρ_k. These circles are called **Gerschgorin circle**. The bounds defined by (3.9.5) being independent are called the Gerschgorin bounds. If any of the Gerschgorin circle is isolated, then it contains exactly one eigenvalue.

Example 3.9.1 Find the eigenvectors and eigenvalues of the matrix

$$A = \begin{bmatrix} 8 & -6 & 2 \\ -6 & 7 & -4 \\ 2 & -4 & 3 \end{bmatrix}$$

Solution: The characteristic equation of matrix A is

$$|A - \lambda I| = 0 \Rightarrow \begin{vmatrix} 8-\lambda & -6 & 2 \\ -6 & 7-\lambda & -4 \\ 2 & -4 & 3-\lambda \end{vmatrix} = -\lambda^3 + 18\lambda^2 - 45\lambda = 0$$

$$\Rightarrow \qquad \lambda(\lambda - 3)(\lambda - 15) = 0$$

$$\Rightarrow \qquad \lambda = 0, 3, 15$$

Therefore, the eigenvalues of A are 0, 3, 15

Suppose $X = [x, y, z]'$ be the eigenvector corresponding to λ. Then we have

$$[A - \lambda I]X = \begin{vmatrix} 8-\lambda & -6 & 2 \\ -6 & 7-\lambda & -4 \\ 2 & -4 & 3-\lambda \end{vmatrix} \begin{bmatrix} x \\ y \\ z \end{bmatrix} = 0 \qquad (3.9.6)$$

For $\lambda = 0$, we get

$$8x - 6y + 2z = 0 \;, \quad -6x + 7y - 4z = 0 \quad 2x - 4y + 3z = 0 .$$

These equations provide us a single linearly independent solution given by (1, 2, 2) so that every non-zero multiple of this vector is an eigenvector corresponding to $\lambda = 0$. Similarly, the eigenvectors corresponding to $\lambda = 3$ and $\lambda = 15$ are the arbitrary non-zero multiples of the vector (2, 1,-2) and (2, 2, 1) respectively, which are obtained from equations (3.9.6). Hence the three eigenvectors are (1,2,2),(2,1,-2), (2,-2,1). An eigenvector $(x, y, z)'$ such that $x^2 + y^2 + z^2 = 1$, is said to be normalized. Thus the normalized eigenvectors of A are (1/3, 2/3, 2/3), (2/3, 1/3, -2/3) and (2/3, -2/3, 1).

3.9.3 Power method

In many engineering problems, it is required to compute the numerically largest eigenvalue and the corresponding eigenvector. In such cases following iterative method is convenient and well suited for machine computations.

Let $X_1, X_2, X_3 \cdots X_n$ be the eigenvectors corresponding to the eigenvalues $\lambda_1, \lambda_2, \lambda_3, \cdots \lambda_n$. Then an arbitrary column vector can be written as $X = m_1 X_1 + m_2 X_2 + \cdots + m_n X_n$ and therefore,

$$AX = m_1 AX_1 + m_2 AX_2 + \cdots + m_n AX_n$$

$$= m_1 \lambda_1 X_1 + m_2 \lambda_2 X_2 + \cdots + m_n \lambda_n X_n$$

Similarly, $$A^2X = m_1\lambda_1^2X_1 + m_2\lambda_2^2X_2 + \cdots + m_n\lambda_n^2X_n$$

and in general,

$$A^rX = m_1\lambda_1^rX_1 + m_2\lambda_2^rX_2 + \cdots + m_n\lambda_n^rX_n$$

If $|\lambda_1| > |\lambda_2| > |\lambda_3| > \cdots > |\lambda_n|$, then λ_1 is the largest root and contribution of the term $m_1\lambda_1^rX_1$, to sum on right hand side increases with r and therefore every time we multiply a column vector by A, it becomes nearer to the eigenvector X_1. We make the largest component of the resulting column vector unity to avoid the factor m_1.

Suppose X be the column vector nearer to the solution and we evaluate AX and write it as $\lambda_1^{(1)}X^{(1)}$ after normalization. This gives us the first approximation to the eigenvalue $\lambda_1^{(1)}$ and eigenvector $X_1^{(1)}$. Similarly we evaluate $AX^{(1)} = \lambda^{(2)}X^{(2)}$ to obtain second approximation and so on. We continue to repeat this process till $\left|X^r - X^{r-1}\right| < \in$, $\in$ being arbitrarily small and can be randomly selected to order to achieve the desired level of accuracy i.e. unless $\left|X^{(r)} - X^{(r-1)}\right|$ becomes negligibly small. Then $\lambda^{(r)}$ will be the largest eigenvalue and $X^{(r)}$ the corresponding eigenvector. This iterative procedure of finding the dominant eigenvalue of a matrix is known as **Rayleigh's** power method. If we rewrite $AX = \lambda X$, *as* $A^{-1}AX = \lambda A^{-1}X$ or $X = \lambda A^{-1}X$, we have $A^{-1}X = \frac{1}{\lambda}X$ .If we use this equation, then the above method provides us the smallest eigenvalue of the matrix A.

Example: 3.9.2: Find the largest eigenvalue and the corresponding eigenvector of matrix

$$A = \begin{bmatrix} 1 & 6 & 1 \\ 1 & 2 & 0 \\ 0 & 0 & 3 \end{bmatrix}$$

Solution: Let the initial eigenvector be $X^{(0)} = [1,0,0]'$ so that

$$AX^{(0)} = \begin{bmatrix} 1 & 6 & 1 \\ 1 & 21 & 0 \\ 0 & 0 & 3 \end{bmatrix}\begin{bmatrix} 1 \\ 0 \\ 0 \end{bmatrix} = \begin{bmatrix} 1 \\ 1 \\ 0 \end{bmatrix} = X^{(1)}, \text{(say)}$$

This implies that the first approximation to the eigenvalue is 1 and the corresponding eigenvector

$$X^{(1)} = X^{(1)} = [1,\ 1,\ 0]'$$

Hence $$AX^{(1)} = \begin{bmatrix} 1 & 6 & 1 \\ 1 & 2 & 0 \\ 0 & 0 & 6 \end{bmatrix}\begin{bmatrix} 1 \\ 1 \\ 0 \end{bmatrix} = \begin{bmatrix} 7 \\ 3 \\ 0 \end{bmatrix} = 7\begin{bmatrix} 1 \\ 0.43 \\ 0 \end{bmatrix}$$

from which we see that $X^{(2)} = [1,\ 0.43,\ 0]'$ and that an approximate eigenvalue is 7.

Repeating the above process, we get

$$3.58\begin{bmatrix}1\\0.51\\0\end{bmatrix};\ 4.06\begin{bmatrix}1\\0.5\\0\end{bmatrix};\ 4\begin{bmatrix}1\\1/2\\0\end{bmatrix};\ 4\begin{bmatrix}2\\1\\0\end{bmatrix};\ 4\begin{bmatrix}1\\1/2\\0\end{bmatrix}$$

Thus the largest eigenvalue is 4 and the corresponding eigenvector is $[1,\ 1/2,\ 0]'$.

Example 3.9.3 Determine the largest eigenvalue and the corresponding eigenvector of the matrix

$$A = \begin{bmatrix}2 & -1 & 0\\ -1 & 2 & -1\\ 0 & -1 & 2\end{bmatrix}.$$

Solution: Let the initial eigenvector be $X^{(0)} = [1,0,0]'$. Then

$$AX^{(0)} = \begin{bmatrix}2 & -1 & 0\\ -1 & 2 & -1\\ 0 & -1 & 2\end{bmatrix}\begin{bmatrix}1\\0\\0\end{bmatrix} = 2\begin{bmatrix}1\\-0.5\\0\end{bmatrix}$$

$\Rightarrow X^{(1)} = [1,-0.5,0]'$ and the approximate eigenvalue is 2.

Repeating the process, we get

$$AX^{(1)} = 2.5\begin{bmatrix}1\\-0.8\\0.2\end{bmatrix};\ AX^{(2)} = 2.8\begin{bmatrix}1\\-1\\0.43\end{bmatrix};\ AX^{(3)} = 3.43\begin{bmatrix}0.87\\-1\\0.54\end{bmatrix};\ AX^{(4)} = 3.41\begin{bmatrix}0.8\\-1\\0.61\end{bmatrix}$$

$$AX^{(5)} = 3.41\begin{bmatrix}.76\\-1\\065\end{bmatrix};\ AX^{(6)} = 3.41\begin{bmatrix}0.74\\-1\\0.67\end{bmatrix}.$$

Hence the largest eigenvalue is 3.41 and the corresponding eigenvector is $[0.74, -1, 0.67]'$.

3.9.4 Jacobi's method

Let A be a given real symmetric matrix. Its eigenvalues are real and there exists a real orthogonal matrix B such that $B^{-1}AB$ is a diagonal matrix D. Jacobi's method consists of diagonalizing A by applying a series of orthogonal transformations $B_1, B_2, \cdots B_r$ so that their product B satisfies the equation $B^{-1}AB = D$. We consider numerically largest non-diagonal element a_{ij} and form a 2×2 sub matrix $A_1 = \begin{bmatrix} a_{ii} & a_{ij} \\ a_{ji} & a_{jj} \end{bmatrix}$, where $a_{ij} = a_{ji}$ which can be easily diagonalized.

Consider an orthogonal matrix $B_1 = \begin{bmatrix} \cos\theta & -\sin\theta \\ \sin\theta & \cos\theta \end{bmatrix}$

Then $B_1^{-1} A_1 B_1 = \begin{bmatrix} \cos\theta & -\sin\theta \\ \sin\theta & \cos\theta \end{bmatrix} \begin{bmatrix} a_{ii} & a_{ij} \\ a_{ji} & a_{ji} \end{bmatrix} \begin{bmatrix} \cos\theta & -\sin\theta \\ \sin\theta & \cos\theta \end{bmatrix}$

$$= \begin{bmatrix} a_{ii}\cos^2\theta + a_{jj}\sin^2\theta + a_{ij}\sin 2\theta & a_{ij}\cos 2\theta + \frac{1}{2}(a_{jj} - a_{ii})\sin 2\theta \\ a_{ij}\cos 2\theta + \frac{1}{2}(a_{jj} - a_{ii})\sin 2\theta & a_{ii}\sin^2\theta + a_{jj}\cos^2\theta - a_{ij}\sin 2\theta \end{bmatrix} \tag{3.9.7}$$

This matrix will reduce to the diagonal form, if $a_{ij}\cos 2\theta + \frac{1}{2}(a_{jj} - a_{ii})\sin 2\theta = 0$ i.e. if,

$$\tan 2\theta = \frac{2a_{ij}}{a_{ii} - a_{jj}} \tag{3.9.8}$$

This equation gives four values of θ, but to get the least possible rotation, we choose $-\frac{\pi}{4} \le \theta \le \frac{\pi}{4}$. Thus the matrix (3.9.7) reduces to a diagonal matrix. As a next step, the largest non-diagonal element, in magnitude, in the new rotated matrix is found and the above procedure is repeated using the orthogonal matrix B_2 and so on a series of such transformations are performed so as to annihilate the non-diagonal elements. After making r transformations, we obtain

$$B_r^{-1} B_{r-1}^{-1} \cdots B_1^{-1} A B_1 B_2 \cdots B_{r-1} B_r = B^{-1} A B \tag{3.9.9}$$

for $r \to \infty$, $B^{-1}AB$ approaches a diagonal matrix whose diagonal elements are the eigenvalues of A. also corresponding columns of $B = B_1 B_2 B_3 \cdots B_r$ are the eigenvectors of A.

Example 3.9.4 Use Jacobi's method to find all the eigenvalue and eigenvector of the matrix

$$A = \begin{bmatrix} 1 & \sqrt{2} & 2 \\ \sqrt{2} & 3 & \sqrt{2} \\ 2 & \sqrt{2} & 1 \end{bmatrix}$$

Solution: Here $a_{11} = 1, a_{33} = 1$ and the largest non-diagonal elements is $a_{13} = 2 = a_{31}$

$$\therefore \qquad \tan 2\theta = \frac{2a_{13}}{a_{11} - a_{33}} = \frac{2 \times 2}{1 - 1} \to \infty \Rightarrow 2\theta = \frac{\pi}{2} \Rightarrow \theta = \frac{\pi}{4}$$

Then $$B_1 = \begin{bmatrix} \cos\theta & 0 & -\sin\theta \\ 0 & 1 & 0 \\ \sin\theta & 0 & \cos\theta \end{bmatrix} = \begin{bmatrix} \frac{1}{\sqrt{2}} & 0 & \frac{-1}{\sqrt{2}} \\ 0 & 1 & 0 \\ \frac{1}{\sqrt{2}} & 0 & \frac{1}{\sqrt{2}} \end{bmatrix}$$

Therefore, $$D_1 = B_1^{-1} A B_1 = \begin{bmatrix} 3 & 2 & 0 \\ 2 & 3 & 0 \\ 0 & 0 & -1 \end{bmatrix}$$

Now largest non-diagonal element is $a_{12} = a_{21} = 2 \text{ and } a_{11} = 3 = a_{22}$ *and*

$$\textit{hence } \tan 2\theta \frac{2a_{12}}{a_{11} - a_{22}} \to \infty \Rightarrow \theta = \frac{\pi}{4}.$$

Therefore, we have

$$B = \begin{bmatrix} \cos\theta & -\sin\theta & 0 \\ \sin\theta & \cos\theta & 0 \\ 0 & 0 & 1 \end{bmatrix} = \begin{bmatrix} \frac{1}{\sqrt{2}} & \frac{-1}{\sqrt{2}} & 0 \\ \frac{1}{\sqrt{2}} & \frac{1}{\sqrt{2}} & 0 \\ 0 & 0 & 1 \end{bmatrix} \Rightarrow D_2 = \begin{bmatrix} 5 & 0 & 0 \\ 5 & 1 & 0 \\ 0 & 0 & -1 \end{bmatrix}$$

The eigenvalues of A are 5, 1,-1 and the corresponding eigenvectors are the columns of the matrix

$$B = B_1 B_2 = \begin{bmatrix} \frac{1}{2} & \frac{-1}{2} & \frac{-1}{\sqrt{2}} \\ \frac{1}{\sqrt{2}} & \frac{1}{\sqrt{2}} & 0 \\ \frac{1}{2} & \frac{-1}{2} & \frac{1}{2} \end{bmatrix}$$

3.9.5 Given's method

The Jacobi's method has the disadvantage that the element annihilated by a transformation, may not remain zero during the subsequent transformations. Given overcome this problem by not disturbing the zeros already formed. But his method leads to tridiagonal matrix instead of diagonal one in the Jacobi's method above. If A be a real symmetric matrix, then Given's method consist of the following steps:

Steps-1 To reduce matrix A to a tridiagonal symmetric matrix.

Consider the matrix
$$A=\begin{bmatrix} a_{11} & a_{21} & a_{31}\\ a_{12} & a_{22} & a_{32}\\ a_{13} & a_{23} & a_{33}\end{bmatrix} \tag{3.9.10}$$

and the orthogonal rotation matrix B in (2,3) plane as

$$B_1=\begin{bmatrix} 1 & 0 & 0\\ 0 & \cos\theta & -\sin\theta\\ 1 & \sin\theta & \cos\theta\end{bmatrix}$$

Then
$$B_1^{-1}A_1B_1=\begin{bmatrix} 1 & 0 & 1\\ 0 & \cos\theta & \sin\theta\\ 0 & -\sin\theta & \cos\theta\end{bmatrix}\begin{bmatrix} a_{11} & a_{21} & a_{31}\\ a_{12} & a_{22} & a_{32}\\ a_{13} & a_{23} & a_{33}\end{bmatrix}\begin{bmatrix} 1 & 0 & 0\\ 0 & \cos\theta & -\sin\theta\\ 1 & \sin\theta & \cos\theta\end{bmatrix}$$

In the resulting matrix, (1, 3) element $= -a_{12}\sin\theta + a_{13}\cos\theta$, which will be zero, if $-a_{12}\sin\theta + a_{13}\cos\theta = 0$ i.e. if $\tan\theta = a_{13}/a_{12}$. With this value of θ, the above transformation gives zeros in (1,3) and (3,1) positions. Again we perform rotation in the (2, 4) plane and put the resulting element (1, 4) equal to zero. This would not affect the zeros obtained earlier. Proceeding in this way, the transformations are applied to the matrix so as to annihilate the elements $(1,3),(1,4),(1,5),\ldots\ldots,(1,n);(2,4),(2,5),\ldots\ldots,(2,n)$ in this order.

Finally we arrive at the tri-diagonal matrix

$$p=\begin{bmatrix} p_1 & q_1 & 0 & 0\cdots 0 & \\ q_1 & p_2 & q_2 & 0\cdots 0 & \\ 0 & q_2 & p_3 & 0\cdots 0 & \\ \cdots & \cdots & & & p_{n-1}q_{n-1}\\ 0\ 0 & 0 & & 0\cdots & q_{n-1}p_n\end{bmatrix} \tag{3.9.11}$$

Step-2 To find the eigenvalues of a tridiagonal matrix

Suppose the resulting tridiagonal matrix after first transformation be

$$\begin{bmatrix} \alpha_{11}-\lambda & \alpha_{12} & 0\\ \alpha_{12} & \alpha_{22}-\lambda & \alpha_{23}\\ 0 & \alpha_{23} & \alpha_{33}-\lambda\end{bmatrix}=0=f_3(\lambda)\text{ Say}$$

$\therefore\ f_0(\lambda)=1$,

$$f_1(\lambda)\alpha_{11}-\lambda=\alpha_{11}-\lambda f_0(\lambda),\ f_2(\lambda)=\begin{vmatrix} \alpha_{11}-\lambda & \alpha_{12}\\ \alpha_{12} & \alpha_{22}-\lambda\end{vmatrix}=(\alpha_{22}-\lambda)f_1(\lambda)-\alpha_{12}^2f_0(\lambda)$$

$$\therefore\quad f_3(\lambda)=(\alpha_{33}-\lambda)\begin{vmatrix} \alpha_{11}-\lambda & \alpha_{12}\\ \alpha_{12} & \alpha_{22}-\lambda\end{vmatrix}-\alpha_{23}\begin{vmatrix} \alpha_{11}-\lambda & 0\\ \alpha_{12} & \alpha_{23}\end{vmatrix}=(\alpha_{33}-\lambda)f_2(\lambda)-\alpha_{23}^2f_1(\lambda)$$

In general, the recurrence formula is

$$f_k(\lambda)=(\alpha_{kk}-\lambda)f_{k-1}(\lambda)-(\alpha_{k-1,k})^2 f_{k-2}(\lambda),2\le k\le n \qquad (3.9.12)$$

The equation $f_k(\lambda)=0$ is the characteristic equation which can be solved by any standard method and hence the roots of (3.9.12) will be the eigenvalue of the given symmetric matrix. The sequence $f_0(\lambda), f_1(\lambda), f_2(\lambda),\cdots f_k(\lambda)$ is called the strum sequence.

Step-3 To find the eigenvectors of the tridiagonal matrix:

If Y be an eigenvector of the tridiagonal matrix P and if $B_1, B_2 \cdots B_j$ the orthogonal matrices employed in reducing the matrix A to the form $P\theta$ then the corresponding eigenvector of A is given by $X = B_1, B_2 \cdots B_j Y$ .The number of rotations required for the Given's method are equivalent to the number of non-tridiagonal elements of the matrix. For a 3×3 matrix, only one rotation is required, where as for a 4×4 matrix, 3 rotations are needed and so on. The amount of computation goes on decreasing from one rotation to the next, as the order of the matrix of computation also starts reducing.

Example 3.9.5 Use Given's method, reduce the matrix $A=\begin{bmatrix}2 & 1 & 3\\ 1 & 4 & 2\\ 3 & 2 & 3\end{bmatrix}$ to the tridiagonal form

Solution: As there is only one non-tridiagonal element $a_{13}(=3)$which has to be reduced to zero, only one rotation is required. To annihilate a_{13}, we define the orthogonal matrix in the (2, 3) plane as

$$B=\begin{bmatrix}1 & 0 & 0\\ 0 & \cos\theta & -\sin\theta\\ 1 & \sin\theta & \cos\theta\end{bmatrix}$$

where θ is given by $\tan\theta=\frac{a_{13}}{a_{12}}=\frac{3}{1}=3\Rightarrow \sin\theta=\frac{3}{\sqrt{10}}$ and $\cos\theta=\frac{1}{\sqrt{10}}$.

$$\therefore \qquad A_1=B^{-1}AB=\begin{bmatrix}2 & 3.16 & 0\\ 3.16 & 4.3 & -1.9\\ 0 & -1.9 & 3.9\end{bmatrix}$$

This is the required tridiagonal from of the given matrix.

3.9.6 House-Holder's method

Householder suggested an alternative procedure for the reduction of a symmetric matrix to the tridiagonal form. This method, though more complicated, require half through as much computation, as the Given's method and it is a substantial improvement on the Given's procedure since it reduces an entire row and column by a single transformation. Here the matrix is reduced to tridiagonal form using elementary orthogonal transformations. Let $A=(a_{ij})_{n\times n}$ be an nth order real symmetric matrix. This method consists in pre and post multiplying A by a real symmetric orthogonal matrix P such that PAP reduces to the tridiagonal form. Let the matrix P be of the form $P=I-2ww', w$ being a column matrix such that

$$ww' = w_1^2+w_2^2+\cdots+w_n^2=1$$

Then $$P' = (I - 2ww')' = I - 2ww' = P$$

and $$P'P = (I - 2ww')'(I - 2ww') = I - 4ww' + 4ww'.ww' = I$$

Thus P is a symmetric orthogonal matrix. Now take w with first $(k-1)$ zero components,

so that $w'_k = \{0, 0, \cdots, 0, \quad x_k, \cdots, x_n\}$

Since $w'_k w_k = 1$, we have $x_k^2 + x_{k+1}^2 + \cdots + x_n^2 = 1$

Then $P_k^{-1} A P_k = P'_k A P_k = P_k A P_k$

We now from successive iterations, we have

$$A_k = P_k A_{k-1} P_k; \; k = 2, 3, 4, \cdots, (n-1).$$

As a first transformation, we find *x's* so that zeros are created in the positions

$(1, 3), (1, 4), \cdots (1, n)$ and $(3, 1), (4, 1), (5, 1) \ldots\ldots (n, 2)$

As a second transformation, we find *x's* so that zeros are created in the positions
$(2, 4), (2, 5) \ldots\ldots (2, n)$ and $(4, 2), (5, 2), \ldots\ldots (n, 2)$.

After $(n-2)$ such transformations, we arrive at a tridiagonal matrix.

Example 3.9.6: Reduce the matrix $A = \begin{bmatrix} 1 & 3 & 4 \\ 3 & 2 & -1 \\ 4 & -1 & 1 \end{bmatrix}$ to the tridiagonal form by Householder's method

Solution: Hence $\sqrt{a_{12}^2 + a_{13}^2} = 5$ and we take $w'_2 = [0, \; x_2, \; x_3]$ so that

$$P = \mathrm{I} - 2w_2 w'_2 = \begin{bmatrix} 1 & 0 & 0 \\ 0 & 1 - 2x_2^2 & -2x_2 x_3 \\ 0 & -2x_2 x_3 & 1 - 2x_3^2 \end{bmatrix}$$

Now (1, 3) element of PAP will become zero only if corresponding element of AP is zero, which implies that $a_{13} - 2p_1 x_3 = 0$ or $4 - 2p_1 x_3 = 0$.

Also $a_{11}^2 + a_{12}^2 + a_{13}^2 = a_{11}^2 + \left(a_{12} - 2p_1 x_2^{(i)}\right)^2 + 0$, by invariant property

Definition 3.9 Two matrices *A* and *B* are said to be similar matrices if there exists a non-singular matrix *P* such that $A = P^{-1} B P$

Theorem 3.9 Schur's theorem: Every square matrix *A* can be written as $U^{-1} T U$ with *T* upper triangular and *U* unitary matrices with $U^H U = I$.

Algorithm 3.9.1 similarity transformation into upper **Heisenberg** (triangular) form using Householder reflections:

Given the matrix A of order n as stored in the first n columns of work array H of order $n\times(n-2)$.

1. For $k=1,2,\cdots,n-2$ do: steps 2 to 14

2. $\beta := signum\left(\mathrm{h}_{k+1,k}\right)\left(\sum_{j=k+1}^{n}\left(h_{jk}\right)^2\right)^{½}$

3. $h_{k+1,k} := h_{k+1,k} + \beta$

4. $\alpha^{-1} := \beta \mathrm{h}_{k+1,k}$

5. For $j=1,2,\cdots n,$ do: steps 6 to 8

6. $\gamma := \left(\sum_{i=k+1}^{n} h_{ik} h_{ji}\right) / \alpha^{-1}$

7. For $i=k+1,k+2,\cdots,n,$ do: steps 8

8. $h_{ji} : h_{ji} - \gamma h_{ik}$

9. For $j=k+1,\mathrm{k}+2,\cdots,\mathrm{n},\mathrm{do}:$ steps 10 to 12

10. $\gamma := \left(\sum_{i=k+1}^{n} h_{ik} h_{ij}\right) / \alpha^{-1}$

11. For $i=k+1,k+2,\cdots,n,$ do: steps 12

12. $h_{ij} := h_{ij} - \gamma h_{ik}$

13. $h_{k+1,n+1} := h_{k+1,k},\ h_{k+1,n+2} := \alpha^{-1}$

14. $h_{k+1},k := -\beta$

15. Stop

3.9.7 QR Method

The QR method consists in constructing the factorization of the matrix $A_1 = A$ in the form $A_1 = Q_1 R_1$, where Q_1 is orthogonal and R_1 is an upper triangular matrix. This implies that

$$Q_1^{-1} A_1 Q_1 = R_1 Q_1 = A_2 \qquad \Rightarrow A_2 = R_1 Q_1 \tag{3.9.13}$$

We now factorize A_2 to obtain $A_2 = Q_2 R_2$, where Q_2 is orthogonal and R_2 is upper triangular matrix. From equation (3.9.15) we obtain

$A_3 = R_2 Q_2$ and so on.

The process is repeated to construct the sequence A_1, A_2, $A_3,\ldots, A_k$, where

$$A_k = Q_k R_k \tag{3.9.14}$$

where Q_k are orthogonal and R_k are upper triangular matrices for all k. When A_1 is nonsingular, the QR factorization is unique. The factorization can be carried out by performing premultiplication upon the given matrix A_1 by orthogonal matrix P of the form $1-2VV^T$ in order to successively reduce the columns of A_1. Thus P_1A_1 contains zeros below the diagonal in its first column, $P_2P_1A_1$ in its second column below the diagonal, and so on. Repeating this procedure with each column of A_1, we obtain an upper triangular matrix, so that we have

$$R = P_{n-1}P_{n-2}P_{n-3},\ldots, P_2P_1A_1 \tag{3.9.15}$$

To complete the construction, we define the orthogonal matrix

$$Q^T = P_{n-1}P_{n-2}P_{n-3},\ldots, P_2P_1 \tag{3.9.16}$$

so that $A_1 = QR$ as required. The sequence $\{A_k\}$ converges either to a triangular matrix with the values of A on its diagonal, or near- triangular matrix from which the eigenvalues can also be easily calculated.

3.10 Singular value decomposition (SVD)

A decomposition of a rectangular matrix like LU decomposition in case of a square matrix is also possible which is called the singular value decomposition (SVD). The SVD is of great importance in matrix theory since it is useful in finding the generalized inverse of a singular matrix and has several image processing applications. Let $A = (a_{ij})_{m\times n}$ be a real matrix with $m \geq n$. Then the matrices A^TA and AA^T are non-negative, symmetric and have identical eigenvalues, say λ_n. Then we can obtain n orthonormal eigenvectors, say X_n of A^TA such that

$$A^TAX_n = \lambda_n X_n \tag{3.10.1}$$

Let Y_n be the n orthonormal eigenvectors of A^TA, then

we have $$A^TY_n = \lambda_n Y_n \tag{3.10.2}$$

The matrix A can be decomposed into the form

$$A = UDV^T \tag{3.10.3}$$

where $$U^TV = V^TV = VV^T = I_n \tag{3.10.4}$$

and $$D = d_{iag}\left(\sqrt{\lambda_1}, \sqrt{\lambda_2}, \cdots, \sqrt{\lambda_n}\right) \tag{3.10.5}$$

The decomposition defined by (3.10.3) is called the singular value decomposition of A. The matrix V consists of n orthonormal eigenvectors X_n of A^TA is a matrix of order n. Thus the eigenvectors of AA^T i.e. Y_n are given by

$$Y_n = \frac{1}{\sqrt{\lambda_n}} AX_n \tag{3.10.6}$$

The matrix D is a diagonal matrix given by

$$D = \begin{bmatrix} \sqrt{\lambda_1} & 0 & 0 \cdots 0 \\ 0 & \sqrt{\lambda_2} & 0 \cdots 0 \\ . \quad . & . \quad . & . \quad . \quad . \\ 0 & 0 & 0 \cdots \sqrt{\lambda_n} \end{bmatrix} \tag{3.10.7}$$

where $\sqrt{\lambda_i}, i = 1,2,3 \cdots \text{n}$ are called the singular values of A and are such that

$$\sqrt{\lambda_1} \geq \sqrt{\lambda_2} \geq \sqrt{\lambda_3} \geq \cdots \geq \sqrt{\lambda_n} \geq 0. \tag{3.10.8}$$

If the rank of A is $r < n$, then

$$\sqrt{\lambda}_{r+1} = \sqrt{\lambda}_{r+2} = \cdots = \sqrt{\lambda}_n = 0 \tag{3.10.9}$$

It can be shown that such a singular value decomposition is unique if λ_i are distinct and the in equally (3.10. 8) is satisfied. For a square matrix A of order n, U, D and V are also square matrices of the same order and the inverse of A can be easily computed, because $A^{-1} = VD^{-1}U^T$ and $D^{-1} = diag\left(\frac{1}{\sqrt{\lambda_1}}, \frac{1}{\sqrt{\lambda_2}}, \cdots \frac{1}{\sqrt{\lambda_n}}\right)$

If any λ_i is zero, then the matrix is singular and for very small λ_i, A is very nearly singular.

Example 3.10.1 Fined the SVD of the matrix $A = \begin{bmatrix} 1 & 2 \\ 1 & 1 \\ 1 & 3 \end{bmatrix}$

Solution: The given matrix is $A = \begin{bmatrix} 1 & 2 \\ 1 & 1 \\ 1 & 3 \end{bmatrix}$

Therefore $A^T = \begin{bmatrix} 1 & 2 & 1 \\ 1 & 1 & 3 \end{bmatrix}$ and $A^T A = \begin{bmatrix} 3 & 6 \\ 6 & 14 \end{bmatrix}$

The eigenvalues of $A^T A$ are $\lambda_1 = 16.64 \, and \, \lambda_2 = 0.36$ and the corresponding eigenvectors are

$X_1 = \begin{bmatrix} 0 & .4033 \\ 0 & .9166 \end{bmatrix} and \, X_2 = \begin{bmatrix} 0 & .9166 \\ 0 & .4033 \end{bmatrix}$. Also $\sqrt{\lambda_1} = 4.080$ and $\sqrt{\lambda_2} = 0.60$. The eigenvectors of AA^T are given by $y_n = \frac{1}{\sqrt{\lambda_n}} \text{Ax}_\text{n}, n = 1,2$ as $y = \begin{bmatrix} 0.5480 \\ 0.3235 \\ 0.7727 \end{bmatrix}$ and $\begin{bmatrix} 0.1833 \\ 0.8555 \\ 0.4849 \end{bmatrix}$

The SVD of A is given by e required decomposition. If any λ_i is zero, then the matrix is singular and for very small λ_i, A is very nearly singular.

Exercises

1. Use Gauss elimination method to solve
   (i) $2x + y + z = 10,\ 3x + 2y + 3z = 18,\ x + 4y + 9z = 16$
   (ii) $5x + 2y + z = 4, 7x + y - 5z = 8, 3x + 7y + 4z = 10$
   (iii) $2x + 2y + z = 12,\ 3x + 2y + 2z = 8,\ 5x + 10y - 8z = 10$
   (iv) $x + 2y - z = 3,\ x - y + 2z = 1,\ 2x - 2y + 3z = 2,\ x - y + z = -1$
   (v) $2x - y = 0,\ -x + 2y - z = 0,\ -y + 2z - u = 0,\ -z + 2u = 1$.
2. Use Gauss-Jordan method to solve
   (i) $5x - 2y + z = 4,\ 7x + y - 5z = 8,\ 3x + 7y + 4z = 10$
   (ii) $10x + y + z = 12,\ x + 10y + z = 12,\ x + y + 10z = 12$
   (iii) $2x - 3y + z = -1,\ x + 4y + 5z = 25,\ 3x - 4y + z = 2$
   (iv) $2x_1 + x_2 + 5x_3 + x_4 = 5,\ x_1 + x_2 - 3x_3 + 4x_4 = -1,\ 3x_1 + 6x_2 - 2x_3 + x_4 = 8,$
   $2x_1 + 2x_2 + 2x_3 - 3x_4 = 2$
   (v) $x_1 + 2x_2 - x_3 = 3,\ x_1 - x_2 + 2x_3 = 1,\ 2x_1 - 2x_2 + 2x_3 = 2,\ x_1 - x_2 + x_3 = -1$
3. Use Crout's (Cholesky's) method to solve
   (i) $10x + y + z = 12,\ 2x + 10y + z = 13,\ 2x + 2y + 10z = 14$
   (ii) $2x - 6y + 8z = 24,\ 5x + 4y - 3z = 2,\ 3x + y + 2z = 16$
   (iii) $5x - 2y + z = 4,\ 7x + y - 5z = 8,\ 3x + 7y + 4z = 10$
   (iv) $2x_1 - x_2 + x_3 = -1,\ 2x_2 - x_3 + x_4 = 1,\ x_1 + 2x_3 - x_4 = -1,\ x_1 + x_2 + 2x_4 = 3$
4. Use Gauss elimination, Gauss-Jordan and L U method to find the inverse of the following matrices

   (i) $\begin{bmatrix} 5 & -2 & 1 \\ 7 & 1 & -5 \\ 3 & 7 & 4 \end{bmatrix}$ (ii) $\begin{bmatrix} 3 & 2 & 4 \\ 2 & 1 & 1 \\ 1 & 3 & 5 \end{bmatrix}$ (iii) $\begin{bmatrix} 1 & 2 & 3 \\ 3 & 2 & 1 \\ 2 & 1 & 3 \end{bmatrix}$

5. Use Jacobi and Gauss-Seidal methods to solve the following equations correct up to three decimal places.
   (i) $10x + 2y + z = 9,\ 2x + 20y - 2z = -44,\ -2x + 3y + 10z = 22$
   (ii) $83x + 11y - 4z = 95,\ 7x + 52y + 13z = 104,\ 3x + 8y + 29z = 71$
   (iii) $5x - y + z = 10,\ 2x + 4y = 12,\ x + y + 5z = -1$
   (iv) $5x + 2y + z = 12,\ x + 4y + 2z - 15,\ x + 2y + 5z = 20$

(v) $54x + y + z = 110,\ 2x + 15y + 6z = 72,\ -x + 6y + 27z = 85$

(vi) $10x_1 - 2x_2 - x_3 - x_4 = 3,\ -2x_1 + 10x_2 - x_3 - x_4 = 15,$
$-x_1 - x_2 + 10x_3 - 2x_4 = 27,\ -x_1 - x_2 - 2x_3 + 10x_4 = -9$

6. Use relaxation method to solve

(i) $3x + 9y - 2z = 11,\ 4x + 2y + 13z = 24,\ 4x - 4y + 3z = -8$

(ii) $10x - 2y - 2z = 6,\ -x + 10y - 2z = 7,\ -x - y + 10z = 8$

(iii) $10x - 2y - 3z = 205,\ -2x + 10y - 2z = 154,\ -2x - y + 10z = 120$

7. Use factorization and partitioning method to find the inverse of the matrices

(i) $\begin{bmatrix} 2 & -2 & 4 \\ 2 & 3 & 2 \\ -1 & 1 & -1 \end{bmatrix}$ (ii) $\begin{bmatrix} 5 & -2 & 1 \\ 7 & 1 & -5 \\ 3 & 7 & 4 \end{bmatrix}$ (iii) $\begin{bmatrix} 1 & .1 & 1 \\ 4 & 3 & -1 \\ 3 & 5 & 3 \end{bmatrix}$ (iv) $\begin{bmatrix} 1 & 3 & 3 & 2 \\ 1 & 4 & 3 & 4 \\ 1 & 3 & 4 & 5 \\ 2 & 5 & 3 & 2 \end{bmatrix}$

8. Find the largest eigenvalue and corresponding eigenvectors of the matrices by power method

(i) $\begin{bmatrix} 10 & -2 & 1 \\ -2 & 10 & -2 \\ 1 & -2 & 10 \end{bmatrix}$ (ii) $\begin{bmatrix} 1 & 3 & -1 \\ 3 & 2 & 4 \\ -1 & 4 & 10 \end{bmatrix}$ (iii) $\begin{bmatrix} 1 & 2 & 3 \\ 0 & -4 & 2 \\ 0 & 0 & 7 \end{bmatrix}$

9. Reduce the following matrices to the tridiagonal from by House - Holder's method

(i) $\begin{bmatrix} 1 & 3 & 4 \\ 3 & 2 & 1 \\ 4 & 2 & 1 \end{bmatrix}$ (ii) $\begin{bmatrix} 2 & -1 & -1 \\ -1 & 2 & -1 \\ -1 & -1 & 2 \end{bmatrix}$ (iii) $\begin{bmatrix} 1 & 1 & \frac{1}{2} \\ 1 & 1 & \frac{1}{4} \\ \frac{1}{2} & \frac{1}{4} & 2 \end{bmatrix}$

10. Find all the eigenvalues and the eigenvectors of the matrices

(i) $\begin{bmatrix} 5 & 0 & 1 \\ 0 & -2 & 0 \\ 1 & 0 & 5 \end{bmatrix}$ (ii) $\begin{bmatrix} 2 & 3 & 1 \\ 3 & 2 & 2 \\ 1 & 2 & 1 \end{bmatrix}$ (iii) $\begin{bmatrix} 1 & 2 & 2 \\ 2 & 1 & 2 \\ 2 & 2 & 1 \end{bmatrix}$ (iv) $\begin{bmatrix} 0 & 1 & 4 \\ 1 & 3 & 1 \\ 2 & 1 & 0 \end{bmatrix}$

by using (i) Jacobi (ii) Given and (iii) House - Holder and (iv) QR methods.

11. Compute the SVD of the matrix (i) $A = \begin{bmatrix} 1 & 2 \\ 2 & 1 \\ 1 & 3 \end{bmatrix}$ and (ii) $A = \begin{bmatrix} 20 & 4 \\ 10 & 14 \\ 5 & 5 \end{bmatrix}$

CHAPTER 4

INTERPOLATION AND FUNCTION APPROXIMATIONS

4.1 Introduction

Often engineers, scientists, organizers and sociologists have to take some decisions concerning the phenomena of which they know only the behavior from experimentally measured values. In certain cases, for example in physics, the fundamental knowledge of the phenomena in question allows us in proposing a precise, deterministic mathematical model which we call the model of knowledge. In many branches of applied mathematics, it is required to express a given data, obtained from observations in from of the law connecting the two variables involved. Such a law inferred by some scheme, is known as empirical law. Several equations of different types can be obtained to express the given data approximately. But the problem is to find the equation of the curve of 'best fit' which may be most suitable for predicting the unknown values. The process of finding such an equation of 'best fit' is known as curve fitting. The method of least squares is probably the best to fit a unique curve to a given data. It is widely used in applications and can be easily implemented on a computer. There are several cases when we have information or data 'y' available at several discrete locations 'x' for example tabulated values of the steam, trigonometric, logarithmic, and other functions etc. In fact, till the use of on-line measurement devices and recorders became popular, experimental results taken in a laboratory were available in a similar form. We may be required to interpolate or extrapolate these data or may at times, be interested in computing slopes or integrals of functions described by them. This chapter is devoted to the discussion of several techniques for doing this.

4.2 Least squares curve-fit and trigonometric approximations

The graphical method and the method of group averages have the obvious drawback of being unable to give a unique curve fit. The principle of least squares, however, provides an elegant procedure of fitting a unique curve to a given data. A French mathematicians Adrian Marie Legendre in 1806 suggested the "Principle of least squares," which states that the curve of best fit is that for which the errors (or residuals) are as small as possible i.e., the sum of the squares of the errors is a minimum. The principle of least squares does not help us to determine the form of appropriate curve which can fit a given data. It only determines the least possible values of the constants in the equation when the form of the curve is known before hand. The selection of the curve is a matter of experience and practical consideration. Here we shall discuss the fitting of various types of curves by the method of least squares.

4.2.1 Straight line fitting

Suppose it is required to fit a straight line

$$y = a + bx \tag{4.2.1}$$

to a given set of observations $(x_1, y_1), (x_2, y_2), \ldots (x_n, y_n)$. Since (4.2.1) passes through the data points (x_i, y_i), we have

$$y_i = a + bx_i \tag{4.2.2}$$

The error e_i between the observed and expected values of $y = y_i$ is defined as

$$e_i = y_i - (a + bx_i),\ i = 1,2,3,\ldots n \tag{4.2.3}$$

Therefore the sum of the squares of theses errors is

$$E = \sum_{i=1}^{n} e_i^2 = \sum_{i=1}^{n} [y_i - (a + bx_i)]^2 \tag{4.2.4}$$

Now for E to be minimum, we must have

$$\frac{\partial E}{\partial a} = 0, \quad \frac{\partial E}{\partial b} = 0$$

There provide us

$$\sum_{i=1}^{n} y_i = na + b\sum_{i=1}^{n} x_i$$

and

$$\sum_{i=1}^{n} x_i y_i = a\sum_{i=1}^{n} x_i + b\sum_{i=1}^{n} x_i^2 \tag{4.2.5}$$

The equations (4.2.5) are called normal equations. Solving these equations for a and b, we obtain

$$a = \frac{1}{n}\left[b\sum_{i=1}^{n} x_i - \sum_{i=1}^{n} y_i\right] \tag{4.2.6}$$

$$b = \frac{n\sum_{i=1}^{n} x_i y_i - \left(\sum_{i=1}^{n} x_i\right)\left(\sum_{i=1}^{n} y_i\right)}{n\sum_{i=1}^{n} x_i^2 - \left(\sum_{i=1}^{n} x_i\right)^2} \tag{4.2.7}$$

Upon substitution the values of a and b in (4.2.1) we obtain the required line of "best fit".

4.2.2 Parabolic curve fitting

Suppose the equation of the parabola to be fit is given by

$$y = a + bx + cx^2 \tag{4.2.8}$$

Let the data points be $(x_i, y_i), i = 1,2,3,\ldots n$. Since (4.2.8) passes through these data points,

we have

$$y_i = a + bx_i + cx_i^2 \tag{4.2.9}$$

The error e_i between the observed and expected values of $y = y_i$ is defined as

$$e_i = y_i - \left(a + bx_i + cx_i^2\right),\ i = 1,2,3,...n \tag{4.2.10}$$

Therefore, the sum of the squares of these errors is

$$\text{E} = \sum_{i=1}^{n} e_i^2 = \sum_{i=1}^{n} \left(y_i - a - bx_i - cx_i^2\right)^2 \tag{4.2.11}$$

For E to be minimum, we have

$$\frac{\partial \text{E}}{\partial a} = 0,\ \frac{\partial \text{E}}{\partial b} = 0,\ \frac{\partial \text{E}}{\partial c} = 0$$

which lead to the normal equations

$$\begin{aligned} \Sigma y_i &= na + b\Sigma x_i + c\Sigma x_i^{\,2} \\ \Sigma x_i y_i &= a\Sigma x_i + b\Sigma x_i^{\,2} + c\Sigma x_i^{\,3} \\ \Sigma x_i^{\,2} y_i &= a\Sigma x_i^{\,2} + b\Sigma x_i^{\,3} + c\Sigma x_i^{\,4} \end{aligned} \tag{4.2.12}$$

Solving equations (4.2.12) for a, b and c, and using these values in equation (4.2.8), we obtain the desired curve of best fit.

4.2.3 Fitting of other curves

1. Power curve: Let the curve be given by

$$y = a x^b \tag{4.2.13}$$

Taking logarithms, we get

$$\log_{10}{}^{y} = \log_{10}{}^{a} + b\log_{10}{}^{x}$$

i.e.

$$Y = A + bX \tag{4.2.14}$$

where

$$X = \log_{10}{}^{x},\ Y = \log_{10}{}^{y}$$

Therefore normal equations for (4.2.14) are

$$\Sigma Y_i = nA + b\Sigma X_i$$

$$\Sigma X_i Y = A\Sigma X_i + b\Sigma X_i^2$$

From these A and b can be determined. Then a can be calculated from $A = \log_{10}{}^{a}$. The values of a and b so obtained are used in (4.2.13) to get required curve of best fit.

2. Exponential curve: Let the curve given by,

$$y = a e^{bx} \tag{4.2.15}$$

Taking logarithms, we get

$$\log_{10}^{y} = \log_{10}^{a} + bx \log_{10}^{e}$$

i.e.

$$Y = A + Bx \tag{4.2.16}$$

where

$$Y = \log_{10}^{y},\ \mathrm{A} = \log_{10}^{a},\ \textit{and}\ \mathrm{B} = b\log_{10}^{e}$$

Here the normal equations are

$$\Sigma Y_i = nA + B \Sigma x_i$$

$$\Sigma x_i Y_i = A \Sigma x_i + B \Sigma x_i^{\ 2}$$

From these A and B can be found and consequently a, b can be calculated and used in (4.2.15) to obtain the desired curve of 'best fit'.

3. Gas equation: The gas equation is given by

$$xy^a = b \left(or\ \ p v^r = k\right) \tag{4.2.17}$$

Taking logarithms, we get

$$\log_{10}^{y} = \frac{1}{a} \log_{10}^{b} - \frac{1}{a} \log_{10}^{x}$$

This is of the form

$$Y = A + BX \tag{4.2.18}$$

where

$$Y = \log_{10} y, \quad A = \frac{1}{\mathrm{a}} \log_{10} b, \quad B = \frac{-1}{\mathrm{a}}, \quad X = \log_{10} x$$

Here also the problem reduces to find a straight line of best fit through the given data.

Example 4.2.1 Fit a second-degree curve to the following data

x:	1.0	1.5	2.0	2.5	3.0	3.5	4.0
y:	1.1	1.3	1.6	2.0	2.7	3.4	4.1

Solution: Let the required curve be $y = a + bx + cx^2$

We put $X = 2x - 5$ so that this equation becomes

$$y = a + bX + cX^2$$

x	y	Y	Xy	X^2	X^2y	X^3	X^4
1.0	−3	1.1	−3.3	9	9.9	−27	81
1.5	−2	1.3	−2.6	4	5.2	−8	16
2.0	−1	1.6	−1.6	1	1.6	−1	1
2.5	0	2.0	0.0	0	0.0	0	0
3.0	1	2.7	2.7	1	2.7	1	1
3.5	2	6.8	6.8	4	13.6	8	16
4.0	3	12.3	12.3	9	36.9	27	81
Total	0	16.2	14.3	28	69.9	0	196

The normal equations are

$$7a+28c=16.2\,,\ 28b=14.3\,,28a+196c=69.9$$

On solving these equations lead to $a=2.07$, b = 0.511, c = 0.061

Therefore,

$$y=2.07+0.511X+0.061X^2\ =2.07+0.511(2x-5)+0.061(2x-5)^2$$

$$=1.04-0.198x+0.244x^2$$

Thus the required second-degree curve (parabola) is

$$y=1.04-0.198x+0.244x^2$$

Example 4.2.2 Fit a straight line to the following data:

Year (x):	1951	1961	1971	1981	1991
Production (y): in thousand tons	8	10	12	10	16

Also find the expected production in 1996.

Solution: Suppose the equation of the required straight line be $y=a+bx$

x	y	xy	x^2
1951	8	15608	3806401
1961	10	19610	3845521
1971	12	23652	3884841
1981	10	19810	3924361
1991	16	31856	3964081
$\Sigma x=9855$	$\Sigma y=56$	$\Sigma xy=110536$	19425205

The normal equations are

$$\Sigma y = na + b\Sigma x$$

$$\Sigma xy = a\Sigma x + b\Sigma x^2$$

imply that

$$5a + 9855b = 56$$

$$9855a + 19425205b = 110536$$

Solving these equations, we get $a = 304.16, b = .16$

$\therefore$ The required straight line is $y = .16\,x$ -304.16

4.2.4 General least square method

Here we propose the following model in terms of the unknown coefficients $c_j, j = 1,2,3,...m$ as

$$y_i = c_1 f_1(x_i) + c_2 f_2(x_i) + \cdots + c_m f_m(x_i),\ i = 1,2,3,\cdots n \qquad (4.2.19)$$

where $x_i \in R^1$ or R^n. The error committed at the *i*th point in approximating the observed value y_i for the expected value y_i^* is

$$e_i = y_i - y_i^*,\ \text{for } i=1,2,3,\ldots n$$

$$= y_i - \sum_{j=1}^{m} c_j f_j(x_i),\ i = 1,2,3,...n$$

This is a system of n equations in $(n+m)$ unknowns c_j, $j = 1,2,3,...m$ and $e_i, i = 1,2,3,...n$ and hence it has infinite number of solutions. Among all these solutions, we define the best solution as the one that minimizes the scalar quantity

$$E = \sum_{i=1}^{n} w_i e_i^2 \qquad (4.2.20)$$

where $w_i \geq 0$ are the weights so that

$$\| y - y^* \|_2 = \sum_{i=1}^{n} (y_i - y_i^*)^2 w_i \qquad (4.2.21)$$

defines the weighted least squares norm. We look for minimizing E with respect to the parameters $\{c_1, c_2, c_3, ..., c_m\}$. The necessary condition in which E is minimum in the space of $\{c_1, c_2, ... c_m\}$ is that

$$\frac{\partial E}{\partial c_k} = 0,\ \text{k} = 1,2,3\ldots n \qquad (4.2.22)$$

Permuting the summation and partial derivative operation and taking the derivative, we obtain

$$2\sum_{i-1}^{n} w_i e_i \frac{\partial e_i}{\partial c_k} = 0, k = 1,2,...,m \tag{4.2.23}$$

Using equation (4.2.19) in (4.2.23), we obtain

$$\sum_{i=1}^{n} w_i \left[y_i - \sum_{j=1}^{m} c_j f_j(x_i) \right] f_k(x_i) = 0, k = 1,2,\ldots m$$

This implies that

$$\sum_{j=1}^{m} c_j \sum_{i=1}^{n} w_i f_j(x_i) f_k(x_i) = \sum_{i=1}^{n} w_i y_i f_k(x_i), \quad k = 1,2,3...m$$

$\Rightarrow$

$$\sum_{j=1}^{m} a_{kj} c_j = b_k, k = 1,2...m \tag{4.2.24}$$

where

$$a_{kj} = \sum_{i=1}^{n} f_j(x_i) f_k(x_i) w_{i,} \quad k = 1,2,...m, j = 1,2,3...m$$

$$b_k = \sum_{i=1}^{n} w_i y_i f_k(x_i), \quad k = 1,2...m \tag{4.2.25}$$

In matrix form we have

$$\begin{bmatrix} a_{11}\, a_{12} \ldots a_{1n} \\ a_{21}\, a_{22} \ldots a_{2n} \\ \vdots \quad \vdots \quad \vdots \\ \mathrm{a}_{m1}\; a_{2m} \cdots a_{mn} \end{bmatrix} \begin{bmatrix} c_1 \\ c_2 \\ \vdots \\ c_m \end{bmatrix} = \begin{bmatrix} b_1 \\ b_2 \\ \vdots \\ b_m \end{bmatrix} \tag{4.2.26}$$

The quantities a_{kj} and b_k defined in (4.2.25) are completely determined because the pairs $(x_{i,} y_i)$ and the function $f_j(x)$ are known in the linear system (4.2.26), it remains only to determine these m unknowns $\{c_{1,} c_2, \ldots, c_m\}$.

Algorithm 4.2.1 Least squares method: Linear model

The n pairs (x_i, y_i) of data points are given along with their weights w_i

1. Define the function $f_j(x)$ composing the model so that $y_i^* = \sum_{j=1}^{m} c_j f_j(x_i)$

2. For $k = 1,2,3,\cdots m$, calculate the terms

$$a_{kj} = \sum_{i=1}^{n} f_j(x_j) f_k(x_i) w_i, j = 1,2,3,\ldots m$$

and

$$b_k = \sum_{i=1}^{n} w_i y_i f_k(x_i)$$

3. Solve the linear system of the symmetric matrix equation

$$\sum_{j=1}^{m} a_{kj} c_{j=b_k}, k = 1,2,3,\ldots m$$

The non-linear models can also be solved by adopting any of the following three classes of method for the solution. Analytic linearization of the model, when it is possible, Newton like method which consists of numerical linearization of the model using direct optimization methods.

4.2.5 Quality of approximation

For the approximation of a real continuous function $y(x), x \in [a,b]$ or a set of numerical data $\{(x_i.y_i), i = 1,\ldots n\}$, by an analytic function $y^*(x)$ or $y(x_i)$ we always like to do a good approximation i.e. the quantity $y - y^*$ must be small in same sense. The distance between the real function y and its model y^* can be measured by the norm $z = \left\| y - y^* \right\|_p$. For the approximation of a set of numerical data $(x_{i,}y_i)$ where $y_i = y(x_i)$, the common norms are

$$L_1 - norm(Laplace\ norm): \left\| y - y^* \right\|_1 = \sum_{i=1}^{n} \left| y_i - y_i^* \right| \quad (4.7.27)$$

$$L_2 - norm(Euclideam\ \text{norm}): \left\| y - y^* \right\|_2 = \sum_{i=1}^{n} \left| y_i - y_i^* \right|^2 \quad (4.2.28)$$

$$L_\infty norm(Laplace - Chebyshev\ \text{minimum norm}): \left\| y - y^* \right\|_\infty = \max_i \left| y_i - y_i^* \right| \quad (4.2.29)$$

By definition, the best approximation y^* of y in the sense of best squares will imply that $\left\| y - y^* \right\|$ will be the minimum in the $L_2 - norm$. In practice, the $L_2 - norm$ defined by (4.2.28) is often inadequate for the following reasons.

The values x and $f_i(x)$ defined by

$$y^* = c_1 f_1(x) + c_2 f_2(x) + \ldots + c_m f_m(x) \quad (4.2.30)$$

may correspond to different physical values or different scales.

Some values can be more precise than the others and, therefore, one should be certain that the approximation y^* will be more influenced by the precise values than by the others, for example, in computing the variance of the error in experimental values y_i .

For the above reasons, we associate to each number i of experimental values a weight w_i and the norm of least-squares becomes the weighted least squares norm , defined by

$$\left\|y-y^*\right\|_2 = \sum_{i-1}^{n}\left(y_i - y_1^*\right)^2 w_i \qquad (4.2.31)$$

where $w_i \geq 0$, $i = 1,2,\cdots n$. In case of approximation of a continuous function defined on the interval [a, b], the discrete summation operator is replaced by the integral operator in the definitions of norm. Thus, we have

$$\begin{aligned} L_1 - norm &: \left\|y-y^*\right\|_1 = \int_a^b \left|y(x)-y^*(x)\right| dx \\ L_2 - norm &: \left\|y-y^*\right\|_2 = \int_a^b \left[y(x)-y^*(x)\right]^2 dx \qquad (4.2.32) \\ L_\infty - norm &: \left\|y-y^*\right\|_\infty = \max_{x\in[a,b]} \left|y(x)-y^*(x)\right| \end{aligned}$$

4.2.6 Approximation by trigonometric polynomials

Scientists and engineers often study physical phenomena, such as light and sound, which have a periodic character. They are described by functions $f(x)$ which are periodic that is

$$f(x+T) = f(x) \qquad (4.2.33)$$

for all x and some fixed number T. The number T is called period of the function $f(x)$. It will be sufficient to consider functions that have period 2π. If the function $g(x)$ has period T, then $f(x) = g(xT/2\pi)$ will be periodic with period 2π, because

$$f(x+2\pi) = g\left(\frac{xT}{2\pi}+T\right) = g\left(\frac{xT}{2\pi}\right) = f(x) \qquad (4.2.34)$$

Henceforth in this section we shall assume that $f(x)$ is a periodic function with period 2π, that is,

$$f(x+2\pi) = f(x), \text{ for all x}. \qquad (4.2.35)$$

Since the only periodic polynomials are the constant functions, one has to use other function classes for the effective approximation of periodic functions, and the trigonometric polynomials offer themselves as an appropriate alternative.

Definition 4.2.1 A series of the form

$$p(x) = \frac{a_0}{2} + \sum_{j=1}^{n}\left(a_j \cos jx + b_j \sin jx\right) \qquad (4.2.36)$$

with $a_0, a_j, j = 1,2,3,\cdots n$ are real or complex constants; is called a trigonometric polynomial of order *n*.

Definition 4.2.2 The function $f(x)$ is said to be piecewise continuous on $[a,b]$, if there exists values $t_0, t_1, \cdots, t_k$ with $a = t_0 < t_1 < t_2 < \cdots < t_k = b$ such that $f(x)$ is continuous on each open interval $t_{i-1} < x < t_i$ for $i = 1,2,3,\cdots k$ and $f(x)$ has left hand and right-hand limits at each of the points t_i.

Definition 4.2.3 Assume that $f(x)$ is periodic with period 2π and that $f(x)$ is piecewise continuous on $[-\pi, \pi]$. The Fourier series $S(x)$ for $f(x)$ is defined as

$$S(x) = \frac{a_0}{2} + \sum_{j=1}^{\infty} \left(a_j \cos jx + b_j \sin jx\right) \tag{4.2.37}$$

where the coefficients a_j and b_j are computed with Euler's formulae

$$a_j = \frac{1}{\pi} \int_{-\pi}^{\pi} f(x) \cos jx \; dx, \; j = 0,1,2,3\cdots$$

and

$$b_j = \frac{1}{\pi} \int_{-\pi}^{\pi} f(x) \sin jx \, dx, \; j = 1,2,3\cdots \tag{4.2.38}$$

This series converges to $f(x)$ if the function satisfies Dirichlet's conditions namely, if $f(x)$ is continuous with a piecewise continuous first order derivatives. Since by Euler's formula

$$e^{ix} = \cos x + i \sin x, i = \sqrt{-1} \tag{4.2.39}$$

Therefore, it is convenient to write trigonometric polynomial of order x as

$$p(x) = \sum_{k=-n}^{n} c_k e^{ikx} \tag{4.2.40}$$

where

$$c_k = (a_k - ib_k)/2, \; c_{-k} = (a_k + ib_k)/2, k = 0,1,2,3\cdots n \tag{4.2.41}$$

The equation (4.2.40) represents a real function if and only if it is its own complex conjugate, that is,

$$\overline{\sum_{k=-n}^{n} c_k e^{ikx}} = \sum_{k=-n}^{n} \bar{c}_k \bar{e}^{ikx} = \sum_{k=-n}^{n} \bar{c}_{-k} e^{ikx}$$

This implies that $p(x)$ is real if and only if $c_k = \bar{c}_{-k}$ *for* all k and hence

$$a_k = 2R_e(c_k), \; b_k = -2 1_m(c_k). \tag{4.2.42}$$

Approximation by trigonometric polynomials is dominated by the Fourier series (4.2.37). Assuming that the series on right-hand side of (4.2.37) is convergent for all $x \in [-\pi, \pi]$, the relation $S(x) = f(x)$ holds at all points $x \in [-\pi, \pi]$ where $f(x)$ is continuous. If $x = a$ is a point of discontinuity of $f(x)$, then

$$S(a) = \left(f(\bar{a}) + f(a^+)\right)/2, \tag{4.2.43}$$

where $f(\bar{a})$ and $f(a^+)$ denote the left and right hand limits, respectively. With this understanding, we obtain the Fourier expansion

$$f(x) = \frac{a_0}{2} + \sum_{k=1}^{\infty}\left(a_k \cos kx + b_k \sin kx\right) = \sum_{k=-\infty}^{\infty} c_k e^{ikx}, \tag{4.2.44}$$

where the Fourier coefficients c_k are defined by (4.2.41). Clearly,

$$\frac{1}{2\pi}\int_0^{2\pi} \overline{e^{ijx}}\, e^{ikx}\, dx = \frac{1}{2\pi}\int_0^{2\pi} e^{i(k-j)x}\, dx = \begin{cases} 1, & k=j \\ 0, & k \neq j \end{cases} \tag{4.2.45}$$

which shows that the functions $1,\ e^{\pm ix}, e^{\pm i2x}, \cdots$ are orthonormal with respect to the scalar or inner product. The partial sum $\sum_{k=-n}^{n} c_k e^{ikx}$ of the Fourier series (4.2.44) for $f(x)$ is the best approximation to $f(x)$ by trigonometric polynomials of order n with respect to the norm

$$\| g \| = \| g \|_2 = \left[\frac{1}{2\pi}\int_0^{2\pi} | g(x)|^2 dx\right]^{1/2} \tag{4.2.46}$$

Further, it can be shown that Parseval's relation holds.

$$\sum_k |c_k|^2 = \frac{1}{2\pi}\int_0^{2\pi} | f(x)|^2 dx \tag{4.2.47}$$

The Fourier coefficients c_k for the function $f(x)$ are used to understand the function $f(x)$. Suppose $f(x)$ is a real 2π periodic function. If we think of $f(x)$ as the position at time x of some object moving on a line, then our 2π-periodic function $f(x)$ describes a periodic motion.

If now $c_k = | c_k | e^{i\theta_k}$, then we can write the Fourier series for $f(x)$ as

$$f(x) \approx 2\sum_{j=0}^{\infty} | c_k | \cos(\theta_k + kx). \tag{4.2.48}$$

Hence, we have represented our periodic motion described by $f(x)$ as a sum or superposition of simple harmonic oscillations. The *kth* such motion $2(c_k)\cos(\theta_k + kx)$ has amplitude $2| c_k |$, frequency k, angular frequency $\frac{k}{2\pi}$ and period or wave length $2\pi / k$, phase angle θ_k. The number $|c_k|$ measures the extent of which a simple harmonic motion of angular frequency 'k' is present in the total motion. The sequence $|c_0|, |c_1|, |c_2|, \cdots$ (or the sequence of their squares) is called the power spectrum or simply the spectrum of $f(x)$. As a consequence of Parseval's relation (4.2.48), the spectrum for $f(x)$ is bounded by $\|f\|_2$, but $f(x)$ may have widely differing behavior depending on just how the "total energy" $\|f\|_2$ is distributed over the spectrum $|c_0|, |c_1|, |c_2|, \cdots$. A "noisy" function will have sizable $|c_k|$ for large k, while for a smooth function, the spectrum will decrease rapidly as k increases. A favorite method of smoothing consists in generating the Fourier coefficients of the given function $f(x)$ from data; filtering these coefficients, this means to suppress certain frequencies, usually the high frequencies, in some manner, and then reconstituting the function as a Fourier series with these purified or filtered coefficients.

It is usually not possible to calculate the Fourier coefficients in (4.2.44) exactly, because the integral cannot be evaluated in closed form or the function $f(x)$ is not known exactly. In other case, numerical integration is used. Here the composite trapezoidal approximation rule of integration

$$\int_{\sigma}^{2\pi} g(x)dx \approx \frac{2\pi}{N} \sum_{n=0}^{N-1} g\left(\frac{2\pi n}{N}\right) \qquad (4.2.49)$$

suffices. The rule can be obtained by replacing the 2π-periodic function $g(x)$ under the integral sign by a piecewise linear interpolate which agrees with $g(x)$ at its equispaced break points $0, \pm\frac{2\pi}{N}, \pm\frac{4\pi}{N}, \pm\frac{6\pi}{N}, \cdots$. We denote by $f_N(k)$ the corresponding approximation to c_k:

$$f_N(k) = \frac{1}{N} \sum_{n=0}^{N-1} f(x_n) \bar{e}^{ix_n}, \; x_n = \frac{2\pi n}{N}, \; n = 0,1,2,3,\cdots N-1. \qquad (4.2.51)$$

The points x_n are called the sampling points and the numbers $f(x_n)$ are the corresponding sample values. The number $\frac{2\pi}{N}$ is called the sampling interval and its reciprocals number $N/2\pi$, is called the sampling frequency. We define the discrete inner product

$$\langle g, h \rangle_N = \frac{1}{N} \sum_{n=0}^{N-1} \bar{h}(x_n) g(x_n) \qquad (4.2.51)$$

Using the product

$$\langle e^{ikx}, e^{ijx} \rangle_N = \begin{cases} 1, & if\, k = j (\text{mod N}) \\ 0, & if\, k \neq j (\text{mod N}) \end{cases} \qquad (4.2.52)$$

We have

$$f_N(j) = < f, e^{ijx} > N \tag{4.2.53}$$

Assuming the Fourier series $\sum c_k e^{ikx}$ converges absolutely to $f(x)$ we conclude that

$$f_N(j) = \left\langle \sum_{k=-\infty}^{\infty} c_k e^{ikx}, e^{ijx} \right\rangle N = \sum_k c_k \left\langle e^{ikx}, e^{ijx} \right\rangle N$$

$$= \sum_{k=j(\text{mod}\, N)} c_k \tag{4.2.54}$$

The approximate Fourier coefficient f_N (j) is made up of the entire exact Fourier coefficients c_k whose corresponding function e^{ikx} cannot be distinguished by the inner product (4.2.51) from the function e^{ijx}. This phenomenon is called *aliasing*. It can be shown that $|c_j| = \theta\left\langle |j|^{-k-1} \right\rangle$ in case $f(x)$ has k-1 continuous derivatives and its kth derivative is piecewise continuous. Combining this with (4.2.54) we find that

$$f_N(j) = c_j + 0\left(N^{-k}\right) \tag{4.2.55}$$

in case $f(x)$ has k-1 continuous derivatives and its kth derivative is piecewise continuous. When we sample a function at N equally spaced points in $[0, 2\pi]$, the aliasing effect prevents us from seeing periodic phenomena in $f(x)$ with frequencies higher than $\left(\frac{N}{2}\right)/2\pi$. Thus, if we wish to observe a certain periodic phenomenon of frequency υ, then we must sample at a frequency atleast as large as 2υ.

For any $m \le n$, the mth order trigonometric polynomial

$$p_m(x) = \sum_{j=-m}^{m} f_N(j)\, e^{ijx} \tag{4.2.56}$$

is the best approximation to $f(x)$ by trigonometric polynomials of order m with respect to the discrete mean-square norm

$$\|g\|_2 = \left(\langle g, g \rangle_N\right)^{1/2} = \frac{1}{N}\left(\sum_{j=0}^{N-1} |g(j2\pi/N)|^2\right)^{\frac{1}{2}} \tag{4.2.57}$$

For $m = n$,

$$p_n(x) = \sum_{j=-n}^{n} f_N(j) e^{ijx} \tag{4.2.58}$$

interpolates $f(x)$ at the sampling points $x_j = \dfrac{2\pi j}{N}$, for all j.

If $f(x)$ is a real function, then we can write the interpolating polynomial, according to (4.2.40), in real form as

$$p_n(x)=\frac{a_0}{2}+\sum_{m=1}^{n}(a_m \cos mx + b_m \sin mx) \tag{4.2.59}$$

with

$$a_m = 2\text{Re } f_N(m) = \frac{2}{N}\sum_{k=0}^{N-1} f(x_k)\cos mx_k$$
$$b_m = -2\,I_m f_N(m) = \frac{2}{N}\sum_{k=0}^{n-1} f(x_k)\sin mx_k \tag{4.2.60}$$

The procedures for the approximation of the best approximation are identical for continuous and discrete cases. However, the continuous approximation is theoretically interesting because of the total absence of smoothing of discrete values.

4.3 Finite differences

The calculus of finite differences deals with the changes that take place in the value of a function due to finite changes in the independent variable. On the other head, in infinitesimal calculus, we study those changes of a function which occur when the independent variable changes continuously in a given interval.

Suppose that the function $y=f(x)$ is tabulated for equally spaced values $(x_i, y_i), i=1,2,3,\cdots n$ such that $x_i = x_0 + ih$. If we are required to recover the values of $f(x)$ or its derivatives for some intermediate values of x in the range $x_0 \le x \le x_n$, following three types of differences are found useful.

4.3.1 Forward differences

The differences $y_1 - y_0$, $y_2 - y_1$, $\cdots, y_n - y_{n-1}$ denoted by Δy_0, $\Delta y_1 \cdots \Delta y_{n-1}$ respectively, are called the first forward differences and the operator Δ is known as forward difference operator. The differences of the first forward differences are called second forward differences and are denoted by $\Delta^2 y_0, \Delta^2 y_1 \cdots etc.$ Thus we have

$$\Delta y_n = y_{n+1} - y_n \tag{4.3.1}$$

$$\Delta^2 y_n = \Delta y_{n+1} - y_n = y_2 - 2y_1 + y_0$$

$$\Delta^3 y_n = \Delta^2 y_{n+1} - \Delta^2 y_n = y_3 - 3y_2 + 3y_1 - y_0$$

and so on. In general,

$$\Delta^p y_n = \Delta^{p-1} y_{n+1} - \Delta^{p-1} y_n \tag{4.3.2}$$

defines the *pth* forward differences. Following table shows how the forward differences of all orders can be formed.

The operator Δ obeys the following laws:

(i) $\Delta[f(x) \pm g(x)] = \Delta f(x) \pm \Delta g(x)$

(ii) $\Delta[cf(x)] = c\Delta f(x)$, c being a constant

Table 4.1 Forward difference table

x	y	Δ	Δ^2	Δ^3	Δ^4	Δ^5
x_0	y_0					
		Δy_0				
x_1	y_1		$\Delta^2 y_0$			
		Δy_1		$\Delta^3 y_0$		
x_2	y_2		$\Delta^2 y_1$		$\Delta^4 y_0$	
		Δy_2		$\Delta^3 y_1$		$\Delta^5 y_0$
x_3	y_3		$\Delta^2 y_2$		$\Delta^4 y_1$	
		Δy_3		$\Delta^3 y_2$		
x_4	y_4		$\Delta^2 y_3$			
		Δy_4				
x_5	y_5					

(iii) $\Delta^m \Delta^n f(x) = \Delta^{m+n} f(x)$, m, n are positive integers

In view of (i) and (ii), Δ is a linear operator, but

$$\Delta[f(x).g(x)] \neq \Delta f(x).\Delta g(x)$$

4.3.2 Backward differences

The differences $y_1 - y_0$, $y_2 - y_1$, $\cdots$, $y_n - y_{n-1}$ when denoted by $\nabla y_1, \nabla y_2, \cdots \nabla y_n$ respectively, are called the first backward differences where ∇ is the backward difference operator. Thus, we have

$$\nabla y_n = y_n - y_{n-1} \tag{4.3.3}$$

$$\nabla^2 y_n = \nabla y_n - \nabla y_{n-1}$$

$$\nabla^3 y_n = \nabla^2 y_n - \nabla^2 y_{n-1}$$

and so on. In general

$$\nabla^p y_n = \nabla^p y_n - \nabla^p y_{n-1} \tag{4.3.4}$$

defines the pth backward difference. Following table shows how the forward differences of all orders can be formed:

Table 4.2: Backward differences table.

x	y	∇	∇^2	∇^3	∇^4	∇^5
x_0	y_0					
		∇y_1				
x_1	y_1		$\nabla^2 y_2$			
		∇y_2		$\nabla^3 y_3$		
x_2	y_2		$\nabla^2 y_3$		$\nabla^4 y_4$	
		∇y_3		$\nabla^3 y_4$		$\nabla^5 y_5$
x_3	y_3		$\nabla^2 y_4$		$\nabla^4 y_5$	
		∇y_4		$\nabla^3 y_5$		
x_4	y_4		$\nabla^2 y_5$			
		∇y_5				
x_5	y_5					

4.3.3 Central differences

The central difference operator δ is defined by the relations

$$\delta y_{\frac{1}{2}} = y_1 - y_0,\ \delta y_{3/2} = y_2 - y_1, \cdots \delta y_{n-\frac{1}{2}} = y_n - y_{n-1} \tag{4.3.5}$$

Similarly, higher order central differences can be defined with values of x and y as in the proceeding two tables, a central difference table can be also formed. Thus, we have table 4.3. It is clear from the three tables that in a definite numerical case, the same numbers occur in the same position whether we use forward, backward or central differences. Thus, we have $\Delta y_0 = \nabla y_1 = \delta y_{1/2}$, $\Delta^3 y_2 = \nabla^3 y_5 = \delta^3 y_{7/2}$ etc. Therefore we noticed that it is only the notation which changes and not the difference. Of all the formulae, those involving central differences are most useful in practice as the coefficients in such formulae decrease much more rapidly.

Table 4.3 Central difference table

x	y	δ	δ^2	δ^3	δ^4	δ^5
x_0	y_0					
		$\delta y_{1/2}$				
x_1	y_1		$\delta^2 y_1$			
		$\delta y_{3/2}$		$\delta^3 y_{3/2}$		
x_2	y_2		$\delta^2 y_2$		$\delta^4 y_2$	
		$\delta y_{5/2}$		$\delta^3 y_{5/2}$		$\delta^5 y_{5/2}$
x_3	y_3		$\delta^2 y_3$		$\delta^4 y_3$	
		$\delta y_{7/2}$		$\delta^3 y_{7/2}$		
x_4	y_4		$\delta^2 y_4$			
		$\delta y_{9/2}$				
x_5	y_5					

4.3.4 Detection of errors

Difference tables can be used to check error in tabular values. Suppose there is an error $\in$ in the entry y_5 of a table. As higher differences are formed this error spreads out and is considerably magnified. The following difference table illustrates the propagation of error.

It is clear from the table 4.4 that:

(i) The error increases with the order of differences.

(ii) The coefficients of the error $(\in)$ in any column are the combinatorial coefficients of $(1+\in)^n$.

(iii) The algebraic sum of the error in any difference column is zero.

(iv) The maximum error in each column occurs opposite to entry containing the error.

These facts enable us to detect error in a difference table.

Table 4.4: Propagation of error

x	y	Δ	Δ^2	Δ^3	Δ^4	Δ^5
x_0	y_0					
		Δy_0				
x_1	y_1		$\Delta^2 y_0$			
		Δy_1		$\Delta^3 y_0$		
x_2	y_2		$\Delta^2 y_1$		$\Delta^4 y_0$	
		Δy_2		$\Delta^3 y_1$		$\Delta^5 y + \in$
x_3	y_3		$\Delta^2 y_2$		$\Delta^4 y_1 + \in$	
		Δy_3		$\Delta^3 y_2 + \in$		$\Delta^5 y_1 - 5\in$
x_4	y_4		$\Delta^2 y_3 + \in$		$\Delta^4 y_1 - \in$	
		$\Delta y_4 + \in$		$\Delta^3 y_3 - 3\in$		$\Delta^5 y_2 + 10\in$
x_5	$y_5 + \in$		$\Delta^2 y_4 - 2\in$		$\Delta^4 y_2 + 6\in$	
		$\Delta y_5 - \in$		$\Delta^3 y_4 + 3\in$		$\Delta^5 y_3 - 10\in$
x_6	y_6		$\Delta^2 y_5 + \in$		$\Delta^4 y_4 - 4\in$	
		Δy_6		$\Delta^3 y_5 - \in$		$\Delta^5 y_4 + 5\in$
x_7	y_7		$\Delta^2 y_6$		$\Delta^4 y_5 + \in$	
		Δy_7		$\Delta^3 y_6$		$\Delta^5 y_5 - \in$
x_8	y_8		$\Delta^2 y_7$		$\Delta^4 y_6$	
		Δy_8		$\Delta^3 y_7$		
x_9	y_9		$\Delta^2 y_8$			
		Δy_9				
y_{10}	y_{10}					

Examples 4.3.1 In the following table locate and correct the error, by using difference table, in which y is a cubic polynomial in x.

x:	0	1	2	3	4	5	6	7
y:	25	21	18	18	27	45	76	123

Solution The difference table is given by

x	y	Δy	$\Delta^2 y$	$\Delta^3 y$
0	25			
		-4		
1	21		1	
		-3		2
2	18		3	
		0		6
3	18		9	
		9		0
4	27		9	
		18		4
5	45		13	
		31		3
6	76		16	
		47		
7	123			

Since y is a polynomial of third of degree, so $\Delta^3 y$ must be constant $i,e,$ the same. The sum of the third order difference being 15, each entry under $\Delta^3 y$ must be $15/5 = 3$. Thus the four entries under $\Delta^3 y$ are in error and can be written as

$$3+(-1),\ 3-(-1),\ 3+(-1),\ 3-(-1)$$

We take $\in = -1$ and find that the entry corresponding to $x = 3$ is in error. Therefore, $y + \in = 18$, this implies that the true values of $y = 18 - \in = 18-(-1) = 19$

4.3.5 Differences of a polynomial

Let $f(x)$ be a polynomial of nth degree in x, so that

$$f(x) = a_0 x^n + a_1 x^{n-1} + a_2 x^{n-2} + \cdots + a_n \tag{4.3.6}$$

Then

$$\begin{aligned}\Delta f(x) &= f(x+h) - f(x) \\ &= a_0\left[(x+h)^n - x^n\right] + a_1\left[(x+h)^{n-1} - x^{n-1}\right] + a_2\left[(x+h)^{n-2} - x^{n-2}\right] + \cdots \\ &= a_0\, nh\, x^{n-1} + a_1' x^{n-2} + \cdots + a_{n-1}' x + a_n' \end{aligned} \tag{4.3.7}$$

where a_1', $a_2', \cdots a_n'$ are the new coefficients. Thus the first difference of a polynomial of nth degree is a polynomial of degree $(n-1)$. Similarly, we have

$$\begin{aligned}\Delta^2 f(x) &= \Delta[f(x+h) - f(x)] = \Delta f(x+h) - \Delta f(x) \\ &= a_0 nh\left[[x+h]^{n-1} - x^{n-1}\right] + a_1'\left[(x+h)^{n-2} - x^{n-2}\right] + \cdots + a_{n-1}' h \\ &= a_0 n(n-1)h^2\, x^{n-2} + a_1'' x^{n-3} + a_2'' x^{n-4} + \cdots + a'' \end{aligned} \tag{4.3.8}$$

Therefore, the second differences represent a polynomial of degree $(n-2)$. Continuing this process for nth differences, we obtain a polynomial of degree zero i.e.

$$\Delta^n f(x) = a_0 n(n-1)(n-2)\cdots 1.h^n = a_0 n!\, h^n \tag{4.3.9}$$

which is a constant. Hence the $(n+1)$th and higher difference of a polynomial of nth degree will be zero. The Converse is also true *i.e.* if the nth differences of a function tabulated at equally spaced intervals are constant, the function is a polynomial of degree *n*. This fact is important in numerical analysis because it enables us to approximate a function by an nth degree polynomial, if its nth order differences become nearly constant.

The difference of polynomials can be conveniently computed with the help of factorial notation. A product of the form

$$x(x-1)(x-2)\cdots(x-r+1) = [x]^r \tag{4.3.10}$$

is called a factorial. In particular,

$$[x] = x,\ [x]^2 = x(x-1),\ [x]^3 = x(x-1)(x-2) \text{ etc.}$$

In general $[x]^n = x(x-1)(x-2)\cdots(x-\overline{n-1})$. In case the interval of differencing is h, then

$$[x]^n = x(x-h)(x-2h)\cdots(x-\overline{n-1}h) \tag{4.3.11}$$

The factorial notation is of special utility in the theory of finite differences and it helps in finding the successive differences of a polynomial directly by simple rule of differentiation.

Similarly, given any difference of a function in factorial notation, we can find the corresponding function by simple integration. Now, we shall show that

$$\Delta^n [x]^n = n! \text{ and } \Delta^{n+1}[x]^n = 0.$$

We have

$$\Delta^n [x]^n = [x+h]^n - [x]^n$$

$$= \{(x+h)(x+h-h)(x+h-2h)\cdots(x+h-\overline{n-1}\,h)\} - x(x-h)(x-2h) - (x-\overline{n-1}\,h)$$

$$= x(x-h)(x-2h)\cdots(x-\overline{n-2}h)\,[x+h-(x-nh+h)] = nh[x]^{n-1}$$

Similarly

$$\Delta^2 [x]^n = \Delta\{n\,h[x]^{n-1}\} = n\,h\,\Delta[x]^{n-1} = n\,h(n-1)h[x]^{n-2}$$

$$= n(n-1)h^2[x]^{n-2}$$

Proceeding in this way, we get

$$\Delta^{n-1}[x]^n = n(n-1)\cdots 2h^{n-1}x$$

$$\Delta^n [x]^n = n(n-1)(n-2)\cdots 2.1\,h^{n-1}(x+h-x) = n!\,h^n$$

Also $$\Delta^{n+1}[x]^n = n!\,h^n - n!\,h^n = 0$$

In particular, when $h = 1$, we have

$$\Delta[x]^n = n[x]^{n-1} \text{ and } \Delta^n [x]^n = n!$$

Thus we conclude that the result of differencing $[x]^r$ is analogous to that of differentiating x^r when $h = 1$. Also every polynomial of degree n can be expressed as a factorial polynomial of the same degree and vice-versa.

Example 4.5.1 Express $f(x) = 2x^3 - 3x^2 + 3x - 10$ in factorial form and hence evaluate $\Delta^3 f(x)$.

Solution: Suppose

$$f(x) = A[x]^3 + B[x]^2 + C[x] + D \tag{4.3.12}$$

Using the method of synthetic division, dividing by x, $x-1$, $x-2$, etc. successively, we have

	x^3	x^2	X	x^0
1	2	−3	3	−10=D
	2	2	−1	
2	2	−1	2 = C	
	–	4		
3	2	3 = B		
	–			
	2 = A			

Hence

$$f(x) = 2[x]^3 + 3[x]^2 + 2[x] - 10 \quad (4.3.13)$$

It can be noted that the coefficient of highest power if x remains unchanged while transforming a polynomial to factorial notation.

∴
$$\Delta f(x) = 6[x]^2 + 6[x] + 2$$
$$\Delta^2 f(x) = 12[x] + 6$$
$$\Delta^3 f(x) = 12$$

Alternatively, we can also express the polynomial $f(x)$ in the factorial notation as follows.

Suppose

$$f(x) = 2x^3 - 3x^2 + 3x - 10 = 2x(x-1)(x-2) + Bx(x-1) + Cx + D$$

Putting $x = 0$, we get D =- 10.

Putting $x = 1$,

$$C + D = 2 - 3 + 3 - 10 = -8 \Rightarrow c = -8 + 10 = 2$$

Putting $x = 2$,

$$2B + 2C + D = 16 - 12 + 6 - 10 \Rightarrow B = 3$$

Hence

$$f(x) = 2x(x-1)(x-2) + 3x(x-1) + 2x - 10$$
$$= 2[x]^3 + 3[x]^2 + 2[x] - 10$$

∴
$$\Delta f(x) = 6[x]^2 + 6[x] + 2$$
$$\Delta^2 f(x) = 12[x] + 6$$
$$\Delta^3 f(x) = 12$$

4.3.6 Difference operators and symbolic relations

In Addition to the operators Δ, ∇ and δ, there are operators E and μ which are defined below:

(i) **Shifting operator:** The shifting operator E is the operation of increasing the argument x by h so that

$$Ef(x) = f(x+h),\ E^2 f(x) = f(x+2h),\ E^3 f(x) = f(x+3h) \text{ so on.}$$

$$E^n f(x) = f(x+nh). \tag{4.3.14}$$

Similarly,

$$\bar{E}^n f(x) = f(x-nh) \tag{4.3.15}$$

Thus

$$E^2 y_0 = y_1,\ Ey_1 = y_2 \cdots Ey_n = y_{n+1}. \tag{4.3.16}$$

Also

$$E^2 y_0 = E(Ey_0) = E(y_1) = y_2,$$

$$E^3 y_0 = E(E^2 y_0) = E(y_2) = y_3,$$

and so on. In general

$$E^n y_0 = y_n. \tag{4.3.17}$$

(ii) **Averaging operator:** The average operator μ is defined as

$$\mu y_n = \frac{1}{2}\left(y_{n+\frac{1}{2}} + y_{n-\frac{1}{2}}\right) \tag{4.3.18}$$

In the difference calculus E is regarded as the fundamental operator and $\Delta, \nabla, \delta, \mu, D = \frac{\mathrm{d}}{\mathrm{dx}}$ can be expressed in terms of E.

We have

$$\Delta y_n = y_{n+1} - y_n = Ey_n - y_n = (E-1)y_n$$

$$\Rightarrow \qquad \Delta \equiv E - 1 \text{ or } E \equiv 1 + \Delta \tag{4.3.19}$$

$$\nabla y_n = y_n - y_{n-1} = y_n - E^{-1} y_n = (1 - E^{-1})y_n$$

$$\Rightarrow \qquad \nabla \equiv 1 - E^{-1} \text{ or } \mathrm{E} = (1 - \nabla)^{-1} \tag{4.3.20}$$

$$\delta y_n = y_{n+\frac{1}{2}} - y_{n-\frac{1}{2}} = E^{1/2} y_n - E^{-1/2} y_n = (E^{1/2} - E^{-1/2})y_n$$

$$\Rightarrow \qquad \delta \equiv E^{1/2} - E^{-1/2} \tag{4.3.21}$$

$$\mu y_n = \frac{1}{2}\left(y_{n+\frac{1}{2}} + y_{n-\frac{1}{2}}\right) = \frac{1}{2}\left(E^{1/2} y_n + E^{-1/2} y_n\right)$$

$$= \left(\frac{E^{1/2} + E^{-1/2}}{2}\right) y_n$$

$\Rightarrow$
$$\mu \equiv \frac{E^{1/2} + E^{-1/2}}{2} \tag{4.3.22}$$

$$(E\nabla) y_n = E(y_n - y_{n-1}) = Ey_n - Ey_{n-1} = y_{n+1} - y_n = \Delta y_n$$

$\Rightarrow$
$$\Delta \equiv E\nabla$$

Also
$$(\nabla E) y_n = \nabla(Ey_n) = \nabla(y_{n+1}) = y_{n+1} - y_n = \Delta y_n$$

$\Rightarrow$
$$\Delta \equiv \nabla E$$

$$(\delta E^{1/2}) y_n = \delta(E^{1/2} y_n) = \delta\left(y_{n+\frac{1}{2}}\right) = y_{n+1} - y_n = \Delta y_n$$

$\Rightarrow$
$$\Delta \equiv \delta E^{1/2}$$

Thus

$$\Delta \equiv E\nabla \equiv \nabla E \equiv \delta E^{1/2} \tag{4.3.23}$$

Again, we have

$$\mu = \frac{1}{2}\left(E^{1/2} + E^{-1/2}\right)$$

$\Rightarrow$
$$\mu^2 \equiv \frac{\left(E^{1/2} + E^{-1/2}\right)^2}{4} = \frac{1}{4}\left[4 + \left(E^{1/2} - E^{-1/2}\right)^2\right]$$

$$= \frac{1}{4}\left[4 + \delta^2\right] = 1 + \frac{\delta^2}{4}$$

$\Rightarrow$
$$\mu \equiv \sqrt{1 + \frac{\delta^2}{4}} \tag{4.3.24}$$

Now

$$Ef(x) = f(x+h) = f(x) + \frac{h}{1!} f'(x) + \frac{h^2}{2!} f''(x) + \cdots \text{ (by Taylor's theorem)}$$

$$= f(x) + \frac{hD}{1!} f(x) + \frac{h^2 D^2}{2!} f(x) + \cdots$$

$$= \left(1 + \frac{hD}{1!} + \frac{h^2D^2}{2!} + \frac{h^3D^3}{3!} + \cdots\right) f(x)$$

$$= e^{hD} f(x)$$

$$\Rightarrow \qquad E = e^{hD} = 1 + \Delta \text{ or } D = \frac{1}{h} \log E = \frac{1}{h} \log(1+\Delta) \qquad (4.3.25)$$

Thus we have established the symbolic relations

(i) $E \equiv 1 + \Delta$ (ii) $\nabla \equiv 1 - E^{-1}$ (iii) $\delta = E^{1/2} - E^{-1/2}$ (iv) $\mu = \frac{1}{2}\left(E^{1/2} + E^{-1/2}\right)$

(v) $\Delta \equiv E\nabla \equiv \nabla E = \delta E^{1/2}$ (vi) $\mu = \sqrt{1 + \delta^2/4}$ (vii) $E \equiv 1 + \Delta = e^{hD}$ (viii) $D = \frac{1}{h}\log(1+\Delta)$

We have seen that using relation (4.3.19), a number of useful identities can be derived. This relation is used to separate the effect of E into that of the powers of Δ and this method of separation is called the method of separation of symbols. This is demonstrated in the following example.

Example 4.3.2: Using the method of separation of symbols prove that

(i) $\Delta^n u_{x-n} = u_x - nu_{x-1} + \frac{n(n-1)}{2} u_{x-2} + \cdots + (-1)^n u_{x-n}$

(ii) $u_1 x + u_2 x^2 + u_3 x^3 + \cdots = \frac{x}{1-x} u_1 \left(\frac{x}{1-x}\right)^2 \Delta u_1 + \left(\frac{x}{1-x}\right)^3 \Delta^2 u_1 + \cdots$

(iii) $u_0 + \frac{u_1 x}{1!} + \frac{u_2 x^2}{2!} + \frac{u_3 x^3}{3!} + \cdots = e^x \left(u_0 + x\Delta u_0 + \frac{x^2}{2!}\Delta^2 u_0 + \frac{x^3}{3!}\Delta^3 u_0 + \cdots\right)$

Solution:

(i) $R.H.S. = u_x - nu_{x-1} + \frac{n(n-1)}{2} u_{x-2} + \cdots + (-1)^n u_{x-n}$

$$= u_x - nE^{-1}u_x + \frac{n(n-1)}{2} E^{-2} u_x + \cdots + (-1)^n E^{-n} u_x$$

$$= \left[1 - nE^{-1} + \frac{n(n-1)}{2} E^{-2} + \cdots + (-1)^n E^{-n}\right] u_x$$

$$= \left(1 - E^{-1}\right)^n u_x = \Delta_n E^{-x} u_x$$

$$= \Delta^n u_{x-n} = L.H.S.$$

Hence the result is proved.

(ii) $L.H.S. = x u_1 + x^2 u_2 + x^3 u_3 + \cdots$

$$= x u_1 + x^2 E u_1 + x^3 E^2 u_1 + \cdots$$

$$= x\left(1 + xE + x^2 E^2 + \cdots\right)u_1$$

$$= x\frac{1}{1-xE}u_1, \text{ using sum of infinite } G.P.$$

$$= x\left[\frac{1}{1-x(1+\Delta)}u_1\right] = x\left[\frac{1}{1-x-x\Delta}\right]u_1$$

$$= \frac{x}{1-x}\left(1-\frac{x\Delta}{1-x}\right)^{-1}u_1$$

$$= \frac{x}{1-x}\left(1+\frac{x\Delta}{1-x}+\frac{x^2\Delta^2}{(1-x)^2}+\cdots\right)u_1$$

$$= \frac{x}{1-x}u_1 + \frac{x^2}{(1-x)^2}\Delta u_1 + \frac{x^3}{(1-x)^3}\Delta^2 u_1 + \cdots$$

$$= R.H.S.$$

Hence result in proved.

(iii) $L.H.S. = u_0 + \frac{x}{1!}u_1 + \frac{u_2}{2!}x^2 + \frac{u_3}{3!}x^3 + \cdots$

$$= u_0 + \frac{x}{1!}Eu_0 + \frac{x^2}{2!}E^2 u_0 + \frac{x^3}{3!}E^3 u_0 + \cdots$$

$$= \left(1 + \frac{xE}{1!} + \frac{x^2E^2}{2!} + \frac{x^3E^3}{3!} + \cdots\right)u_0$$

$$= e^{xE}u_0 = e^{x(1+\Delta)}u_0 = e^x e^{\Delta} x_0$$

$$= e^x\left(a + \frac{x\Delta}{\angle 1} + \frac{x^2\Delta^2}{\angle 2} + \frac{x^3\Delta^3}{\angle 3} + \cdots\right)u_0$$

$$= x^x\left[u_0 + \frac{x}{\angle 1}\Delta u_0 + \frac{x^2}{\angle 2}\Delta^2 u_0 + \frac{x^3}{\angle 3}\Delta^3 u_0 + \cdots\right]$$

$$= R.H.S.$$

Hence the result is proved.

Example 4.3.3 Prove with usual notations that

(i) $hD = \log(1+\Delta) = -\log(1-\nabla) = \sinh^{-1}(\mu\delta)$

(ii) $\Delta = \frac{1}{2}\delta^2 + \delta\sqrt{(1+\delta^2/4)}$

(iii) $\Delta^3 y_2 = \nabla^3 y_5$

Solution:

(i) We know that $e^{hD} = E = 1 + \Delta$

$$hD = \log(1+\Delta)$$

Also $$hD = \log E = -\log(E^{-1}) = -\log(1-\nabla)$$

Again $$\mu = \frac{1}{2}\left(E^{1/2} + E^{-1/2}\right) \text{and } \delta = E^{1/2} - E^{-1/2}$$

$\therefore$ $$\mu\delta = \frac{1}{2}\left(E^{1/2} + E^{-1/2}\right)\left(E^{1/2} - E^{-1/2}\right)$$

$$= \frac{1}{2}\left(E - E^{-1}\right) = \frac{1}{2}\left(e^{hD} - e^{-hD}\right)$$

$$= \sinh(hD)$$

$\Rightarrow$ $$hD = \sinh^{-1}(\mu\delta)$$

Thus $$hD = \log(1+\Delta) = -\log(1-\nabla) = \sinh^{-1}(\mu\delta)$$

$$\frac{1}{2}\delta^2 + \delta\sqrt{1+\frac{\delta^2}{4}} = \frac{1}{2}\left(E^{1/2} + E^{-1/2}\right)^2 + \left(E^{1/2} - E^{-1/2}\right)\left[\sqrt{1+(E^{1/2} - E^{-1/2})^2/4}\right]$$

$$= \frac{1}{2}\left(E + E^{-1} - 2\right) + \left(E^{1/2} - E^{-1/2}\right)\left[\left(\sqrt{E + E^{-1} + 2}\right)/2\right]$$

$$= \frac{1}{2}\left(E + E^{-1} - 2\right) + \left(E^{1/2} - E^{-1/2}\right)\left(E^{1/2} + E^{-1/2}\right)/2$$

$$= \frac{1}{2}\left[\left(E + E^{-1} - 2\right) + \left(E - E^{-1}\right)\right] = \frac{1}{2}(2E - 2)$$

$$= E - 1 = \Delta,$$

Hence the result is proved.

(ii) $\Delta^3 y_2 = (E-1)^3 y_2 = \left(E^3 - 3E^2 + 3E - 1\right)y_2 = y_5 - 3y_4 + 3y_3 - y_2$

$\Delta^3 y_5 = \left(1 - E^{-1}\right)^3 y_5 = \left(1 - 3E^{-1} + 3E^{-2} - E^{-3}\right)y_5 = y_5 - 3y_4 + 3y_3 - y_2$

$\therefore$ $\Delta^3 y_2 = \Delta^3 y_5$

4.4 Interpolation

Suppose (x_i, y_i), $i = 0,1,2,3,4,\cdots n$ be the set of $(n+1)$ given data points of the function $y = f(x)$. The process of finding the value of y corresponding to any value of $x = x_i$ between x_0 and x_n, is called interpolation. Thus interpolation is the technique of estimating the value of a function for any intermediate value of the independent variable whereas the process of computing the value of the function outside the given range is called extrapolation. If the function $f(x)$ is known explicitly, then the value of y corresponding to any value of x can easily be found. Conversely, if the form of $f(x)$ is not known (as is the case in most applications), it is very difficult to determine the exact form of $f(x)$ with the help of tabulated values (x_i, y_i). In such cases, it is required to find a simpler function, say $\phi(x)$, such that $f(x)$ and $\phi(x)$ agree at the set of tabulated points. The function $\phi(x)$ is a called interpolating or smoothing function and any other value may be calculated from $\phi(x)$. If $\phi(x)$ is polynomial, then the process is called polynomial interpolation and $\phi(x)$ is called the interpolating polynomial. Similarly, different types of interpolation arise depending on whether $\phi(x)$ is a finite trigonometric series, series of Bessel functions, etc. However, in this chapter we shall restrict ourselves to polynomial interpolation only. The approximation of an unknown function by means of a polynomial is justified as a consequence of a famous theorem due to Weierstrass (1885), which states that if there is a function $y = f(x)$, which is continuous in $a \le x \le b$, then we can approximate it by a polynomial, $p_n(x)$, of degree $n(\in)$ such that $|f(x) - p_n(x)| < \in$, $a \le x \le b$. Thus, we can, in principle, fit a 'good' polynomial to a set of data.

The study of interpolation is based on the calculus of finite differences. Here we derive some important interpolation formulae by means of forward backward and central differences of a function, which are often employed in engineering and scientific investigations.

4.4.1 Newton's forward interpolation formula

Let (x_i, y_i), $i = 0,1,2,\cdots n$ be the set of $(n+1)$ data values of the function $y = f(x)$, which are equispaced so that $x_i = x_0 + ih$, $i = 0,1,2,3,\cdots,n$. Suppose it is required to evaluate $f(x)$ for $x = x_0 + ph$, where p is any real number. Then

$$y_p = f(x_0 + ph) = E^p f(x_0) = (1+\Delta)^p y_0, (\because E = 1+\Delta,\ y_0 = f(x_0))$$

$$= \left[1 + p\Delta + \frac{p(p-1)}{2!}\Delta^2 + \frac{p(p-1)(p-2)}{3!}\Delta^3 + \cdots\right] y_0, \text{ by using binomial theorem.}$$

$$= y_0 + p\Delta y_0 + \frac{p(p-1)}{2!}\Delta^2 y_0 + \frac{p(p-1)(p-2)}{3!}\Delta^3 y_0 + \cdots \tag{4.4.1}$$

If $y = f(x)$ is a polynomial of nth degree, then $\Delta^{n+1} y_0$ and higher order differences will be zero. Hence (4.4.1) becomes

$$y_p = y_0 + p\Delta y_0 + \frac{p(p-1)}{2!}\Delta^2 y_0 + \frac{p(p-1)(p-2)}{3!}\Delta^3 y_0 + \cdots + \frac{p(p-1)\cdots(p-\overline{n-1})}{n!}\Delta^n y_0 \quad (4.4.2)$$

It is called Newton's forward interpolation formula. This formula is used for interpolating the values of y near the beginning of a set of tabulated values and extrapolating values of y a little backward i.e. to the left of y_0 .

Newton's backward interpolation formula

Let $(x_i, y_i), i = 0,1,2,3\cdots n$ be the set of tabulated values of the function $y = f(x)$. Suppose it is required to evaluate $f(x)$ for $x = x_n + ph$, where p being a real number. Then we have

$$y_p = f(x_n + ph) = E^p f(x_n) = (1-\nabla)^{-p} y_n, \left(\because E^{-1} = 1 - \nabla, y_n = f(x_n)\right)$$

$$= \left[1 + p\nabla + \frac{p(p+1)}{2!}\nabla^2 + \frac{p(p+1)(p+2)}{3!}\nabla^3 + \cdots\right] y_n, \text{ using binomial theorem.}$$

$$= y_n + p\nabla y_n + \frac{p(p+1)}{2!}\nabla^2 y_n + \frac{p(p+1)(p+2)}{3!}\nabla^3 y_n + \cdots \quad (4.4.3)$$

If $y = f(x)$ is a polynomial of degree n then $\nabla^{n+1} y_n$ and higher order differences vanish. Hence (4.4.3) becomes

$$y = y_n + p\nabla y_n + \frac{p(p+1)}{2\,!}\nabla^2 y_n + \cdots + \frac{p(p+1)(p+2)\cdots(p+n-1)}{n\,!}\nabla^n y_n \quad (4.4.4)$$

It is called Newton's backward interpolation formula as (4.4.4) contains y_n and backward differences of y_n. This formula is used for interpolating the values of y near the end value of a set of tabulated values and also for extrapolating values of y a little ahead i.e. to the right of y_n.

4.4.2 Errors in polynomial interpolation

Let the function $y = f(x)$, defined by the $(n+1)$ data points $(x_i, y_i), \mathrm{i} = 0,1,2.\cdots n$ be continuous and differentiable $(n+1)$ times. Let $\phi_n(x)$ be an approximating polynomial of degree not exceeding n, to $y = f(x)$ such that

$$\phi_n(x_i) = f(x_i) = f(x_i) = y_i, \mathrm{i} = 0,1,2,3,\cdots,\mathrm{n}. \quad (4.4.5)$$

Since the expression $y(x) - \phi_n(x)$ vanishes at $x = x_i$, for all $i = 0,1,2,\cdots x$ so we taken

$$y(x) - \phi_n(x) = A\pi_{n+1}(x) \quad (4.4.6)$$

where

$$\pi_{n+1}(x) = (x - x_0)(x - x_1)\cdots(x - x_n) \quad (4.4.7)$$

and A is to be determined such that equation (4.4.6) holds for any intermediate value of x, say $x = x'$, satisfying $x_0 x' < x_n$. Then clearly, we obtain

$$A = \frac{y(x') - \phi_n(x')}{\pi_{n+1}(x')} \tag{4.4.8}$$

We define a function $g(x)$ such that

$$g(x) = y(x) - \phi_n(n) - A\pi_{n+1}(x) \tag{4.4.9}$$

where A is given by (4.4.8). Then, we have

$$g(x_0) = g(x_1) = \cdots = g(x_n) = g(x') = 0.$$

This means $g(x)$ vanishes $(n+2)$ times in the interval $x_0 \le x \le x_n$. Consequently, by the repeated application of Rolle's Theorem, $g'(x)$ must vanish $(n+1)$ times, $g''(x)$ must vanish n times, etc. in the interval $x_0 \le x \le x_n$. In particular, $g^{n+1}(x)$ must vanish once in this interval. Let this point be $x = \xi$, $x_0 < \xi < x_n$. Differentiating (4.4.9) $(n+1)$ times with respect to x and putting $x = \xi$, we get

$$0 = y^{(n+1)}(\xi) - A.(n+1)!$$

so that

$$A = \frac{y^{n+1}(\xi)}{(n+1)!} \tag{4.4.10}$$

Now equations (4.4.8) and (4.4.10) provide us

$$y(x') - \phi_n x' = \frac{y^{n+1}(\xi)}{(n+1)!}\pi_{n+1}(x')$$

Thus, in general, we have

$$y(x) - \phi_n(n) = \frac{\pi_{n+1}(x)}{(n+1)!} y^{(n+1)}(\xi), x_0 < \xi < x_n \tag{4.4.11}$$

which is the required expression for the error. This formula is almost useless in practical computations but extremely useful in theoretical work in different branches of numerical analysis, because $g(x)$ is generally, unknown and hence we do not have any information about $y^{n+1}(x)$. In particular, we will use (4.4.11) to determine errors in Newton's interpolation formulae below. The formula (4.4.11) in case of Newton's interpolating forward formula can be rewritten as

$$y(x) - y_n(x) = \frac{(x - x_0)(x - x_1)\cdots(x - x_n)}{(n+1)!} y^{n+1}(\xi), \; x_0 < \xi < x_n \; y_p = y_n(x) \tag{4.4.12}$$

Upon applying Taylor's series expression to $y(x+h)$, we have

$$y(x+h) = y(x) + hy'(x) + \frac{h^2}{2}y''(x) + \cdots$$

$$\Rightarrow \qquad y'(x) = \frac{y(x+h) - y(x)}{h} = \frac{1}{h}\Delta y(x), \text{ on neglecting } h^2 \text{ and higher powers of } h.$$

$$\Rightarrow \qquad D\,y(x) = \frac{1}{h}\Delta y(x),\ D = d/dx.$$

Therefore $D \equiv \frac{1}{h}\Delta$ and so $D^{n+1} = \frac{1}{h^{n+1}}\Delta^{n+1}$. We obtain

$$y^{n+1}(x) \approx \frac{1}{h^{n+1}}\Delta^{n+1} y(x) \qquad (4.4.13)$$

Equation (4.4.12) can therefore be written as

$$y(x) - y_n(x) = \frac{p(p-1)(p-2)\cdots(p-n)}{(n+1)!}\Delta^{n+1} y(\xi) \qquad (4.4.14)$$

in which form it is suitable for computations .

Similarly, it can be shown that the error in Newton's backward interpolation formula (4.4.4) may be written as

$$y(x) - y_n(x) = \frac{p(p+1)(p+2)\cdots(p+n)}{\angle n+1}\nabla y^{n+1}(\xi) \qquad (4.4.15)$$

where $x_0 < \xi < x_n$ and $x = x_n + ph$

Example 4.4.1: The population of a town in the decennial census was as given below. Estimate the population in the years 1955 and 1985.

Year (x):	1951	1961	1971	1981	1991
Population (y): (in thousands)	46	66	81	93	101

Solution: The difference table is

x	Y	Δ	Δ^2	Δ^3	Δ^4
1951	46				
		20			
1961	66		-5		
		15		2	
1971	81		-3		-3
		12		-1	
1981	93		-4		
		8			
1991	101				

(i) Here $x = 1955$ and interpolation is desired at the beginning of the table so we use Newton's forward interpolation formula (4.4.2). We take $h = 10$, $x_0 = 1951$ so that $x = x_0 + ph$ provides us $p = \frac{1955-1951}{10} = \frac{4}{10} = 0.4$. Then

$$y(1955) = 46 + 0.4(20) + \frac{0.4(0.4-1)}{6}(-5) + \frac{0.4(0.4-1)(0.4-2)}{6}(2)$$

$$+\frac{0.4(0.4-1)(0.4-2)(0.4-3)}{24}(-3) = 54.85 \text{ thousands}$$

(ii) Here x = 1985 and interpolation is required at the end of the table and hence we use Newton's backward interpolation formula (4.4.4). We take h = 10, $x_n = 1991$ so that $x = x_n + ph$ gives us $p = \frac{1985-1991}{10} = -0.6$. Then

$$y(1985) = 101 - (0.6)8 + \frac{-0.6(-0.6+1)}{2}(-4) + \frac{-0.6(-0.6+1)\,(-0.6+2)}{6}(-1)$$

$$+\frac{-0.6(-0.6+1)(-0.6+2)(-0.6+3)}{24}(-3) = 96.84 \text{ thousands.}$$

4.5 Central difference interpolation

In this section, we shall discuss and develop central difference formulae which are best suited for interpolation near the middle of a tabulated set. The most important central difference formulae are due to Stirling, Bessel and Everett. Gauss's formulae are also of interest from theoretical stand-point only.

4.5.1 Gauss's forward and backward interpolation formulae

Let us assume Gauss forward interpolation formula, which uses differences lie on the solid line in Table 4.5.1below, of the form

$$y_p = y_0 + G_1\Delta y_0 + G_2\Delta^2 y_{-1} + G_3\Delta^3 y_{-1} + G_4\Delta^4 y_{-2} + \cdots \tag{4.5.1}$$

where $G_1, G_2, G_3, \cdots$ are to be determined. By Newton's forward difference formula, we have

$$y_p = E^p y_0 = (1+\Delta)^p y_0$$

$$= y_0 + p\Delta y_0 + \frac{p(p-1)}{2\,!}\Delta^2 y_0 + \frac{p(p-1)(p-2)}{3\,!}\Delta^3 y_0 + \cdots \tag{4.5.2}$$

Now

$$\Delta^2 y_{-1} = \Delta^2 E^{-1} y_0 = \Delta^2(1+\Delta)^{-1} y_0 = \Delta^2\left(1-\Delta+\Delta^2-\Delta^3+\cdots\right)y_0$$

$$= \Delta^2 y_0 - \Delta^3 y_0 + \Delta^4 y_0 - \Delta^5 y_0 + \cdots$$

$$\Delta^3 y_{-1} = \Delta^3 y_0 - \Delta^4 y_0 + \Delta^5 y_0 - \Delta^6 y_0 + \cdots$$

$$\Delta^4 y_{-2} = \Delta^4 E^{-2} y_0 = \Delta^4 (1-\Delta)^{-2} y_0 = \Delta^4 \left(1 - 2\Delta + 3\Delta^2 - 4\Delta^3 + \cdots\right) y_0$$

$$= \Delta^4 y_0 - 2\Delta^5 y_0 + 3\Delta^6 y_0 - 4\Delta^7 y_0 + \cdots$$

Table 4.5.1 Central difference table

X	y	Δ	Δ^2	Δ^3	Δ^4	Δ^5	Δ^6
x_{-3}	y_{-3}						
		Δy_{-3}					
x_{-2}	y_{-2}		$\Delta^2 y_{-3}$				
		Δy_{-2}		$\Delta^3 y_{-3}$			
x_{-1}	y_{-1}		$\Delta^2 y_{-2}$		$\Delta^4 y_{-3}$		
		Δy_{-1}		$\Delta^3 y_{-2}$		$\Delta^5 y_{-3}$	
x_0	y_0		$\Delta^2 y_{-1}$		$\Delta^4 y_{-2}$		$\Delta^6 y_{-3}$
		Δy_0		$\Delta^3 y_{-1}$		$\Delta^5 y_{-2}$	
x_1	y_1		$\Delta^2 y_0$		$\Delta^4 y_{-1}$		
		Δy_1		$\Delta^3 y_0$			
x_2	y_2		Δ^2_1				
		Δy_2					
x_3	y_3						

and so on. Thus (4.5.1) becomes

$$y_p = y_0 + G_1 \Delta y_0 + G_2\left(\Delta^2 y_0 - \Delta^3 y_0 + \Delta^4 y_0 - \Delta^5 y_0 + \cdots\right)$$
$$+ G_3\left(\Delta^3 y_0 - \Delta^4 y_0 + \Delta^5 y_0 - \Delta^6 y_0 + \cdots\right)$$
$$+ G_4\left(\Delta^4 y_0 - 2\Delta^5 y_0 + 3\Delta^6 y_0 - 4\Delta^7 y_0 + \cdots\right) + \cdots \quad (4.5.3)$$

Comparing (4.5.2) and (4.5.3), we obtain

$$G_1 = p, \quad G_2 = \frac{p(p-1)}{2\,!}, \quad G_3 = \frac{(p+1)p(p-1)}{3\,!}, \quad G_4 = \frac{(p+1)p(p-1)(p-2)}{4\,!} .etc.$$

Hence the Gauss forward interpolation formula is given by

$$y_p = y_0 + p\Delta y_0 + \frac{p(p-1)}{2\,!}\Delta^2 y_{-1} + \frac{(p-1)p(p-1)}{3\,!}\Delta^3 y_{-1} + \frac{(p-1)p(p-1)(p-2)}{4\,!}\Delta^4 y_{-2} + \ldots \quad (4.5.4)$$

This formula is used to interpolate the values of y for $p(0 < p < 1)$ measured forwardly from origin.

The Gauss's Backward Interpolation formula uses the differences which lie on the dashed line in Table 4.5.1 and can therefore be assumed of the form

$$y_p = y_0 + G_1'\Delta y_{-1} + G_2'\Delta^2 y_{-1} + G_3'\Delta^3 y_{-2} + G_4'\Delta^4 y_{-2} + \cdots \quad (4.5.5)$$

where $G_1', G_2', G_3', G_4' \cdots$ have to determined . Following the above procedure and comparing with Newton's backward difference formula, we obtain

$$G_1' = p,\ G_2' = \frac{p(p+1)}{2!},\ G_3' = \frac{(p+1)p(p-1)}{3!},\ G_4' = \frac{(p+2)(p+1)p(p-1)}{4!}\ etc.$$

Therefore Gauss backward interpolation formula is given by

$$y_p = y_0 + p\Delta y_{-1} + \frac{p(p+1)}{2!}\Delta^2 y_{-1} + \frac{(p+1)p(p-1)}{3!}$$

$$\Delta^3 y_{-2} + \frac{(p+2)(p+1)p(p-1)}{4!}\Delta^4 y_{-2} + \cdots \quad (4.5.6)$$

This formula is used interpolate the values of y for a negative value of plying between -1 and 0 Gauss' formulas are not of much practical use, however these have theoretical significance.

4.5.2 Stirling's formula

If we take mean of the Gauss forward and backward formula (4.5.4) and (4.5.6), we obtain

$$y_p = y_0 + p\frac{\Delta y_{-1} + \Delta y_0}{2} + \frac{p^2}{2}\Delta^2 y_{-1} + \frac{p(p^2-1)}{3!}\frac{\Delta^3 y_{-1} + \Delta^3 y_{-2}}{2} + \frac{p^2(p^2-1)}{4!}\Delta^4 y_{-2} + \cdots \quad (4.5.7)$$

This is known as Striling's formula. It uses the average of the differences along broken doted line in table 4.5.1. In the central differences notation, Stirling formula (4.5.7) takes the form

$$y_p = y_0 + p\mu\delta y_0 + \frac{p^2}{2!}\delta^2 y_0 \frac{p(p^2-1^2)}{3!}\mu\delta^3 y_0 + \frac{p^2(p^2-1^2)}{4!}\delta^4 y_0 + \cdots \quad (4.5.8)$$

since

$$\frac{1}{2}(\Delta y_0 + \Delta y_{-1}) = \frac{1}{2}(\delta y_{1/2} + \delta y_{-1/2}) = \mu\delta y_0,\ \frac{1}{2}(\Delta^3 y_{-1} + \Delta^3 y_{-2}) = \frac{1}{2}\left(\delta^3 y_{\frac{1}{2}} + \delta^3 y_{-\frac{1}{2}}\right) = \mu\delta^3 y_0 \text{ etc.}$$

4.5.3 Bessel's formula

This is a very useful formula for practical interpolation, and it uses the differences as shown in the following table where brackets mean that the average has to be taken

$$\begin{array}{lllllll} \vdots & \vdots \\ x_{-1} & y_{-1} \\ x_0 & \begin{pmatrix} y_0 \\ y_1 \end{pmatrix} \Delta y_0 & \begin{pmatrix} \Delta^2 y_{-1} \\ \Delta^2 y_0 \end{pmatrix} & \Delta^3 y_{-1} & \begin{pmatrix} \Delta^4 y_{-2} \\ \Delta^4 y_{-1} \end{pmatrix} & \Delta^5 y_{-2} & \begin{pmatrix} \Delta^6 y_{-3} \\ \Delta^6 y_{-2} \end{pmatrix} \\ x_1 \\ \vdots & \vdots \end{array}$$

Therefore, we have assume Bessel's formula of the form

$$y_p = \frac{y_0 + y_1}{2} + B_1 \Delta y_0 + B_2 \frac{\Delta^2 y_{-1} + \Delta^2 y_0}{2} + B_3 \Delta^3 y_{-1} + B_4 \frac{\Delta^4 y_{-2} + \Delta^4 y_{-1}}{2} + \cdots$$

$$= y_0 + \left(B_1 + \frac{1}{2}\right)\Delta y_0 + B_2 \frac{\Delta^2 y_{-1} + \Delta^2 y_0}{2} + B_3 \Delta^3 y_{-1} + B_4 \frac{\Delta^4 y_{-2} + \Delta^4 y_{-1}}{2} + \cdots \quad (4.5.9)$$

The Newton's forward differences interpolation formula is given by

$$y_p = y_0 + p\Delta y_0 + \frac{p(p-1)}{2!}\Delta^2 y_0 \frac{p(p-1)(p-2)}{3!}\Delta^3 y_0 + \frac{p(p-1)(p-2)(p-3)}{4!}\Delta^4 y_0 + \cdots \quad (4.5.10)$$

Comparing (4.5.9) after simplifying the differences, we obtain

$$B_1 + \frac{1}{2} = p,\ \mathrm{B}_2 = \frac{p(p-1)}{2!},\ \mathrm{B}_3 = \frac{p(p-1)\left(p - \frac{1}{2}\right)}{3!},\ \mathrm{B}_4 = \frac{(p+1)p(p-1)(p-2)}{4!} \text{ etc.}$$

Thus the Bessel formulae (4.5.9) becomes

$$y_p = y_0 + p\Delta y_0 + \frac{p(p-1)}{2!}\frac{\Delta^2 y_{-1} + \Delta^2 y_0}{2} + \frac{p(p-1)\left(p - \frac{1}{2}\right)}{3!}\Delta^3 y_{-1}$$

$$+ \frac{(p+1)p(p-1)(p-2)}{4!}\frac{\Delta^4 y_{-2} + \Delta^4 y_7}{2} + \cdots \quad (4.5.11)$$

In central differences notation, (4.5.11) can be written as

$$y_p = y_0 + p\delta y_{\frac{1}{2}} + \frac{p(p-1)}{2!}\mu\delta^2 y_{\frac{1}{2}} + \frac{\left(p - \frac{1}{2}\right)p(p-1)}{3!}\delta^3 y_{\frac{1}{2}}$$

$$+ \frac{(p+1)p(p-1)(p-2)}{4!}\mu\,\delta^4 y_{\frac{1}{2}} + \cdots \quad (4.5.12)$$

since

$$\frac{1}{2}\left(\Delta^2 y_{-1}+\Delta^2 y_0\right)=\mu\delta^2 y_{\frac{1}{2}},\quad \frac{1}{2}\left(\Delta^4 y_{-2}+\Delta^4 y_{-1}\right)=\mu\delta^4 y_{\frac{1}{2}} \text{ etc.}$$

4.5.4 Everett's formula

This formula is extensively used and involves only even differences on and below the central line as shown below.

$$\begin{array}{cccccl} x_0 & y_0 & \Delta^2 y_{-1} & \Delta^4 y_{-2} & \Delta^6 y_{-3}\cdots & \text{central line} \\ & — & — & — & — & \\ x_1 & y_1 & \Delta^2 y_0 & \Delta^4 y_{-1} & \Delta^6 y_{-2} & \end{array}$$

Hence this formula has the form

$$y_p = E_0 y_0 + E_2\Delta^2 y_{-1} + E_4{}^4 y_{-2} + \cdots + F_0 y_1 + F_2\Delta^2 y_0 + F_4\Delta^4 y_{-1} + \cdots \qquad (4.5.13)$$

where coefficients E_0, F_0, E_2, F_2, E_4, F_4,$\cdots$ can be determined by the earlier method as used in the proceeding cases, we obtain .

$$E_0 = 1-p = q, \qquad F_0 = p$$

$$E_2 = q\left(q^2-1^1\right), \qquad F_2 = p\left(p^2-1^2\right)$$

$$E_4 = \frac{q\left(q2-1^2\right)\left(q^2-2^2\right)}{5!}, \qquad F_4 = \frac{p\left(p^2-1^2\right)\left(p^2-2^2\right)}{5!} \text{ etc.}$$

$$y_p = q\,y_0 + \frac{q\left(q^2-1^2\right)}{3!}\Delta^2 y_{-1} + \frac{q\left(q^2-1^2\right)\left(q^2-2^2\right)}{5!}\Delta^4 y_{-2} + \cdots$$

$$+\,py_1 + p\frac{p^2-1^2}{3!}\Delta^2 y_0 + \frac{p\left(p^2-1^2\right)\left(p^2-2^2\right)}{5!}\Delta^4 y_{-1} + \cdots \qquad (4.5.14)$$

where $q = 1 - p$. There is a close relationship between Bessel's formula and Everett's formula and one can be deduced from the other by suitable rearrangements. It is also interesting to observe that Bessel's formula truncated after third differences is Everett's formula truncated after second differences.

4.5.5 Choice of an interpolation formula

As for as practical interpolation is concerned, we have to see which formula yields most accurate results in a particular problem. The coefficients in the central difference formulae are smaller and converge faster than those in Newton's formulae. After a few terms, the coefficients in the String's formula decrease more rapidly than those of the Bessel's formula and the coefficients of Bessel's formula decrease more rapidly than those of Newton's formula. As such, whenever possible, central difference formula should be used in preference to Newton's formulae. The right choice of an interpolation formula however, depends on the position of the interpolated value in the given data.

(i) To find a tabulated value near the beginning of the table use Newton's forward formula.

(ii) To find a value near the end of the table, use Newton's backward formula.

(iii) To find an interpolated value near the centre of the table, use either Stiring's or Bessel's or Everett's formula.

The central difference table is

X	P	y_p	Δy_p	$\Delta^2 y_p$	$\Delta^3 y_p$	$\Delta^4 y_p$	$\Delta^5 y_p$
1939	–3	12					
			3				
1949	–2	15		2			
			5		0		
1959	–1	20		2		3	
			7		3		–10
1969	0	27		5		7	
			12		–4		
1979	1	39		1			
			13				
1989	2	52					

If the interpolation is required for $\frac{-1}{2} \le p \le \frac{1}{4}$, prefer Stirling's s formula. If interpolation is required for $\frac{1}{4} \le p \le \frac{3}{4}$, then use Bessel's or Everette's formula. But in the case where a series of calculations have to be performed, it would be inconvenient to use both these formulae, and a choice must be made between them. The choice depends on the order of the highest difference that could be neglected so that contributions from it and further differences would be less than half a unit in the last decimal place. If this highest difference is of even order, Stirling's 's formula is recommended, if it is of even order, Bessel's formula might be preferred. Even the estimation of the maximum value of a difference of any order in an interpolation formula is also not difficult.

Gauss backward formula is

$$y_p = y_0 + p\Delta y_{-1} + \frac{(p+1)p}{2!}\Delta^2 y_{-1} + \frac{(p+1)p(p-1)}{3!}\Delta^3 y_{-2}$$

$$+ \frac{(p+2)(p+1)p(p-1)}{4!}\Delta^4 y_{-2} + \frac{(p+2)(p+1)p(p-1)(p-2)}{5!}\Delta^5 y_3 + \cdots$$

we have

$$y(1974)=27+(0.5)7+\frac{(1.5)(0.5)}{2}+\frac{(1.5)(.5)(-.5)}{6}(3)+\frac{(2.5)(1.5)(-.5)}{24}(-7)$$

$$+\frac{(2.5)(1.5)(0.5)(-.5)(-1.5)}{120}(-10)$$

$$=27+3.5+1.875-0.1875+0.2743-0.1172 \;=32.345 \text{ thousands.}$$

Example 4.5.1: Use Gauss's forward formula to evaluate y_{30}, given that

$$y_{21}=18.4708,\ y_{25}=17.8144,\ y_{29}=17.1070,\ y_{33}=16.3432 \text{ and } y_{37}=15.5154.$$

Solution: Taking $x_0=29,\ h=4,\ x=30,$ we have $p=\frac{x-x_0}{h}=\frac{30-29}{4}=0.25$

The difference table is given below:

x	p	y_p	Δy_p	$\Delta^2 y_p$	$\Delta^3 y_p$	$\Delta^4 y_p$
21	–2	18.4708				
			–0.6564			
25	–1	17.8144		–0.0510		
			–0.7074		–0.0054	
29	0	17.1070		–0.0564		–0.0022
			–0.7638		–0.0076	
33	1	16.3432		–0.0640		
			–0.8278			
37	2	15.5154				

Use Gauss forward difference formula

$$y_p=y_0+p\Delta y_0+\frac{p(p-1)}{\angle 2}\Delta^2 y_{-1}+\frac{(p+1)p(p-1)}{\angle 3}\Delta^3 y_{-1}+\frac{(p+1)p(p-1)(p-2)}{\angle 4}\Delta^4 y_{-2}+\cdots$$

we get $y_{30}=17.1070+(.25)(-0.7638)+\frac{(0.25)(-0.75)}{2}(-0.0564)+\frac{(1.25)(.25)(-0.75)}{6}$

$$(-0.0076)+\frac{(1.25)(.25)(-0.75)(-1.75)}{24}(-0.0022)$$

$$=17.1070-0.19095+0.00529+0.0003-0.00004=16.9216 \text{ approximately.}$$

Example 4.5.2: Use Gauss backward formula to find the population of a town for the year 1974 given that

year:	1939	1949	1959	1969	1979	1989
Population (in thousands):	12	15	20	27	39	52

Solution: Here we take $x_0 = 1969$, h = 10, $x = 1974$, $x = 1974$ so that $p \dfrac{1974-1969}{10} = 0.5$

Example 4.5.3 Find the value of e^x when (a) $x = 0.644$ using Stirling, Bessel and Everett's formulae (b) $x = 0.638$ by using Stirling and Bessel's formulae; from the following data values.

x:	0.61	0.62	0.63	0.64	0.65	0.66	0.67
$y = e^x$:	1.840431	1.858928	1.877610	1.896481	1.915541	1.934792	1.954237

The difference table is

x	$y = e^x$	Δ	Δ^2	Δ^3	Δ^4
0.61	1.840431				
		0.018497			
0.562	1.858928		0.000185		
		0.018682		0.000004	
0.63	1.877610		0.000189		–0.00004
		0.018871		0.0000	
0.64	1.896481		0.000189		0.000002
		0.019060		0.000002	
0.65	1.915541		0.000191		0.000001
		0.019251		0.000003	
0.66	1.934792		0.000194		
		0.019445			
0.67	1.954237				

(a) Here $p = \dfrac{0.644-0.64}{0.01} = 0.4$

Stirling Formula:

$$y(0.644) = 1.896481 + (0.4)\frac{0.018871+0.019060}{2} + \frac{0.16}{2}(0.000189)$$

$$= 1.896481 + 0.0075862 + 0.00001512$$

$$= 1.904082$$

Bessel Formula:

$$y(0.644) = 1.896481 + 0.4(0.019060) + \frac{0.4(0.4-1)}{2}\frac{0.000189 + 0.000191}{2}$$

$$= 1.896481 + 0.0076240 - 0.00002280$$

$$= 1.904082$$

Everette's Formula :

$$y(0.644) = 0.6(1.896481) + \frac{0.6(0.36-1)}{2}(0.000189)$$

$$+ 0.4(1.915541) + \frac{0.4(0.16-1)}{2}(0.000191)$$

$$= 1.1378886 - 0.000012096 + 0.7662164 - 0.000010696$$

$$= 1.904082$$

The third difference contribution to both Stirling's and Bessel's formulae is neglible and in all the cases the value obtained is correct to six decimal places.

(b) We have noted that Stirling's formula gives the most accurate result for $-\frac{1}{4} \le p \le \frac{1}{4}$, and Bessel's formula is most efficient for $-\frac{1}{4} \le p \le \frac{3}{4}$. In order to use these formulae; therefore, we have to choose x_0 so that p satisfies the approximate inequality.

(i) To use Stirling's formula, we choose $x_0 = 0.64$ and $x_n = 0.638$ so that $p = -0.2$. Hence, we have

$$y(0.638) = 1.896481 - 0.2\frac{0.018871 + 0.019060}{2} + \frac{0.04}{2}(0.000189)$$

$$= 1.896481 - 0.0037931 + 0.0000038$$

$$= 1.892692$$

which is correct to the last decimal place.

(ii) For Bessel's formula, we choose $x_0 = 0.63, x_n = 0.638$ so that $p = 0.8$. Hence we obtain

$$y(0.638) = 1.877610 + 0.8(0.018871) + \frac{0.8(0.8-1)}{2}(0.000189)$$

$$= 1.877610 + 0.0150968 - 0.0000151$$

$$= 1.892692$$

4.6 Interpolation with unevenly spaced points

The various interpolation formulae derived so far possess have the disadvantage of being applicable only to equally spaced values of the argument. It is therefore desirable to develop interpolation formulae for unequally spaced values of the argument. In this section we shall discuss four such formulae: (i) Lagrange's interpolation formula, (ii) Hermit's interpolation formulae; (iii) Newton's divided difference formula; and (iv) Aitkin's method of interpolation by iteration.

4.6.1 Lagrange's interpolation formula

Let $y(x)$ be continuous and $(n+1)$ times differentiable in the interval (a,b). Suppose $(x_i, y_i), i = 0,1,2,\cdots n$ be the set of data values of $y = f(x)$, which are not necessarily equally spaced. Since there are $(n+1)$ pairs of values of x and y, we can represent $f(x)$ by a polynomial of degree n in x, say $L_n(x)$, such that

$$L_n(x_i) = f(x_i) = y_i, \text{i} = 0,1,2,\cdots \text{n} \tag{4.6.1}$$

Now suppose that

$$L_n(x) = a_0 + a_1 x + a_2 x^2 + \cdots + a_n x^n \tag{4.6.2}$$

be the desired polynomial of nth degree satisfying interpolation conditions (4.6.1). Substituting conditions (4.6.1) in (4.6.2), we have

$$\left.\begin{array}{l} y_0 = a_0 + a_1 x_0 + a_2 x_0^2 + \cdots + a_n x_0^n \\ y_1 = a_0 + a_1 x_1 + a_2 x_1^2 + \cdots + a_n x_1^n \\ y_2 = a_0 + a_1 x_2 + a_2 x_2^2 + \cdots + a_n x_2^n \\ \cdots \quad \cdots \quad \cdots \quad \cdots \\ y_n = a_0 + a_1\, x_\text{n} + a_2 x_2^2 + \ldots + a_n x_n^n \end{array}\right\} \tag{4.6.3}$$

The system of equations (4.6.3) will have a solution if

$$\begin{vmatrix} 1 & x_0 & x_0^2 \cdots & x_0^\text{n} \\ 1 & x_1 & x_1^2 \cdots & x_1^\text{n} \\ \vdots & \vdots & \vdots \cdots & \vdots \\ 1 & x_\text{n} & x_\text{n}^2 & x_\text{n}^\text{n} \end{vmatrix} \neq 0 \tag{4.6.4}$$

This is called 'Vandermonde's determinant'.

$$\Rightarrow \qquad (x_0 - x_1)(x_0 - x_2)\cdots(x_0 - x_n)(x_1 - x_2)\cdots(x_1 - x_n)\cdots(x_{n-1} - x_n) \neq 0$$

Eliminating $a_0, \text{a}_1, a_2 \cdots a_n$ from equations (4.6.2) and (4.6.3), we obtain

$$\begin{vmatrix} L_n(x) & 1 & x & x^2 \cdots x^n \\ y_0 & 1 & x_0 & x_0^2 \cdots x_0^n \\ y_1 & 1 & x_1 & x_1^2 \cdots x_1^n \\ \vdots & \vdots & \vdots & \vdots \cdots \vdots \\ y_n & 1 & x_n & x_n^2 \cdots x_n^n \end{vmatrix} = 0 \tag{4.6.5}$$

which shows that $L_n(x)$ is a linear combination of $y_0, y_1, y_2, \cdots y_n$.

Therefore, we have

$$L_n(x) = \sum_{i=0}^{n} l_i(x) y_i \tag{4.6.6}$$

where $l_i(x)$ are polynomials in x of degree n. Because $L_n(x_j) = y_j$, for $j = 0,1,2,\cdots n$, equation (4.6.6) gives us

$$l_i(x_j) = \begin{cases} 0, & i \neq j \\ 1, & i = j \end{cases}$$

Hence $l_i(x)$ may be written as

$$l_i(x) = \frac{(x-x_0)(x-x_1)(x-x_2)\cdots(x-x_{i-1})(x-x_{i+1})\cdots(x-x_n)}{(x_i-x_0)(x_i-x_1)(x_1-x_2)\cdots(x_i-x_{i-1})(x_i-x_{i+1})\cdots(x_i-x_n)} \tag{4.6.7}$$

We define

$$\pi_{n+1}(x) = (x-x_0)(x-x_1)(x-x_2)\cdots(x-x_{i-1})(x-x_i)(x-x_{i+1})\cdots(x-x_n)$$

so that

$$\pi'_{n+1}(x_i) = \frac{d}{dx}[\pi_{n+1}(x)]_{x=x_i}$$

We define

$$\pi_{n+1}(x) = (x-x_0)(x-x_1)(x-x_2)\cdots(x-x_{i-1})(x-x_i)\,(x-x_{i+1})\cdots(x-x_n)$$

so that

$$\Pi'_{n+1}(x_i) = \frac{d}{dx}[\pi_{n+1}(x)]_{x=x_i}$$

$$= (x_i-x_0)(x_i-x_1)(x_i-x_2)\cdots(x_i-x_{i-1})(x_i-x_{i+1})\cdots(x_i-x_n)$$

Then, we have

$$l_i(x) = \frac{\pi_{n+1}(x)}{(x-x_i)\pi'_{n+1}(x)} \tag{4.6.8}$$

Hence

$$L_n(x)=\sum_{i=0}^{n}\frac{\pi_{n+1}(x)}{(x-x_i)\pi'_{n+1}(x)}y_i \tag{4.6.9}$$

This is called Lagrange's interpolation formula. The coefficients defined by (4.6.8) are called Lagrange's interpolation coefficients. Interchanging x and y in (4.6.9), we obtain

$$L_n(y)=\sum_{i=0}^{n}\frac{\pi_{n+1}(y)}{(y-y_i)\pi'_{n+1}(y_i)}x_i \tag{4.6.10}$$

which is useful for inverse interpolation.

Uniqueness: Let, if possible $\overline{L}_n(x)$ be a polynomial, different from $L_n(x)$, of degree n such that $\overline{L}_n(x_i)=y_i, i=0,1,2,\cdots n$. Then the polynomial $p_n(x)$ defined by

$$p_n(x)=L_n(x)-\overline{L}_n(x)$$

must vanish at $(n+1)$ data points $x_i, i=0,1,2,\cdots n$. Hence we have

$$p_n(x)\equiv 0$$

which implies that $L_n(x_i)=\overline{L}_n(x_i)$ and hence the Lagrange's interpolation polynomial is unique.

The Lagrange's formula can be applied whether the values x_i are equally spaced or not. It easy to remember, but difficult to apply. This formula can also be used to split the given function into partial fractions.

Error in Lagrange's Interpolation formula

We know that the errors estimate in polynomial interpolation is given by the equation (4.4.11) namely,

$$y(x)-\phi_n(x)=\frac{\pi_{n+1}(x)}{(n+1)!}y^{(n+1)}(\xi),\ \ x_0<\xi<x_n$$

Therefore replacing $\phi_n(x)$ with $L_n\ (x)$ and $y(x)$ with $f(x)$ here, we get

$$f(x)-L_n(x)=\frac{\pi_{n+1}(x)}{(n+1)!}y^{(n+1)}(\xi),\ \ x_0<\xi<x_n$$

The quantity

$$E_L=\max_{[x_0,x_n]}|\ R_n(x)\ |,\ \ R_n(x)=|\ f(x)-L_n(x)\ |$$

may be taken as an error estimate of error. If we assume that

$$\left|\ y^{n+1}(\xi)\ \right|\le M_{n+1},\ \ x_0\le\xi\le x_n$$

then

$$E_L \le \frac{M_{n+1}}{(n+1)!} \underset{[x_0, x_n]}{Max} | \pi_{n+1}(x) | \quad (4.6.11)$$

4.6.2 Hermite's interpolation formula

We now derive an interpolation formula called Hermite's interpolation formula in which both the function and its first derivative values are to be assigned at each point of interpolation.

Let $(x_i, y_i, y_i'), i = 0,1,2,3,\cdots n$ be the set of $(n+1)$ data points. It is required to find a polynomial of least degree, say $H_{2n+1}(x)$, such that

$$H_{2n+1}(x_i) = y_i, H'_{2n+1}(x_i) = y_i', \ i = 0,1,2,\cdots n \quad (4.6.12)$$

where primes denote differentiation with respect to *x*. The polynomial $H_{2n+1}(x)$ is called Hermite's interpolation polynomial. We have here $(2n+2)$ conditions and therefore the number of coefficients to the determined is $(2n+2)$ and the degree of the polynomial is $(2n+1)$. Anologous to Lagrange's interpolation formula (4.6.9), we take

$$H_{2n+1}(x) = \sum_{i=0}^{n} u_i(x) y_i + \sum_{i=0}^{n} v_i(x) y_i' \quad (4.6.13)$$

where $u_i(x)$ and $v_i(x)$ are polynomial in *x* of degree $(2n+1)$. Using condition (4.6.12), we obtain

$$\left.\begin{array}{l} u_i(x_j) = \begin{cases} 1, & \text{if } i = j \\ 0, & \text{if } i \neq j \end{cases} ; v_i(x) = 0, \text{ for all } i \\ u_i'(x) = 0, \text{ for all } i; v_i'(x_j) = \begin{cases} 1, & \text{if } i = j \\ 0, & \text{if } i \neq j \end{cases} \end{array}\right\} \quad (4.6.14)$$

Since $u_i(x)$ and $\vartheta_i(x)$ are polynomials in *x* of degree $(2n+1)$, we have

$$u_i(x) = A_i(x) [l_i(x)]^2, \ v_i(x) = B_i(x) [l_i(x)]^2 \quad (4.6.15)$$

where $l_i(x)$ are given by (4.6.8) and $A_i(x)$, $B_i(x)$ are both linear functions of *x*. We therefore write

$$\left.\begin{array}{l} u_i(x) = (a_i x + b_i) [l_i(x)]^2 \\ \text{and} \quad v_i(x) = (c_i x + d_i) [l_i(x)]^2 \end{array}\right\} \quad (4.6.16)$$

Using conditions (4.6.14) and (4.6.16), we obtain

$$a_i x_i + b_i = 1, c_i x_i + d_i = 0, \ a_i + 2l_i'(x_i) = 0, c_i = 1 \quad (4.6.17)$$

These equations imply that

$$a_i = -2l_i'(x_i), b_i = 1 + 2x_i l'(x_i), c_i = 1; d_i = -x_i \quad (4.6.18)$$

Hence equations (4.6.17) become

$$\left.\begin{aligned} u_i(x) &= [-2xl_i'(x_i)+1+2x_il_i'(x_i)]\,[l_i(x_i)]^2 \\ &= [1-2(x-x_i)l_i'(x_i)]\,[l_i(x)]^2 \\ \text{and} \qquad v_i(x) &= (x-x_i)[l_i(x)]^2 \end{aligned}\right\} \tag{4.6.19}$$

Using (4.6.19) in (4.6.13), we obtain

$$H_{2n+1}(x)=\sum_{i=0}^{n}[1-2\,(x-x_i)\,l_i'(x_i)]\,[l_i(x)]^2\,y_i+\sum_{i=0}^{n}(x-x_i)[l_i(x)]^2\,y_i' \tag{4.6.20}$$

which is the required Hermite interpolation formula.

Example: 4.6.1 Using Lagrange's interpolation formula, find the form of the function $y(x)$ for the data:

x:	0	1	3	4
y:	−12	0	12	24

Solution: Since $y = 0$ for $x = 1$, therefore $x = 1$ is a factor of $y(x)$.

Let $y(x) = (x-1)R(x)$. Then $R(x) = \dfrac{y}{x-1}$.

Thus we have

X	0	3	4
$R(x)$	12	6	8

Applying Lagrange's formula, we find that

$$R(x)=\frac{(x-3)(x-4)}{(-3)(-4)}(12)+\frac{(x-0)(x-4)}{(3-0)(3-4)}(6)+\frac{(x-0)(x-3)}{(4-0)(4-3)}(8)$$

$$=(x-3)(x-4)-2x(x-4)+2x(x-3)$$

$$=x^2-5x+12$$

Hence the required polynomial is

$$y(x)=(x-1)(x^2-5x+12)$$

Example4.6.2 Determine the Hermite polynomial of degree 5 which fits the following data and hence find an approximate value of $\log(2.7)$. Compare the value with Lagrange's formula.

x	2.0	2.5	3.0
$y=\log x$	0.69315	0.91629	1.09861
$y'=\dfrac{1}{x}$	0.5	0.4000	0.33333

Solution: Hence we use Lagrange's method to compute the polynomials

$$l_0(x), l_1(x) \text{ and } l_2(x) \text{ where}$$

$$l_0(x) = \frac{(x-2.5)(x-3.0)}{(-0.5)(-1.0)} = 2x^2 - 11x + 15$$

Similarly $\qquad l_1(x) = -(4x^2 - 20x + 24), \quad l_2(x) = 2x^2 - 9x + 10$

Therefore, we have

$$l_0'(x) = 4x - 11, \; l_1'(x) = -8x + 20, \; l_2'(x) = 4x - 9$$

Hence

$$l_0'(x_0) = -3, \; l_1'(x_1) = 0, \; l_2'(x_2) = 3$$

Thus

$$u_0(x) = (6x-11)(2x^2 - 11x + 15)^2, \; v_0(x) = (x-2)(2x^2 - 11x + 15)^2$$

$$u_1(x) = (4x^2 - 20x + 24)^2, \; \vartheta 1(x) = (x-2.5)(4x^2 - 20x + 24)^2$$

$$u_2(x) = (19-6x)(2x^2 - 9x + 10)^2, \; v_2(x) = (x-3)(2x^2 - 9x + 10)^2$$

Now the required Hermite's polynomial is given by

$$H_5(x) = (6x-11)(2x^2 - 11x + 15)^2 (0.69315)$$

$$+(4x^2 - 20x + 24)^2 (0.91629) + (19-6x)\,(2x^2 - 9x + 10)^2\,(1.09861)$$

$$+(x-2)(2x^2 - 11x + 15)^2 (0.5) + (x-2.5)(4x^2 - 20x + 24)^2 (0.4)$$

$$+(x-3)(2x^2 - 9x + 10)^2 (0.33333)$$

Putting $x = 2.\,7$ and simplifying, we get

$$\log(2.7) = H_5(2.7) = 0.993252$$

which is correct to six decimal places.

Also the Lagrange's polynomial is obtained as

$$L_2(x) = (2x^2 - 11x + 15)(0.69315) - (4x^2 - 20x + 24)$$

$$(0.91629) + (2x^2 - 9x + 10)\,(1.09861)$$

$$= -0.08164x^2 + 0.81366x - 0.60761$$

Therefore $\log(2.7) = L_2(2.7) = 0.9941164$

The actual value of $\log(2.7) = 0.9932518$.

Thus Hermite's polynomial gives us more accurate result then Lagrange's polynomial.

4.7 Divided differences

The Lagrange's formula has the drawback that if another interpolation value were inserted, then the interpolation coefficients are required to be recalculated. This labour of re-computing the interpolation coefficients is saved by using an interpolation polynomial which has the property that a polynomial of higher degree may be derived from it by simply adding new terms. Newton's general interpolation formula is one such formula which employs what are called divided differences. Before deriving this formula, we shell first define these differences. Let $(x_i, y_i), i = 0,1,2,3,\cdots n$ by set of $(n+1)$ tabulate points. the divided difference of orders $1,2,3,\cdots,n$ are defined by the relations

$$\left.\begin{aligned} [x_0 x_1] &= \frac{y_1 - y_0}{x_1 - x_0},\ [x_1, x_2] = \frac{y_2 - y_1}{x_2 - x_1}, [x_2, x_3] = \frac{y_3 - y_2}{x_3 - x_2} \text{ etc.} \\ [x_0, x_1, x_2] &= \frac{[x_1, x_2] - [x_0, x_1]}{x_2 - x_0}, \\ [x_0, x_1, x_2] &= \frac{[x_1, x_2] - [x_0, x_2]}{x_2 - x_0}, \\ &\cdots\cdots\cdots\cdots\cdots\cdots \\ &\cdots\cdots\cdots\cdots\cdots\cdots \\ [x_0, x_1, x_2, \ldots, x_n] &= \frac{[x_1, x_2, x_3 \ldots, x_n] - [x_0, x_1, x_2 \ldots x_{n-1}]}{x_n - x_0}. \end{aligned}\right\} \quad (4.7.1)$$

Clearly, it can be seen that the divided differences are symmetrical in their arguments i.e. independent of the order of the arguments.

$$[x_0, x_1] = \frac{y_0}{x_0 - x_1} + \frac{y_1}{x_1 - x_0} = [x_1, x_0]$$

$$[x_0, x_1, x_2] = \frac{y_0}{(x_0 - x_1)(x_0 - x_2)} + \frac{y_1}{(x_1 - x_0)(x_1 - x_2)} + \frac{y_2}{(x_2 - x_0)(x_2 - x_1)}$$

$$= [x_1, x_2, x_0] \text{ or } [x_2, x_0, x_1]$$

and so on.

$$[x_0, x_1, \ldots x_n] = \frac{y_0}{(x_0 - x_1)(x_0 - x_2)\ldots(x_0 - x_n)} +$$

$$\frac{y_1}{(x_1, x_0)(x_1 - x_2)\ldots(x_1 - x_n)} + \cdots + \frac{y_n}{(x_n - x_0)(x_n - x_1)\ldots(x_n - x_{n-1})} \quad (4.7.2)$$

Suppose the arguments are equally spaced so that

$$x_1 - x_0 = x_2 - x_1 = \cdots = x_n - x_{u-1} = h.$$

Then , we obtain

$$[x_0, x_1] = \frac{y_1 - y_0}{x_1 - x_0} = \frac{1}{4}\Delta y_0$$

$$[x_0, x, x_2] = \frac{[x_1, x_2] - [x_0, x_1]}{x_2 - x_0} = \frac{1}{2h}\left[\frac{\Delta y_1}{h} - \frac{\Delta y_0}{h}\right] = \frac{1}{2h^2}\Delta^2 y_0 = \frac{1}{h^2 2!}\Delta^2 y_0.$$

and so on, in general, we have

$$[x_0, x_1, x_2 \cdots, x_n] = \frac{1}{h^n n!}\Delta^n y_0 \qquad (4.7.3)$$

Thus the nth divided differences of a polynomial of the nth degree are constant.

4.7.1 Newton's general interpolation formula

Let $y_0, y_1, y_2 \cdots y_n$ be the values of $y = f(x)$ corresponding to the argument $x_0, x_1, x_2, \cdots x_n$. Then from the definition of divided differences, we have

$$[x, x_0] = \frac{y - y_0}{x, -x_0} \Rightarrow y = y_0 + (x - x_0)[x, x_0] \qquad (4.7.4)$$

$$[x, x_0 x_1] = \frac{[x, x_0] - [x_0, x_1]}{x - x_1} \Rightarrow [x, x_0] = [x_0, x_1] + (x - x_1)[x, x_0, x_1]$$

Using this value of $[x, x_0]$, in (4.7.4), we get

$$y = y_0 + (x - x_0)[x_0, x1] + (x - x_0)(x - x_1)\,[x, x_0, x_1] \qquad (4.7.5)$$

Also
$$[x, x_0, x_1, x_2] = \frac{[x, x_0, x_1] - [x_0, x_1, x_2]}{x - x_2}$$

$$\Rightarrow [x, x_0, x_1] = [x_0, x_1, x_2] + (x - x_2)[x, x_0, x_1, x_2]$$

Upon substituting this in (4.7.5), we get

$$y = y_0 + (x - x_0)[x_0, x_1] + (x - x_0)(x - x_1)[x_0, x_1, x_2] + (x - x_0)(x - x_1)(x - x_2)[x, x_0, x_1, x_2]$$

Proceeding in this manner, we get

$$y = f(x) = y_0 + (x - x_0)[x_0, x_1] + (x - x_0)(x - x_1)[x_0, x_1, x_2] + \cdots$$
$$+ (x - x_0)(x - x_1)\cdots(x - x_n)[x_0, x_1, x_2 \cdots x_n] \qquad (4.7.6)$$

This is called Newton's general interpolation formula with divided difference.

The process of reciprocal differences will be useful to a continued faction approximation hich agrees with a tabulated function $f(x)$ at the points $x_0, x_1 x_2, \cdots x_n$. We define the quantities $\phi_0[x_0], \phi_1[x_0, x_1], \phi_2[x_0 x_1 x_2], \cdots$ called the reciprocal difference by

$$\phi_0[x_0] = f(x), \ \phi_1[x_0, x_1] = \frac{x - x_0}{\phi_0[x] - \phi_0[x_0]} = \frac{x - x_0}{f(x) - f(x_0)} \tag{4.7.7}$$

$$\phi_2[x_0, x_1, x_2] = \frac{x - x_1}{\phi_1(x_0 x) - \phi_1[x_0, x_1]}$$

and so on. Then the procedure outlined in the derivation of Newton's divided difference formula, leads to the general formula

$$\phi_0[x] = f(x_0) + \frac{x - x_0}{\phi_1[x_0, x_1]} + \frac{x - x_1}{\phi_2[x_0, x_1 x_2]} + \frac{x - x_2}{\phi_3[x_0, x_1 x_2 x]} + \cdots \tag{4.7.8}$$

which is required continued fraction approximation to the given set of tabulated values.

Example 4.7.1 Using Newton's divided difference formula defines $f(x)$ as a polynomial in x from the following data.

x	-1	0	3	6	7
$f(x)$	3	-6	39	822	1611

Solution: the divided difference table is

x	$f(x)$	1st	2nd	3rd	4th
-1	3				
		-9			
0	-6		-6		
		15		5	
3	39		41		1
		261		13	
6	822		132		
		789			
7	1611				

Then the formula (4.7.6) leads to

$$f(x) = 3 + (x+1)(-9) + x(x+1)(6) + x(x+1)(x-3)(5) + x(x+1)(x-3)(x-6)$$

$$= x^4 - 3x^3 + 5x^2 - 6$$

which is the required polynomial .

4.7.2 Interpolation by iteration

We now describe other method due to Aitken, which has the advantage of being very easily programmed for a digital computer. Suppose we are given a set of $(n+1)$ data points $(x_i, y_i), i = 0,1,2,3,\cdots n$ of $y = f(x)$, where the value of x need not necessarily be equally spaced. Then to find the value of y corresponding to any given value of x, we proceed iteratively as follows: We obtain a first approximation to y by considering the first two points only. Then obtain its second approximation by considering the first there points and so on, The different interpolation are polynomials denoted by $\Delta(x)$ with suitable subscripts then at the first stage of approximation we have

$$\Delta_{01}(x) = y_0 + (x - x_0)[x_0, x_1] = \frac{1}{x_1 - x_0}\begin{vmatrix} y_0 & x_0 - x \\ y_1 & x_1 - x \end{vmatrix}$$

Similarly, we can form $\Delta 02(x), \Delta_{03}(x), \cdots$

Next Δ_{012} is formed by considering first three points as

$$\Delta_{012}(x) = \frac{1}{x_2 - x_1}\begin{vmatrix} \Delta_{01} & x_1 - x \\ \Delta_{02} & x_2 - x \end{vmatrix}$$

and similarly $\Delta 013(x), \Delta_{014}(x)$, etc are formed. At the nth stage of approximation, we obtain

$$\Delta_{012\ldots n}(x) = \frac{1}{x_n - x_{n-1}}\begin{vmatrix} \Delta_{0123\cdots\overline{n-1}}(x) & x_n - x \\ \Delta_{0123\cdots\overline{n-2n}}(x) & x_n - x \end{vmatrix}$$

the computation can be conveniently arranged in the table below

Table 4.7.1 Aitken's scheme

x	y				
x_0	y_0				
		$\Delta_{01}(x)$			
x_1	y_1		$\Delta_{012}(x)$		
		$\Delta_{02}(x)$		$\Delta_{0123}(x)$	
x_2	y_2		$\Delta_{013}(x)$		$\Delta_{01234}(x)$
		$\Delta_{03}(x)$		$\Delta_{01234}(x)$	
x_3	y_3		$\Delta_{014}(x)$		
		$\Delta_{04}(x)$			
x_4	y_4				

A modifications of this scheme suggested by Neville is given in table 4.7.2 which is particularly suited for iterated inverse interpolation

Table 4.7.2 Neville's scheme

x	y				
x_0	y_0				
		$\Delta_{01}(x)$			
x_1	y_1		$\Delta_{012}(x)$		
		$\Delta_{12}(x)$		$\Delta_{0123}(x)$	
x_2	y_2		$\Delta_{123}(x)$		$\Delta_{1234}(x)$
		$\Delta_{23}(x)$		$\Delta_{1234}(x)$	
x_3	y_3		$\Delta_{124}(x)$		
		$\Delta_{24}(x)$			
x_4	y_4				

An obvious advance of Aitken's method is that it gives a good idea of the accuracy of the result at any stage.

Example 4.7.3 : Use Aitken's method to complete $\log_{10} 301$ from the data

x	300	304	305	307
$\log_{10}{}^{x}$	2.4771	2.4829	2.4843	2.4871

Also compare the result with those of Lagrange's and Newton's divided difference formulae.

Solution: Aitken's scheme is given by

x	$\log_{10}{}^{x}$			
300	2.4771			
		2.47855		
304	2.4829		2.47858	
		2.47854		2.47860
305	2.4843		2.47857	
		2.47853		
307	2.4871			

Hence $\log_{10} 301 = 2.4786$

Using Lagrange's interpolation formula, we get

$$\log_{10} 301 = \frac{-3(-4)(-6)}{-4(-5)(-7)}(2.4771) + \frac{1(-4)(-6)}{4(-1)(-3)}(2.4829) + \frac{1(-3)(-6)}{5.1(-2)}(2.4843)$$

$$+\frac{-(-3)(-4)}{7(3)(2)}(2.4871) = 1.2739 + 4.9658 - 4.471 + 0.7106 = 2.4786$$

which is same as above. The divided difference table is

x	$\log_{10} x$	1st	2nd
300	2.4771		
		0.00145	
304	2.4829		0.00001
		0.00140	
305	2.4843		0.00000
		0.00140	
307	2.4871		

Using Netwon's divided formula (4.7.6), we get $\log_{10}301$=2.4771+0.00145+(-3)(-0.00001)=2.4786, which is again the same as obtain above. It is clear that the arithmetic method is simpler than Lagrange's method. The Aitken's scheme has the advantage to given better estimate at each stage in addition the simplest arithmetic.

4.8 Inverse interpolation

For a given a set of values of x and y we have been so for finding the value of y corresponding to a certain value of x. the process of finding of the value of x for a certain value of y is called inverse interpolation. When the values of x are equally spaced, the iterative method should be employed.

4.8.1 Lagrange's method

This procedure is similar to Lagrange's interpolation formula (4.6.9) with the only difference that x is assumed to be expressible as a polynomial in y. Lagrange's formula is merely a relation between two variable ether of which many be taken as the independent variable-Therefore, on interchanging x and by in the Lagrange's formula , we obtain

$$x = \sum_{i-0}^{n} \frac{\pi_{n+1}(y)}{(y - y_{-1})\, \pi'_{n+1}(y_i)} x_i \tag{4.8.1}$$

when $\pi_{n+1}(y) - (y - y_0)(y - y_1)\cdots(y - y_{-i})(y - yi + 1)\cdots(y - y_n)$

The formula (4.8.1) is used for inverse interoperation. It is important to note that inverse interpolation is, in general, meaningful only if the function is single-valued in the interval. The Aitken's and Neville's schemes can also be used to perform inverse interpolation process by interchanging x and y as in the case of Lagrange's formula. Both schemes lead to same result ultimately. In practice, however Neville's scheme should be preferred for the simple reason that in this scheme those points which are nearest to exact are used for interpolation at $x - x_r$.

4.8.2 Iterative method

Consider the Newton's forward difference formula

$$y_p = y_0 + p\Delta y_0 + \frac{p(p-1)}{2!}\Delta^2 y_0 + \frac{p(p-1)(p-2)}{3!}\Delta^3 y_0 + \cdots \tag{4.8.3}$$

This leads to

$$p = \frac{1}{\Delta y_0}\left[y_p - y_0 - \frac{p(p-1)}{2!}\Delta y_0 - \frac{p(p-1)(p-2)}{3!}\Delta^3 y_0 \cdots\right] \tag{4.8.4}$$

Neglecting the second and higher order differences, we obtain

$p_1 = (y_p - y_0)/\Delta y_0$ as first approximation to p

In order to obtain the second approximation, we retain the term with second order differences in (4.8.4) and replace p by p_1 and obtain

$$p_2 = \frac{1}{\Delta y_0}\left[y_p - y0 - \frac{p_1(p_1 - 1)}{2!}\Delta^2 y_0\right]$$

Similarly, the third order approximation is given by

$$p_3 = \frac{1}{\Delta y_0}\left[y_p - y_0 - \frac{p_2(p_2 - 1)}{2!}\Delta^2 y_0 - \frac{p_2(p_{2-1})(p_2 - 2)}{3!}\Delta^3 y_0\right]$$

and so on. This process is continued till two successive approximation of p agree with each other to the desired level of a accuracy.

This technique can equally be applied by starting with any other interpolation formula. Thus method is a power full iterative procedure for finding the roots of an equation to a good degree of accuracy

Example 4.8.1 find the value of x for y= 3000 form the following data by using iterative method

x:	I0	15	20
y:	1754	2648	3564

Solution : Here we take $x_0 = 10,\ h = 5$. The difference table is

x	y	Δy	$\Delta^2 y$
10	1754		
		894	
15	2648		22
		916	
20	3564		

Hence $y_p = 3000,\ y_0 = 1754$, therefore the successive approximations to p are

$$p_1 = \frac{1}{894}[3000 - 1754] = 1.39$$

$$p_2 = \frac{1}{894}\left[3000 - 1754 - \frac{1.39(1.39-1)}{2} \times 22\right] = 1.387$$

$$p_3 = \frac{1}{894}\left[3000 - 1754 - \frac{1.387(1.387-1)}{2} \times 22\right] = 1.3871$$

We, therefore, take $p = 1.387$ correct to three decimal places. Hence the value of x (corresponding to y=3000)= $x_0 + ph = 10 + 1.387 \times 5 = 16.935$.

Example 4.8.2 : Using inverse interpolation , find the really root of $x^3 + x - 3 = 0$, close to x 1.2.

Solution : the difference table is

x	$y(= x^3 + x - 3)$	Δy	$\Delta^2 y$	$\Delta^3 y$	$\Delta^4 y$
1.0	-1				
		0.431			
1.1	-0.569		0.066		
		0.497		0.006	
1.2	-0.072		0.072		0.000
		0.569		0.006	
1.3	0.497		0.078		
		0.647			
1.4	1.144				

Clearly the root of the given equation lies between 1.2 and 1.3. Taking $x_0 = 1.2$ and using Stirling's formula we get

$$0 = -0.072 + x.\frac{0.569+0.497}{2} + \frac{x^2}{2}(0.072) + \frac{x(x^2-1)}{6}\frac{0.006+0.006}{2} \quad (\because y = 0)$$

or

$$0 = -0.072 + 0.532x0.36x^2 + 0.001x^3$$

$$x = \frac{0072}{0.532} - \frac{0.036}{0.532}x^2 - \frac{0.001}{0.532}x^3$$

Therefore first approximation for $x^{(2)} = \dfrac{0.072}{0.0532} = 0.1353$

Using this in the expression for x above we get

$$x^{(2)} = 0.1353 - 0.067(0.1353)^2 - 18797(0.1353) = 0.134$$

Hence the desired root $= 1.2 + 0.1 \times .134 = 1.2134$

4.9 Spline interpolation

It is a topic of recent origin with important applications in numerical differentiation, integration, numerical solution of two point boundary value problems and integral equations. The theory of cubic spline is explained in detail in the book by Ahelberg, Nilson, and Walsh and we restrict ourselves here to the discussion of interpolator cubic splines and some of their important properties. We consider the problem of interpolation between given data points (x_i, y_i), $i = 0,1,2,\cdots N$, where $a = x_0 < x_, < x_2 < \cdots < x_n = b$, by means of smooth polynomial curve. Spline fitting is one of the recent and increasingly important techniques for the solution, whose name is derived formed draftsman's device to draw a curve through these points in such a way that not only the curve itself, but also its slope and curvature are continuous functions. A commonly used cubic spline $S(x)$ is defined by the following properties:

(i) $S(x_i) = y_i$, $i = 0,1,2,3\cdots N$

(ii) $S(x)$, $S'(x)$ and $S''(x)$ are continuous on $[a, b]$, and

(iii) $S(x)$ is a cubic polynomial, say $s_i(x)$, in each subinterval (x_{i-1}, x_i), $i = 1,2,3\cdots N$

From condition (iii) we obtain,

$$S(x) = \begin{cases} s_1(x), & x_0 \le x \le x_1 \\ s_2(x), & x_1 \le x \le x_2 \\ \cdots\cdots\cdots\cdots & \\ \cdots\cdots\cdots\cdots & \\ s_N(x), & x_1 \le x \le x_N \end{cases} \tag{4.9.1}$$

The interpolation condition (i) gives us

$$S(x_0) = s_1(x_0) = y_{0,} \;\; S(x_1) = s_2(x_1) = y_1, \;\; S(x_2) = s_3(x_2) = y_2, \cdots$$

$$S(x_n) = s_n(x_n) = y_n \tag{4.9.2}$$

The continuity conditions (ii) yield relations of the type (4.9.2)

$$s_1(x_1) = s_2(x_1),\ s_1'(x_1) = s_2'\ (x_1),\ s_1''(x_1) = s_2'(x_1) \tag{4.9.3}$$

etc. In the spline theory, the abscissae x_i are called Knots. According to condition

(iii) $s_i(x)$ is a cubic polynomial in (x_{i-1}, x_i) and hence we can write

$$s_i''(x) = \frac{1}{h_i}[(x_i - x)M_{i-1} + (x - x_{i-1})M_i] \tag{4.9.4}$$

where $h_i = x_i - x_{i-1}$, $M_{i-1} = S''(x_{i-1})$ and $M_i = S''(x_i)$.

Integrating (4.9.4) twice, we get

$$s_i(x) = \frac{1}{h}\left[\left(\frac{x_1 - x}{6}\right)^3 M_i + \frac{(x - x_{i-1})^3}{6} M_i\right] + c_i(x_i - x) + d_i(x - x_{i-1}) \tag{4.9.5}$$

where c_i and d_i are constants to be determined. Using conditions $s_i(x_{i-1}) = y_{i-1}$ and $s_i(x_i) = y_i$, we obtain

$$c_i = \frac{1}{h_i}\left[y_{i-1} - \frac{h_i^2}{6} M_{i-1}\right] \text{ and } d_i = \frac{1}{h_i}\left[y_i - \frac{h_i^2}{6} \mathrm{M_i}\right]$$

Using these values of c_i and d_i in (4.9.5), we get

$$s_i(x) = \frac{1}{h_i}\left[\frac{(x_i - x)^3}{6} M_{i-1} = \frac{(x - x_{i-1})^3}{6} M_i + \left(y_{i-1} - \frac{h_i^2}{6} M_{i-1}\right)(x_i - x) + \left(y - \frac{h_i^2}{6} M_i\right)(x - x_{i-1})\right. \tag{4.9.6}$$

On differentiating (4.9.6) w. r .t x, we get obtain

$$s_i''(x) = \frac{1}{h_i}\left[\frac{-3(x_i - x)^2}{6} M_{i-1} + \frac{3(x - x_{i-1})^2}{6} M_i - \left(y_{i-1} - \frac{h_i^2}{6} M_{i-1}\right) + \left(y_i - \frac{h_i^2}{6} M_i\right)\right].$$

which on setting $x = x_i$ leads to

$$s_i'(x_i^-) = \frac{h_i}{2} M_i - \frac{1}{h_i}\left(y_{i-1} - \frac{h^2}{6} M_{i-1}\right) + \left(y_i - \frac{h_i^{\,2}}{6} M_i\right)$$

$$= \frac{1}{\mathrm{h_i}}(y_i - y_{i-1}) + \frac{h_i}{6} M_{i-1} + \frac{h_i}{3} M_i, i = 1,2,3,\cdots N \qquad (4.9.7)$$

To obtain the right hand derivative, we write drown the equation of the cubic spline in the subinterval (x_i, x_{i+1}) setting $i = i+1$ in (4.9.7)

$$s_{i+1}(x) = \frac{1}{hi+1}\left[\frac{(x_{i+1} - x)^3}{6} M_i + \frac{(x - x_i)^3}{6} M_{i+1} + \left(y_i - \frac{h^2{}_{i+1}}{6} M_i\right)(x_{i+1} - x)\right.$$

$$\left. + \left(y_{i+1} - \frac{h^2{}_{1+1}}{6} M_{i+1}\right)(\mathrm{x - x_i})\right] \qquad (4.9.8)$$

where $h_{i+1} = x_{i+1} - x_i$. Differentiating (4.9.8) and setting $x = x_i$, we get

$$s_{i+1}'(x_i^+) = \frac{1}{h_{i+1}}(y_{i+1} - y_i) - \frac{h_{i+1}}{3} M_i - \frac{h_{i+1}}{6} M_{i+1}, i = 0,1,2,\cdots N-1 \qquad (4.9.9)$$

Equating (4.9.7) and (4.9.9), we obtain

$$\frac{h_i}{6} M_{i-1} + \frac{1}{3}(h_i + h_{i+1}) M_i + \frac{h_{i+1}}{6} M_{i+1} = \frac{y_{i+1} - y_i}{h_{i|+1}} - \frac{y_i - y_{-1}}{h_i}, \mathrm{i = 1,2,3,\cdots N\text{-}1} \qquad (4.9.10)$$

For equal intervals, we have $h_i = h_{i+1} = h$, the relation (4.9.10) simplifiers to

$$M_{i-1} + 4M_i + M_{i+1} = \frac{6}{h^2}(y_{i+1} - 2y_i + y_{i-1}), \mathrm{i = 1,2,3\cdots N\text{-}1} \qquad (4.9.11)$$

The equations (4.9.10) have some special significance. If M_o and M_N are known, then the system can be written

$$\left.\begin{aligned}
&2(h_1+h_2)M_1+h_2M_2=6\left(\frac{y_2-y_1}{h_2}-\frac{y_1-y_0}{h_1}\right)-h_1M_0\\
&h_2M_1+2(h_2+h_3)M_2+h_3M_3=6\left(\frac{y_3-y_2}{h_3}-\frac{y_{2-y_1}}{h_2}\right)\\
&h_3M_2+2(h_3+h_4)M_3+h_4M_4=6\left(\frac{y_4-y_3}{h_4}-\frac{y_3-y_2}{h_3}\right)\\
&\cdots\cdots\cdots\cdots\cdots\cdots\cdots\cdots\cdots\cdots\cdots\cdots\\
&h_{N-1}M_{N-2}+2(h_{N-1}+h_N)M_{N-1}=6\left(\frac{y_{N-}y_{N-1}}{h_N}-\frac{y_{N-1}-y_{N-2}}{h_N}\right)-h_NM_n
\end{aligned}\right\} \qquad (4.9.12)$$

Such systems are called tridiagonal systems and can be solved efficiently and accretion by Gauss-elimination method. When M_i are known, equation (4.9.6) gives the required cubic spline in the subinterval (x_{i-1}, x_i). In order to find unique interpolating splines two further relations called "end conditions" are required because (4.9.11) constitute (N-1) equations in (N+1) unknowns $y_{0,} y_1, y_2 \cdots y_N$. There are several ways in which the end conditions may be imposed sothat. different types of cubic splines are obtained Some of these are listed below.

(a) We assume that $S(x)$ is liner in the intervals $(-\infty, x_0)$ and (x_n, ∞). This means that $M_0 = M_N = 0$ and the system (4.9.11) then takes the form

$$\begin{bmatrix} 4 & 1 & 0\cdots0 & 0 & 0 \\ 1 & 4 & 1\cdots & 0 & 0\ 0 \\ \cdots & \cdots & \cdots\cdots\cdots & \cdots & \\ 0 & 0 & 0\cdots1 & 4 & 1 \\ 0 & 0 & 0\cdots0 & 1 & 4 \end{bmatrix} \begin{bmatrix} M_1 \\ M_2 \\ \vdots \\ M_{N-2} \\ M_{N-1} \end{bmatrix} = \frac{6}{h^2} \begin{bmatrix} \delta^2 y_1 \\ \delta^2 y_2 \\ \vdots \\ \delta^2 y_{N-2} \\ \delta^2 y_N - 1 \end{bmatrix} \qquad (4.9.13)$$

where δ is the central difference operator. A cubic spline so defined is called a natural cubic spline, which converges to the order $0(h^3)$.

(b) We may impose the first derivative values at the end point, *viz*

$$m_0 = y_0, \mathrm{m_N} = y_N \qquad (4.9.14)$$

Using these and conditions in equations (4.9.7) and (4.9.9), respectively, we obtain two equations which when combined with (4.9.11) completes the system

(c) In many application, it is convenient to prescribe the second derivative at x_0 and x_N so that we have

$$M_0 = y_0'', \mathrm{M_N} = y_N'' \tag{4.9.15}$$

The resulting system will be a tridiagonal system and the cubic spline so defined has fourth order of convergence and hence is preferable to the natural cubic spline in practical applications.

(d) We may also choose

$$M_0 = M_1, \mathrm{M_{N\text{-}1}} = M_N \tag{4.9.16}$$

and in this case we get parabola's in the end intervals.

The cubic spline has played a document role in practical investigations because of the minimum property given by the following theorem.

Theorem 4.9.1 (Minimum property of cubic spline)

Assume that $f \in C^2[a,b]$ and $S(x)$ is the unique cubic spine interpolant for $f(x)$ which passes through $\{(x_k, f(x_k))\}_{k=0}^{N}$ and satisfies the clamped end conditions $s'(a) = f'(a)$ and $s'(b) = f'(b)$. Then $\int_a^b [S''(x)]^2\,dx \le \int_a^b [f''(x)]^2\,dx$.

Proof $\because \int_a^b S''(x)[f''(x) - S''(x)]\mathrm{d}x = |S''(x)[f'(x) - S'(x)]|_a^b - \int_a^b S'''(x)\,[f'(x) - S'(x)]\,dx$

$$= -\int_a^b S'''(x)[f'(x) - S'(x)]dx$$

Since $S'''(x) = 6s_{k,3}$ on the subinterval $[x_k, x_{k+1}]$ it follows that

$$\int_{x_k}^{x_{k+1}} s'''(x)[f'(x) - s'(x)]dx = 6s_{k,3}[f(x) - s(x)]\Big|_{x_k}^{x_{k+1}}$$

$$= 0 \text{ for } k = 0,1,2,3\cdots N-1$$

Hence $\int_a^b s''(x)[f''(x) - s''(x)]dx = 0$, this implies that $\int_a^b s''(x)f''(x)dx = \int_a^b [s''(x)]^2\,dx$

Since $0 \le [f''(x) - S''(x)]^2$, we get

$$0 \le \int_a^b [f''(x) - s''(x)]^2\,dx$$

$$= \int_a^b [f''(x)]^2 dx - 2\int_a^b f''(x)s''(x)dx + \int_a^b [s(x)]^2 dx$$

$$= \int_a^b [f''(x)]^2 dx - \int_a^b [s(x)]^2 dx$$

which implies that

$$\int_a^b [s''(x)]^2 dx \le \int_a^b [f''(x)]^2 dx$$

Hence the proof

In other worlds Therefore 4.9.1 implies that there exists an interpolating cubic spline $S(x)$, which is the smoothest interpolating function to a set of data points (x_i, y_i), in the $l = 0,1,2,3,\cdots N$, where $a = x_0 < x_1 < x_2 < \cdots < x_n = b$, in the that of all $Z \in C^2[a,b]$, S minimizes the integral $\int_a^b [Z''(x)]^2 dx$. Since $Z''(x)$ is a good approximation to the curvature of the curve $Y = f(x)$, $S(x)$ is said be smoothest interpolating function.

Algorithm 4.9.1 (Cubic spline) To construct and evaluate a cubic spline interpolation $s(x)$ for $N+1$ data points (x_i, y_i) $i = 0,1,2\cdots n$. Provision is made for the following choices of end point constraints

(i) "clamped cubic spline" : specify $S'(x_0)$ and $S'(x_N)$

(ii) Natural cubic spline : $S''(x) = 0 = S''(x_N)$

(iii) Extrapolate $S''(x)$ to the end points

(iv) $S''(x)$ is constant near the endpoint

(v) Specify $S''(x)$ at each end point

Error analysis of cubic spline

In the interval (x_{i-1}, x_i), the cubic spline $S(x)$ is given by

$$S(x) = \frac{1}{h^2}\left[m_{i-1}(x_i - x)^2(x - x_{i-1}) - m_i(x - x_{i-1})^2(x_i - x)\right]$$

$$+\frac{1}{h^3}\left[y_{i-1}(x_i - x)^2\{2(x - x_{i-1}) + h\} + y_i(x - x_{i-1})^2\{2(x_i - x) + h\}\right]$$

If we substitute $x = x_{i-1} + \theta h, 0 \le \theta \le 1$, the equation (4.9.17) transforms to

$$S(x) = \bar{S}(\theta) = \theta h(1-\theta)^2 m_{i-1} - \theta^2(1-\theta)m_i h + (2\theta+1)(1-\theta)^2 y_{i-1} + \theta^2(3-2\theta)y_i \qquad (4.9.18)$$

Let the truncation error be given by

$$e(x) = y(x) - S(x) \tag{4.9.19}$$

But $$y(x) = y(x_{i-1} + \theta h) = \bar{y}(\theta) = y_i - (1-\theta)\text{h}y_i + \frac{(1-\theta)^2}{2} y_i'' + \cdots \tag{4.9.20}$$

by Taylor's series at $x = x_i$. Therefore

$$\bar{e}(\theta) = \bar{y}(\theta) - \bar{S}(\theta) = y_i - (1-\theta)hy_i' + \frac{(1-\theta)^2 h^2}{2} y_i'' \cdots$$

$$-\left[\theta h(1-\theta)^2 m_{i-1} - \theta^2(1-\theta)m_i h + (2\theta+1)(1-\theta)^2 y_{i-1} + \theta^2(3-2\theta)y_i\right]$$

$$= y_i(1-\theta)hy_i' + \frac{(1-\theta)^2 h^2}{2} y'' - \frac{(1-\theta)^3 h^3}{6} y_i''' + \frac{(1-\theta)^4 h^4}{6} y^{iv} + \cdots$$

$$-\left[\theta h(1-\theta)^2\left(y_{i-1}' - \frac{1}{180}h^4 y_{i-1}^{(v)} + \cdots\right) - \theta^2(1-\theta)h\left(y_i' - \frac{1}{180}h^4 y_i^{(v)} + \cdots\right)\right.$$

$$\left. + (2\theta+1)(1\text{-}\theta)^2\left(y_i - hy_i' + \frac{h^2}{2}y_i'' - \frac{h^3}{6}y_i''' + \cdots\right) + \theta^2(3-2\theta)y_i\right] \tag{4.9.21}$$

Since $S'(x_i) = y'(x_i) - \frac{1}{180}h^4 y^{(v)}(x_i) + o(h^6)$. Expanding y_{i-1}' in Taylor's series and simplifying, the equation (4.9.21) becomes

$$\bar{e}(\theta) = \frac{h^4}{24}\theta^2(1-\theta)^2 y_i^{(iv)} + o(h^5) \tag{4.9.22}$$

which shows that the cubic spline given by (4.9.17) has fourth order of convergence

Example 4.9.1: For the following data , finding the cubic splines and evaluate $y(1.5)$ and $y'(3)$,

x:	1	2	3	4
y:	1	2	5	11

Solution: Since the data is equispaced with, $h = 1$ and $\text{N} = 3$, form (4.9.11), we obtain

$$M_{i-1} + 4M_i + M_{i+1} = 6(y_{i-1} - 2y_i + y_{i+1}), 1 = 1.2.$$

Therefore, $$M_0 + 4M_i + M_2 = (6y_o - 2y_1 + y_2) = 12$$

$$M_i + 4M_2 + M_3 = 6(y_1 - 2y_2 + y_3) = 18$$

Since $M_0 = 0,\ M_3 = 0$, therefore

$\Rightarrow \qquad 4M_1 + M_2 = 12, \mathrm{M}_2 + 4M_2 = 18$

Thus

$$s(x) = \frac{(x_{i+1} - x)^3}{6h} M_i + \frac{(x - x_i)^3}{6h} M_{i+1} + \frac{x_{i+1} - x}{h}\left(y - \frac{h^2}{6} M_i\right) + \frac{x - x_i}{h}\left(y_{i+1} - \frac{h^2}{6} m_{i+1}\right)$$

gives the cubic spline as

$$s(x) = \begin{cases} \frac{1}{3}\left(x^3 - 3x^2 + 5x\right), 1 \le \mathrm{x} \le 2 \\ \frac{1}{3}\left(x^3 - 3x^2 + 5x\right), 2 \le \mathrm{x} \le 3 \\ \frac{1}{3}\left(-2x^3 + 24x^2 - 76x + 81\right), 3 \le \mathrm{x} \le 4 \end{cases}$$

Therefore $\qquad y(1.5) = -s(1.5) = \frac{11}{8}$

Also $\qquad y'(3) = s'(3) = \frac{14}{3}.$

Example 4.9.2: Fit a natural cubic spline to the data

x	1	2	3
y	-8	-1	18

and compute $y(1.5)$ and $y'(1)$

Solution: Here $N = 2, m_o M_N = 0$ and hence the recurrence relation (4.9.11)

Given $M_1 = 18$ For the interval $1 \le x \le 2$, the cubic spline obtained gives from (4.9.6)' is given by

$$S(x) = 3(x-1)^3 - 8(2-x) - 4(x-1)$$

Therefore $\qquad y(1.5) = S(1.5) = \frac{-45}{8}$ and $y'(1) = S'(1) = 4$

The tabulated function is $y = x^3 - 9$, and hence the actual values of y(1.5) and $y'(1)$ are respectively $\frac{-45}{8}$ and 3.

Example 4.9.3 Given the data points below, find

$x:$	0	1	2	3
$y:$	0	0.5	2	1.5

(a) clamped cubic spline with the conditions $S'(0) = 0.2$ and $\mathrm{S}'(3) = -1$

(b) Natural cubic spline with $S''(0) = o$ and $S''(3) = 0.$

(c) Extrapolated cubic spline

(d) Parabolically terminated cubic spline $S''(x) = 0$ on $(0,1)$ and on $[x_2, x_3]$

(e) Curvature-adjusted cubic spline with $S''(0) = -0.3$ and $S''(3) = 3.3$

Solution: Here h=1 and N=3

(a) the system of with equations (4.9.7) and (4.9.9) in this case leads to

$$\begin{pmatrix} 3.5 & 1.5 \\ 1.0 & 3.5 \end{pmatrix} \begin{pmatrix} M_1 \\ M_2 \end{pmatrix} = \begin{pmatrix} 5.1 \\ -10.5 \end{pmatrix} \Rightarrow m_1 = 2.52 \text{ and } M_2 = -3.72$$

Now using (4.9.11) we get

$$M_0 = -0.36, M_1 = 0.36$$

Therefore, equation (4.9.6) provides us

$$S_0(x) = 0.48x^3 - 0.18x^2 + 0.2x, \ 0 \le x \le 1$$

$$S_1(x) = -1.04(x-1)^3 + 1.26(x-1)^2 + 1.28(x-1) + 0.5 \quad , 1 \le x \le 2$$

$$S_2(x) = 0.68\,(x-2)^3 - 1.86(x-2)^2 = 0.68(x-2) + 2.0 \quad , 2 \le x \le 3$$

(b) The equation (4.9.11) provides in this case gives us

$$\begin{pmatrix} 4.0 & 1.0 \\ 1.0 & 4.0 \end{pmatrix} \begin{pmatrix} M_1 \\ M_2 \end{pmatrix} = \begin{bmatrix} 6.0 \\ -12.0 \end{bmatrix}$$

Since $M_0 = S''(0) = 0$ and $M_3 = S''(3) = 0,$

$$\Rightarrow M_1 = 2.4 \text{ and } M_2 = -3.6$$

The equation (4.9.6) leads to the cubic spline

$$S_0(x) = 0.4x^3 + 0.1x \ , 0 \le x \le 1$$

$$S_1(x) = -(x-1)^3 + (1.2)(x-1)^2 + (1.3)(x-1) + 0.5, \ \le x \le 2$$

$$S_2(x) = 0.6\,(x-2)^3 - 1.8(x-2)^2 + 0.7(x-2) + 2.0, \ 2 \le x \le 3$$

(b) In this case we extrapolate $S''(x)$ to the end points to obtain the linear system

$$\begin{pmatrix} 6.0 & 0.0 \\ 0.0 & 6.0 \end{pmatrix}\begin{pmatrix} m_1 \\ m_2 \end{pmatrix} = \begin{pmatrix} 6.0 \\ -12.0 \end{pmatrix} \Rightarrow m_1 = 1, m_2 = -2$$

Now using equation (4.9.11), we get

$$M_0 = 4, \ M_3 = -5$$

The resulting cubic spline form (4.9.6) is written as

$$S_0(x) = -0.5x^3 + 2.0x^2 - x, 0 \le x \le 1$$

$$S_1(x) = -0.5(x-1)^3 + 0.5(x-1)^3 + 0.5(x-1)^2 + 1.5(x-1) + 0.5, \ 1 \le x \le 2$$

$$S_2(x) = -0.5(x-2)^3 - (x-2)^2 + (x-2) + 2. \ ,2 \le x \le 3$$

(d) Here, we obtain the linear system of equations

$$\begin{pmatrix} 5.0 & 1.0 \\ 1.0 & 5.0 \end{pmatrix}\begin{pmatrix} M_1 \\ M_2 \end{pmatrix} = \begin{pmatrix} 6.0 \\ -12.0 \end{pmatrix} \Rightarrow M_1 = 1.75 \text{ and } M_2 = -2.75$$ since $S''(x) = 0$ on the subinterval at each end imply that $M_0 = M_1 = 1.75$ and $M_3 = M_2 = -2.75$. then the equation (4.9.6) leads to

$$S_0(x) = 0.875 \ x^2 - 0.37x, \qquad 0 \le x \le 1$$

$$S_1(x) = -0.75 \ (x-1)^3 + 0.875(x-1)^2 + 1.375(x-1) + 0.5, \qquad 1 \le x \le 2$$

$$S_2(x) = -1.375 \ (x-2)^2 + 0.875(x-2) + 2.0, \qquad 2 \le x \le 3$$

which is the required parabolic cubic spline

(e) In this case, from equation (4.9.11) on using $M_0 = S''(o) = -0.3$ and $M_3 = S''(3) = 3.3$, we have

$$\begin{pmatrix} 4.0 & 1.0 \\ 1.0 & 4.0 \end{pmatrix}\begin{pmatrix} M_1 \\ M_2 \end{pmatrix} = \begin{pmatrix} 6.3 \\ -15.3 \end{pmatrix} \Rightarrow M_1 = 2.7 \text{ and } M_2 = -4.5$$

Then the equation (4.9.6) provides us

$$S_0(x) = 0.5x^3 - 0.15x^2 + 0.15x, \qquad 0 \le x \le 1$$
$$S_1(x) = -1.2(x-1)^3 + 1.35(x-1)^2 + 1.35(x-1) + 0.5, \quad 1 \le x \le 2$$
$$S_2(x) = 1.3(x-2)^3 - 2.25(x-2)^2 + 0.45(x-2) + 2.0, \quad 2 \le x \le 3$$

This is the required cubic spline.

Exercises

1. Fit a straight line by the method of least squares to the data :

X :	1	2	3	4	5
Y:	14	27	40	55	68

2. Fit a straight line to the following data:

Year (x):	1951	1961	1971	1981	1991
Production (y):	8	10	12	10	16

3. The result of measurement of electric resistance R of a copper bar at various temperatures t^0C are given below :

t:	19	25	30	36	40	45	50
R:	76	77	79	80	82	83	85

Find a relation R= a+bt, a ,b being constants to be determined.

4. Fit a least square geometric curve $y = ax^6$ to the data:

x:	1	2	3	4	5
y:	0.5	2	4.5	8	12.5

5. Use method of least squares to fit a relation of the form $y = ab^x$ to the data :

x:	2	3	4	5	6
y:	144	172.8	207.4	248.8	298.5

6. The voltage v across a capacitor at time t seconds is given by the following table:

t:	0	2	4	6	8
v:	150	63	28	12	5.6

Use the method of least squares to fit a curve of the form $v = ae^{kt}$ to this data.

7. The readings for x and y given below are known to follow a law of the form

y = a +bx.

x:	-4	1	2	3
y:	4	6	10	8

Find the best possible values of a and b, by the method of least squares.

8. If V(km/hr)and R (kg/tonne) are related by a relation of the type $R = a + bv^2$, find by the method of least squares, a and b with the help of following data:

V:	10	20	30	40	50
R:	8	10	15	21	30

9. Find the parabola of the from $y = a + bx + cx^2$ which fits most closely with the observations:

x:	–3	–2	–1	0	1	2	3
y:	4.63	2.11	0.67	0.09	0.63	2.15	4.58

10. Find the missing term in the following table:

x:	0	1	2	3	4
y:	1	3	9	?	81

Explain why the result differs from $3^3 = 27$?

11. Find the cubic polynomial which takes the values

x:	0	1	2	3
$f(x)$:	1	2	1	10

Hence or otherwise evaluate f(4).

12. Use Newton's interpolation formula to find y when $x = 1.85$ and 2.4 from the data

x:	1.7	1.8	1.9	2.0	2.1	2.2	2.3
$y = e^x$	5.474	6.050	6.686	7.389	8.166	9.025	9.974

13. Given sin $45^0 = 0.7071$, sin $50^0 = 0.7660$, sin $55^0 = 0.8192$, sin $60^0 = 0.8660$, find sin 52^0, using Newton's forward formula.

14. Apply Newton's back word difference formula to the data below to obtain a polynomial of degree 4 in x:

x:	1	2	3	4	5
y:	1	–1	1	–1	1

15. Given $u_1 = 40$, $u_3 = 45, u_5 = 54$, find u_2 and u_4.

16. Given $y_0 = 3$, $y_1 = 12$, $y_2 = 81$, $y_3 = 200$, $y_4 = 100$, $y_5 = 8$, without forming the difference

table , find $\Delta^5 y_0$.

17. Use Stirling's formula to interpolate the value of $y = e^x$ at x=1.91 from the data

x:	1.7	1.8	1.9	2.0	2.1	2.2
$y = e^x$	5.4739	6.0496	6.6859	7.3891	8.1662	9.0250

18. Use Stirling's formula to find u_{32} form the data

$u_{20} = 14.035$, $u_{25} = 13.674$, $u_{30} = 13.257$, $u_{35} = 12.734$, $u_{40} = 12.089$, $u_{45} = 11.309$.

19. Using Gauss's forward formula, find the value of $f(32)$ given that
$f(25) = 0.2707, f(30) = 0.3027, f(35) = 0.3386, f(40) = 0.3794$

20. Using Gauss's backward formula, find the value of $\sqrt{12.516}$ given that

$\sqrt{12500} = 111.803399, \sqrt{12510} = 111.848111, \sqrt{12520} = 111.892806, \sqrt{1230} = 111.937483$.

21. Evaluate sin (0.197) form the following table :

x:	0.15	0.17	0.19	0.21	0.23
$\sin x$:	0.14944	0.16918	0.18886	0.20846	0.22798

22. If $y(1) = -3, y(3) = 9, y(4) = 30$ and $y(6) = 132$, find the four point Lagrange's interpolation polynomial that takes the same values as y at the given points.

23. Evaluate $\sqrt{155}$ using Lagrange's interpolation formula from the data:

x:	150	152	154	156
$y = \sqrt{x}$:	12.247	12.329	12.410	12.490

24. Express the following functions as sum of partial fractions by using Lagrange's interpolation formula.

(i) $\dfrac{3x^2 + x + 1}{(x-1)(x-2)(x-3)}$ (ii) $\dfrac{x^2 + x - 3}{x^3 - 2x^2 - x + 2}$ (iii). $\dfrac{x^2 + 6x - 1}{(x-1)(x+1)(x-4)(x-6)}$

25. Using Hermit's interpolation formula estimate the value of log (3.2) from the following table :

x	:	3	3.5	4.0
$y = \log_x$	:	1.09861	1.25276	1.38629
$y' = \dfrac{1}{x}$	:	0.33333	0.28571	0.25000

26. Find the Hermit polynomial of third degree approximating the function $y(x)$ such that

$y(x_0)=1, y'(x_0)=y'(x_1)=0.$

27. Obtain the natural cubic spline which agrees with $y(x)$ at the set of data points:

x:	2	3	4
y:	11	49	12

Hence compute $y(2.5)$ and $y'(2)$.

28. Determine the cubic spline valid in the interval $\lfloor x_{i-1}, x_1 \rfloor$ for the following data:

(a)

X	6.2	6.5
$y = x \log x$	11.3119	14.1014

(b)

X	6.2	6.4	6.6
$y = x \log x$	11.3119	11.8803	12.4549

(c)

x	1.3	1.4	1.5
$y = \tan x$	3.6021	5.7979	14.1014

29. A function $f(x)$ is defined by

$$f(x)=\begin{cases} 1+x, 0 \le x \le 3 \\ 1+x+(x-3)^3, 3 \le x \le 4 \end{cases}$$

Show that $f(x)$ is a cubic spline in [0,4].v

30. Apply reciprocal differences to cover the function $f(x)=\dfrac{1}{1+x^2}$ from the following data:

x:	0	1	2	3	4	5
$f(x)$:	1	$\frac{1}{2}$	$\frac{1}{5}$	$\frac{1}{10}$	$\frac{1}{17}$	$\frac{1}{26}$

CHAPTER 5

NUMERICAL DIFFERENTIATION AND INTEGRATION

5.1 Introduction

The process, by which we can find the derivative of a function at some assigned value of the independent variable, when we are given a set of values of that function, is called "numerical differentiation". The process of evaluating a definite integral from a set of tabulated values of the integral is known as "numerical integration". If this process is applied to the integration of a function of a single variable, then it is called "quadrature". The problems of numerical differentiation are solved by first approximating the function by an interpolating formula and then differentiating this formula as many times as desired. In case the values of the argument are equally spaced, and we desire to find the derivatives of the function at a point near the beginning (end) of a set of tabular values, we use Newton-Gregory forward (backward) formula. In order to find the derivative at a point near the middle of the table, we should use a central different formula. For calculating the derivatives of a function whose argument values are unequally spaced, we should use Newton's divided difference formula to represent the function. Representing the integrand by an interpolation formula and then integrating at between the given limits solves the problems of numerical integration, like that of numerical differentiation. While using these formulae, it must be observed that the table of values defines the function at these points only and does not completely define the function and hence the function may not be differentiable at all. As such the process of numerical differentiation should be used only if the tabulated values are such that the differences of same order are constants. Otherwise errors are bound to creep in which go on increasing as derivatives of higher order are found. This is due to the fact that difference between the actual function $f(x)$ and the approximating polynomial $\phi(x)$ may be small at the data points but $f'(x)-\phi'(x)$ may be large.

5.2 Numerical differentiation

The general method for deriving the numerical differentiation formula is to differentiate the interpolating polynomial. We illustrate the derivation with Newton's forward formula only because the method of derivation with regard to other formulae being the same. Consider the function $y = f(x)$ which is tabulated for the values $x_i (= x_0 + ih)$, $i = 0,1,2,3,\dots n$. Then the Newton's forward difference formula is given by

$$y = y_0 + u\Delta y_0 + \frac{u(u-1)}{2!}\Delta^2 y_0 + \frac{u(u-1)(u-2)}{3!}\Delta^3 y_0 + \dots \qquad (5.2.1)$$

where $x = x_0 + uh$.

Then

$$\frac{dy}{dx}=\frac{dy}{du}\frac{du}{dx}=\frac{1}{h}\frac{d}{du}\left[y_0+u\Delta y_0+\frac{u^2-u}{2!}\Delta^2 y_0+\frac{u^3-3u^2+2u}{3!}\Delta^3 y_0+\ \ldots\right]$$

$$=\frac{1}{h}\left[\Delta y_0+\frac{2u-1}{2!}\Delta^2 y_0+\frac{3u^2-6u+2}{3!}\Delta^3 y_0+\ldots\right] \tag{5.2.2}$$

$$\frac{d^2y}{dx^2}=\frac{1}{h}\frac{d}{du}\left(\frac{dy}{dx}\right)=\frac{1}{h}\frac{d}{du}\left[\frac{1}{h}\left(\Delta y_0+\frac{2u-1}{2!}\Delta^2 y_0+\frac{3u^2-6u+2}{3!}\Delta^3 y_0+\ldots\right)\right]$$

$$=\frac{1}{h^2}\left[\Delta^2 y_0+\frac{6u-6}{3!}\Delta^3 y_0+\frac{12u^2-36u+22}{4!}\Delta^4 y_0+\ldots\right] \tag{5.2.3}$$

The formulae (5.2.2) and (5.2.3) are used for computing the values of $\frac{dy}{dx}$ and $\frac{d^2y}{dx^2}$ for non-tabulated values of x respectively. For tabular values of x, these formulae take simple forms, for by setting $x = x_0$, we obtain $u = 0$ and hence expressions (5.2.2) and (5.2.3) give us

$$\left[\frac{dy}{dx}\right]_{x=x_0}=\frac{1}{h}\left[\Delta y_0-\frac{1}{2}\Delta^2 y_0+\frac{1}{3}\Delta^3 y_0-\frac{1}{4}\Delta^4 y_0+\ldots\right] \tag{5.2.4}$$

and

$$\left[\frac{d^2y}{dx^2}\right]_{x=x_0}=\frac{1}{h^2}\left[\Delta^2 y_0-\Delta^3 y_0+\frac{11}{12}\Delta^4 y_0-\frac{5}{6}\Delta^5 y_0+\ldots\right] \tag{5.2.5}$$

The higher derivates may be computed from the formulae, which can be obtained by successive differentiation. Alternatively, we know that $1+\Delta = E = e^{hD}$ which implies that

$$D=\frac{1}{h}\log(1+\Delta)=\frac{1}{h}\left[\Delta-\frac{1}{2}\Delta^2+\frac{1}{3}\Delta^3-\frac{1}{4}\Delta^4+\ldots\right] \tag{5.2.6}$$

and

$$D^2=\frac{1}{h^2}\left[\Delta-\frac{1}{2}\Delta^2+\frac{1}{3}\Delta^3-\frac{1}{4}\Delta^4+\ldots\right]^2=\frac{1}{h^2}\left[\Delta^2-\Delta^3+\frac{11}{12}\Delta^4+\ldots\right] \tag{5.2.7}$$

and so on higher order derivatives formulae can be obtained by applying these identities to y_0.

In a similar way, different formulae can be derived by interpolation formulae

(a) Newton's backward difference formula

$$\frac{dy}{dx}=\frac{1}{h}\left[\nabla y_n+\frac{2u+1}{2!}\nabla^2 y_n+\frac{3u^2+6u+2}{3!}\nabla^3 y_n+\ldots\right] \tag{5.2.8}$$

and

$$\frac{d^2y}{dx^2}=\frac{1}{h^2}=\left[\nabla^2 y_n+\frac{6u+6}{3!}\nabla^3 y_n+\frac{12u^2+36u+22}{4!}\nabla^4 y_n+\ldots\right] \tag{5.2.9}$$

for non-tabular values of x, where $x = x_n + uh$. For tabular values, we obtain.

$$\left[\frac{dy}{dx}\right]_{x=x_n} = \frac{1}{h}\left[\nabla y_n + \frac{1}{2}\nabla^2 y_n + \frac{1}{3}\nabla^3 y_n + ...\right] \quad (5.2.10)$$

$$\left[\frac{d^2y}{dx^2}\right]_{x=x_n} = \frac{1}{h^2}\left[\nabla^2 y_n + \nabla^3 y_n + \frac{11}{12}\nabla^4 y_n + \frac{5}{6}\nabla^5 y_n + ...\right] \quad (5.2.11)$$

Alternatively, we also know that $1-\nabla = E^{-1} = e^{-hD}$ which implies that

$$\Rightarrow \quad D = -\frac{1}{h}\log(1-\nabla) = \frac{1}{h}\left[\nabla + \frac{1}{2}\nabla^2 + \frac{1}{3}\nabla^3 + \frac{1}{4}\nabla^4 + ...\right] \quad (5.2.12)$$

and

$$D^2 = \frac{1}{h^2}\left[\nabla + \frac{1}{2}\nabla^2 + \frac{1}{3}\nabla^3 + \frac{1}{4}\nabla^4 + ...\right]^2 = \frac{1}{h^2}\left[\nabla^2 + \nabla^3 + \frac{11}{12}\nabla^4 + ...\right] \quad (5.2.13)$$

and so on. Applying these identities to y_n, we get formula similar to (5.2.11)

(b) Stirling's formula

$$y = y_0 + \frac{u}{1!}\left(\frac{\Delta y_0 + \Delta y\text{-}1}{2}\right) + \frac{u^2}{2!}\Delta^2 y_{-1} + \frac{u(u^2-1^2)}{3!}\left(\frac{\Delta^3 y_{-1} + \Delta^3 y_{-2}}{2}\right) + \frac{u^2(u^2-1^2)}{4!}\Delta^4 y_{-2} + ...$$

where $x = x_0 + uh$.

Differentiating, we get

$$\frac{dy}{dx} = \frac{1}{h}\left[\left(\frac{\Delta y_0 + \Delta y_{-1}}{2}\right) + u\Delta^2 y_{-1} + \frac{3u^2-1}{3!}\left(\frac{\Delta^3 y_{-1} + \Delta^3 y_{-2}}{2}\right) + \frac{4u^3-2u}{4!}\Delta^4 y_{-2} + ...\right] \quad (5.2.14)$$

At $x = x_0$, $u = 0$, we get

$$\left(\frac{dy}{dx}\right) = \frac{1}{h}\left[\left(\frac{\Delta y_0 + \Delta y_{-1}}{2}\right) - \frac{1}{6}\left(\frac{\Delta^3 y_{-1} + \Delta^3 y_{-2}}{2}\right) + \frac{1}{30}\left(\frac{\Delta^5 y_{-2} + \Delta^5 y_{-3}}{2}\right) + ...\right] \quad (5.2.15)$$

Similarly,

$$\left(\frac{d^2y}{dx^2}\right)_{x=x_0} = \frac{1}{h^2}\left[\Delta^2 y_{-1} - \frac{1}{12}\Delta^4 y_{-2} + \frac{1}{90}\Delta^6 y_{-3} + ...\right] \quad (5.2.16)$$

We can similarly also use other interpolation formulae for computing the derivatives.

Example 5.2.1 Given that

x:	1.0	1.1	1.2	1.3	1.4	1.5	1.6
y:	7.989	8.403	8.781	9.129	9.451	9.750	10.031

Find $\frac{dy}{dx}$ and $\frac{d^2y}{dx^2}$ at (*i*) x=1.1 and (*ii*) x=1.6

Solution: The difference table is

x	y	Δ	Δ^2	Δ^3	Δ^4	Δ^5	Δ^6
1.0	7.989						
		0.414					
1.1	8.403		– 0.036				
		0.378		0.006			
1.2	8.781		– 0.030		– 0.002		
		0.348		0.004		0.002	
1.3	9.129		– 0.026		0.000		– 0.001
		0.322		0.004		0.001	
1.4	9.451		– 0.023		0.001		
		0.299		0.005			
1.5	9.750		– 0.018				
		0.281					
1.6	10.031						

(i) Here h=0.1 and $x_0 = 1.1$. Therefore, using formulae (5.2.4) and (5.2.5) to compute derivatives at x=1.1 we get.

$$\left[\frac{dy}{dx}\right]_{x=1.1} = \frac{1}{0.1}\left[0.378 - \frac{1}{2}\,(-0.03) + \frac{1}{3}\,(0.004) - \frac{1}{4}\,(0) + \frac{1}{5}\,(0.001)\right] = 3.946$$

$$\left[\frac{d^2y}{dx^2}\right]_{x=1.1} = \frac{1}{(0.1)^2}\left[-0.03 - (0.004) + \frac{11}{12}(0) - \frac{5}{6}\,(0.001)\right] = 3.545$$

(ii) Here x_n=1.6 and h=0.1. Therefore upon using formulae (5.2.10) and (5.2.11), we get

$$\left(\frac{dy}{dx}\right)_{x=1.6} = \frac{1}{0.1}\left[0.281 + \frac{1}{2}\,(-0.018) + \frac{1}{3}\,(0.005) + \frac{1}{4}\,(-0.001) + \frac{1}{5}\,(-0.001) + \frac{1}{6}(-0.003)\right] = 2.727$$

$$\left[\frac{d^2y}{dx^2}\right]_{x=1.6} = \frac{1}{(0.1)^2}\left[-0.018 + 0.005 + \frac{11}{12}(-0.001) + \frac{5}{6}\,(-0.001) + \frac{137}{180}(-0.003)\right] = -1.703$$

Example 5.2.2 Given that

x:	1.0	1.2	1.4	1.6	1.8	2.0	2.2
y:	2.7183	3.3201	4.0552	4.9530	6.0496	7.3891	9.0250

Compute $\frac{dy}{dx}$ and $\frac{d^2y}{dx^2}$ at (*i*) $x = 1.2$, (*ii*) $x = 1.6$ and (*iii*) $x = 2.2$

Solution: The difference table is

x	y	Δ	Δ^2	Δ^3	Δ^4	Δ^5	Δ^6
1.0	2.7183						
		0.6018					
1.2	3.3201		0.1333				
		0.7351		0.0194			
1,4	4.0552		0.1627		0.0067		
		0.8978		0.0361		0.0013	
1.6	4.9530		0.1988		0.0080		0.0001
		1.0966		0.0441		0.0014	
1.8	6.0496		0.2429		0.0094		
		1.3395		0.0535			
2.0	7.3891		0.2964				
		1.6359					
2.2	9.0250						

At $x = 1.2,\ x_0 = 1,2,\ y_0 = 3.3201$ and $h = 0.2$. Using formulae (5.2.4) and (5.2.5), we obtain

$$\left[\frac{dy}{dx}\right]_{x=1.2} = \frac{1}{0.2}\left[0.7351 - \frac{1}{2}(0.1627) + \frac{1}{3}(0.0361) - \frac{1}{4}(0.0080) + \frac{1}{5}(0.0014)\right] = 3.3205$$

$$\left[\frac{d^2y}{dx^2}\right]_{x=1.2} = \frac{1}{0.04}\left[0.1627 - 0.0361 + \frac{11}{12}(0.0080) - \frac{5}{6}(0.0014)\right] = 3.318$$

(i) At x=1.6 $x_0 = 1.6$ h=0.2. Upon making use of formulae (5.2.15) and (5.2.16), we get

$$\left[\frac{dy}{dx}\right]_{x=1.6} = \frac{1}{0.2}\left[\frac{0.8978 + 1.0966}{2} - \frac{1}{6}\left(\frac{0.0361 + 0.0441}{2}\right) + \frac{1}{30}\left(\frac{0.0013 + 0.0014}{2}\right)\right] = 4.9530$$

$$\left[\frac{d^2y}{dx^2}\right]_{x=1.6} = \frac{1}{0.04}\left[0.1988 - \frac{1}{12}(0.0080) + \frac{1}{90}(0.0001)\right] = 4.9525$$

(ii) At -x =2.2, $x_n = 2.2$, $y_n = 9.0250$ and h =0.2. The formulae (5.2.10) and (5.2.11) lead to

$$\left[\frac{dy}{dx}\right]_{x=2.2} = \frac{1}{0.2}\left[1.6359 + \frac{1}{2}(0.2964) + \frac{1}{3}(0.0535) + \frac{1}{4}(0.0094) + \frac{1}{5}(0.0014)\right] = 9.0228$$

$$\left[\frac{d^2y}{dx^2}\right]_{x=2.2} = \frac{1}{0.04}\left[0.2964 + 0.0535 + \frac{11}{12}(0.0094) + \frac{5}{6}(0.0014)\right] = 8.992$$

In example 5.2.2, the tabulated function is e^x and hence it is easy to see that the error is considerably more in the case of second derivatives. This is due to the reason that although the tabulated function and its approximating polynomial would agree at the set of data points, their slopes at these points may vary considerably. Numerical differentiation is, therefore, an unsatisfactory process and should be used only in rare cases.

Example 5.2.3: Prove that $\frac{dy}{dx} = \frac{1}{h}\left(\delta y - \frac{1}{24}\delta^3 y + \frac{3}{640}\delta^5 y + ...\right)$ and

$$\frac{d^2y}{dx^2} = \frac{1}{h^2}\left(\delta^2 y - \frac{1}{12}\delta^4 y + \frac{1}{90}\delta^6 y + ...\right)$$

Solution: We know that $\delta = E^{1/2} - E^{-1/2} = e^{hD/2} - e^{-hD/2} = 2\sinh\frac{hD}{2}$, where $D = \frac{d}{dx}$.

This implies that

$$D = \frac{2}{h}\sinh^{-1}\frac{\delta}{2} = \frac{2}{h}\left[\frac{\delta}{2} - \frac{\delta^3}{6\times 8} + \frac{3}{40\times 32}\delta^5 - \cdots\right] = \frac{1}{h}\left[\delta - \frac{\delta^3}{24} + \frac{3}{640}\delta^5 - \cdots\right] \quad (5.217)$$

Since by Maclaurins expansion we have $\sinh x = x - \frac{x^3}{6} + \frac{3}{40}x^5 - \cdots$

Therefore $Dy = \frac{1}{h}\left[\delta - \frac{\delta^3}{24} + \frac{3}{640}\delta^5 - \cdots\right]y = \frac{1}{h}\left[\delta y - \frac{1}{24}\delta^3 y + \frac{3}{640}\delta^5 y - \cdots\right]$ (5.2.18)

From (5.2.17), we have

$$D^2 = \frac{1}{h^2}\left[\delta^2 - \frac{1}{12}\delta^4 + \frac{1}{90}\delta^6 - \cdots\right]$$

which implies that

$$D^2 y = \frac{1}{h^2}\left[\delta^2 y - \frac{1}{12}\delta^4 y + \frac{1}{90}\delta^6 y - \cdots\right]$$

This is the required result.

Example 5.2.4 A slider in a machine moves along a fixed straight rod. Its distance *x* cm. along the rod is given below for various values of the time *t* seconds. Find the velocity of the slider and its acceleration

when *t*= 0.3 second.

t=	0	0.1	0.2	0.3	0.4	0.5	0.6
x=	30.13	31.62	32.87	33.64	33.95	33.81	33.24

Solution: The difference table is

t	x	Δ	Δ^2	Δ^3	Δ^4	Δ^5	Δ^6
0	30.13						
		1.49					
0.1	31.62		−0.24				
		1.25		−0.24			
0.2	32.87		−0.48		0.26		
		0.77		0.02		−0.27	
0.3	33.64		−0.46		−0.01		0.29
		0.31		0.01		0.02	
0.4	33.95		−0.45		0.01		
		−0.14		0.002			
0.5	33.81		−0.43				
		−0.57					
0.6	33.24						

As the derivatives are required near the middle of the table, we use Stirling's formulae (5.2.15) and (5.2.16) with h=0.1 and t_0=0.3, to obtain

$$\left[\frac{dx}{dt}\right]_{x=0.3} = \frac{1}{0.1}\left[\frac{0.31+0.77}{2} - \frac{1}{6}\left(\frac{0.01+0.02}{2}\right) + \frac{1}{30}\left(\frac{0.02-0.27}{2}\right) - \cdots\right] = 5.33$$

$$\left[\frac{d^2x}{dt^2}\right]_{x=0.3} = \frac{1}{0.01}\left[-0.46 - \frac{1}{12}(-0.01) + \frac{1}{90}(0.29) - \cdots\right] = -45.6$$

Hence, the required velocity is 5.33 cm / sec and acceleration is −45.6 cm/sec ^{2}.

5.3 Errors in numerical differentiation

There are two types of errors, viz. Truncation errors and round off errors, which generally occur in the numerical computation of derivatives. The truncation error is caused by replacing the tabulated function by means of an interpolation polynomial. This error can usually be estimated by a formula of error estimation in polynomial interpolation. However, the truncation error in any numerical differentiation formula can easily be estimated as below.

Suppose that the tabulated function is such that its differences of a certain order are small and that the tabulated function is well approximated by the polynomial. We consider, for example, Stirling's formula (5.2.15), which can be written in the form

$$\left[\frac{dy}{dx}\right]_{x=x_0} = \frac{\Delta y_{-1} + \Delta y_0}{2h} + T_1 = \frac{y_1 - y_{-1}}{2h} + T_1 \qquad (5.3.1)$$

where T_1 the truncation error, is given by

$$T_1 = \frac{1}{6h}\left|\frac{\Delta^3 y_{-2} + \Delta^3 y_{-1}}{2}\right| \quad (5.3.2)$$

Similarly, formula (5.2.16) leads to

$$\left[\frac{d^2y}{dx^2}\right]_{x=x_0} = \frac{1}{h^2}\,\Delta^2 y_{-1} + T_2 \quad (5.3.3)$$

where

$$T_2 = \frac{1}{12h^2}\,\left|\Delta^4 y_{-2}\right| \quad (5.3.4)$$

The round-off error on the other hand, is inversely proportional to h in case of first order derivatives and inversely proportional to h^2 in the case of second order derivatives and so on . Thus the round-off error increases as h decreases. In case of Stirling's formula (5.3.1), the round-off error does not exceed $2\in/2h = \in/h$, where $\in$ the maximum error in the value is y_i. On the other hand the formula (5.2.15) viz.

$$\left[\frac{dy}{dx}\right]_{x=0.x_0} = \frac{\Delta y_{-1} + \Delta y_0}{2h} - \frac{\Delta^3 y_{-2} + \Delta^3 y_{-1}}{12h} + \cdots = \frac{y_{-2} - 8y_{-1} + 8y_1 - y_2}{12h} + \ldots \quad (5.3.5)$$

has the maximum rounding error $\frac{18\in}{12h} = \frac{3\in}{2h}$, where as the formula (5.2.16)

$$\left[\frac{d^2y}{dx^2}\right]_{x=x_0} = \frac{\Delta^2 y - 1}{h^2} + \ldots = \frac{y_{-1} - 2y_0 + y_1}{h^2} + \ldots \quad (5.3.6)$$

has the maximum rounding error $\frac{4\in}{h^2}$. This shows that in case of higher derivatives, the round-off error incases rather rapidly .

Example 5.3.1 Estimate the errors in the values of $\frac{dy}{dx}$ *and* $\frac{d^2y}{dx^2}$ at x=1.6 for the data given in example 5.2.2.

Solution: Form equation (5.3.2), we have

$$\text{Truncation error} = \frac{1}{6h}\left|\frac{\Delta^3 y_{-1} - \Delta^3 y_0}{2}\right| = \frac{1}{6(0.2)}\ \frac{00361 + 0.0441}{2} = 0.03342$$

Also from equation (5.3.5), we have

$$\text{Round-off error} = \frac{3\in}{2h} = \frac{3(0.5)10^{-4}}{2(0.2)} = 0.00038,$$

Here $\in < 0.00005 = 0.5 \times 10^{-4}$.

$\therefore$ Total error =0.03342+ 0.00038 = 0.0338

From Stirling's formula (5.3.1) with first order differences, we get

$$\left(\frac{dy}{dx}\right)_{x=1.6} = \frac{\Delta y_{-1} + \Delta y_o}{2h} = \frac{0.8978 + 1.0966}{0.4} = \frac{1.9944}{0.4} = 4.9860$$

The exact value is 4.9530 as the tabulated function is e^x so that the error in the above solution is (4.9.860–4.9530)= 0.0330, which agrees with the total error obtained above. Using (5.3.6), we obtain

$$\left(\frac{d^2y}{dx^2}\right)_{x=1.6} = \frac{\Delta^2 y - 1}{h^2} = \frac{0.1988}{0.04} = 4.9700$$

so that error = 4.9700 – 4.9530 = 0.0170.

Also, the truncation error $= \frac{1}{12h^2}\left|\Delta^4 y_{-2}\right| = \frac{1}{12(0.04)}0.0080 = 0.01667$

and the round-off error $= \frac{4\in}{h^2} = \frac{4\times\ 0.5\times 10^{-4}}{0.04} = 0.0050$.

Hence, total error in $\left(\frac{d^2y}{dx^2}\right)_{x=1.6} = 0.0167 + 0.0050 = 0.0217$

5.4 Maximum and minimum values of a tabulated function

Consider the Newton's forward difference formula

$$y = y_0 + p\Delta y_0 + \frac{p(p-1)}{2}\Delta^2 y_0 + \frac{p(p-1)(p-2)}{6}\Delta^3 y_0 + \ldots \quad (5.4.1)$$

Differentiating with respect to p we get

$$\frac{dy}{dp} = \Delta y_0 + \frac{2p-1}{2}\Delta^2 y_0 + \frac{3p^2 - 6p + 2}{6}\Delta^3 y_0 + \ldots \quad (5.4.2)$$

We know that for maxima or minima $\frac{dy}{dp} = 0$. Hence, terminating the right hand side for simplicity, after the third order differences and equating it to zero, we obtain

$$c_0 + c_1 p + c_2 p^2 = 0 \quad (5.4.3)$$

where

$$\left.\begin{aligned} c_0 &= \Delta y_0 - \frac{1}{2}\Delta^2 y_0 + \frac{1}{3}\Delta^3 y_0 \\ c_1 &= \Delta^2 y_0 - \Delta^3 y_0 \\ c_2 &= \frac{1}{2}\Delta^3 y_0 \end{aligned}\right\} \quad (5.4.4)$$

The equation (5.4.3) being a quadratic, can be solved for p and the corresponding values of x are then being found from $x = x_0 + ph$, at which y is maximum or minimum.

Example 5.4.1 Find the minimum values of y from the table

x:	3	4	5	6	7	8
y:	0.205	0.240	0.259	0.262	0.250	0.224

The difference table is

x	y	Δ	Δ^2	Δ^3	Δ^4	Δ^5
3	0.205					
		0.035				
4	0.240		–0.016			
		0.019		0.000		
5	0.259		–0.016		0.001	
		0.003		0.001		–0.001
6	0.262		–0.015		0.000	
		–0.012		0.001		
7	0.250		–0.014			
		–0.026				
8	0.224					

Taking $x_0 = 3$, and $h = 1$, the Newton's forward difference formula gives us

$$y = 0.205 + p(0.035) + \frac{p(p-1)}{2}(-0.016) \tag{5.4.5}$$

Differentiating this with respect to p, we get

$$\frac{dy}{dp} = 0.035 + \frac{2p-1}{2}(-0.016)$$

For y to be minimum, $\frac{dy}{dp} = 0$, which implies that $p = 2.6875$

$\therefore$

$$x = x_0 + ph = 3 + 2.6875\,(1) = 5.6875$$

Using the values in equation (5.4.5) we get

$$y_{\min} = 0.205 + 2.6875(0.035) + \frac{1}{2}(2.6875)(1.6875)(-0.016) = 0.2628$$

Example 5.4.2 Find maximum value of $f(x)$ in the range of x from the following table of values

x:	60	75	90	105	120
$f(x)$:	28.2	38.2	43.2	40.9	37.7

Solution: The difference table is

x	$y = f(x)$	Δ	Δ^2	Δ^3	Δ^4
60	28.2				
		10.0			
75	38.2		−5.0		
		5.0		−2.3	
90	43.2		−7.3		8.7
		−2.3		6.4	
105	40.9		−0.9		
		−3.2			
120	37.7				

From Stirling's formula, we have

$$y = f(x) = y_0 + \frac{p}{1!}\left(\frac{\Delta y_0 + \Delta y_{-1}}{2}\right) + \frac{p^2}{2!}\Delta^2 y_{-1} + \frac{p(p^2-1)}{3!}\left(\frac{\Delta^3 y_{-1} + \Delta^3 y_{-2}}{2}\right) + \frac{p^2(p^2-1)}{4!}\Delta^4 y_{-2} + \cdots \quad (5.4.6)$$

where $x = x_0 + ph$ and $y_0 = f(x_0), x_0 = 90$ and h=15.

Therefore, we have $\qquad \frac{dy}{dx} = \frac{dy}{dp}.\frac{dp}{dx} = \frac{1}{h}\frac{dy}{dp}$

In order to have $y = f(x)$ maximum, we solve the equation

$$\frac{dy}{dx} = f'(x) = 0 \Rightarrow \frac{dy}{dp} = 0 \Rightarrow f'(p) = 0.$$

i.e. $$\frac{\Delta y_0 + \Delta y_{-1}}{2} + \frac{2p}{2!}\Delta^2 y_{-1} + \frac{3p^2-1}{3!}\frac{\Delta^3 y_{-1} + \Delta^3 y-2}{2} + \frac{4p3-24}{4!}\Delta^4 y_{-2} = 0$$

$\Rightarrow$ $$17.4p^3 + 12.3p^2 - 96.3p + 12.10 = 0$$

Solving this equation by Newton-Raphson method, we have $p = 0.128126$

$\therefore$ $$x = 90 + 15p = 90 + 15\,(0.128126) = 91.92189$$

Using these values of x and p in (5.4.6), we get

$$y_{max} = f(91.92189) = 43.2641$$

which in the required maximum value of $f(x)$.

Examples 5.4.3 From the following table find maximum value of y correct to two decimal places

x:	1.2	1.3	1.4	1.5	1.6
y:	0.9320	0.9636	0.9855	0.9975	0.9996

Solution: The difference table is

x	Y	Δ	Δ^2	Δ^3	Δ^4
1.2	0.9320				
		0.0316			
1.3	0.9636		−0.0097		
		0.0219		−0.0002	
1.4	0.9855		−0.0099		0.0002
		0.0120		0.0000	
1.5	0.9975		−0.0099		
		0.0021			
1.6	0.9996				

Let $x_0 = 1.2$. Thus from the Newton's forward difference formula, we have

$$0=0.0316 + \frac{2p-1}{2}(-0.0097)$$

$$\Rightarrow \quad p = 3.8$$

$$\therefore \quad x = x_0 + ph = 1.2 + (3.8)(0.1) = 1.58$$

For this values of x, Newton's backward difference formula at $x_n = 1.6$, gives us

$$y(1.58)=0.9996 - 0.2\,(0.0021)+ \frac{-0.2\,(-0.2+1)}{2}(-0.0099)$$

$$= 0.9996 - 0.0004 + 0.0008 = 1.0$$

which is the required maximum value.

5.5 Numerical integration

Numerical integration is primary tool used by engineers and scientists to obtain approximate answers for definite integrals that cannot be solved analytically. The process of evaluating a definite integral from a set of tabulated values of the integrand $y = f(x)$ is called numerical integration. The general problem of numerical integration may be stated as follows:

Given a set of data points $(x_i, y_i), i = 0,1,2,3,...n$ of a function $y = f(x)$, where $f(x)$ is not known explicitly, it is required to compute the value of the definite integral

$$I = \int_a^b y\,dx \tag{5.5.1}$$

As in the case of numerical differentiation, we here again replace $y=f(x)$, by an interpolating polynomial $\phi(x)$ in order to obtain an approximate value of the definite integral. Thus different integration formulae can be obtained depending upon the type of the interpolation formula used.

Here we derive a general formula for numerical integration by using Newton's forward difference formula. Let the interval $[a, b]$ be divided into n– equal subintervals such that

$$a = x_0 < x_1 < x_2 < \ldots < x_n = b, \text{ with } x_n = x_0 + nh,\ h = \frac{b-a}{n}$$

Hence, the integral (5.5.1) becomes

$$I = \int_{x_0}^{x_n} y\,dx \tag{5.5.2}$$

Using Newton's forward difference formula, we get

$$I = \int_{x_0}^{x_n}\left[y_0 + p\,\Delta y_0 + \frac{p(p-1)}{2!}\,\Delta^2 y_0 + \frac{p(p-1)(p-2)}{3!}\,\Delta^3 y_0 + \ldots\right]dx, \text{ where } x = x_0 + ph.$$

$$= h\int_0^n\left(y_0 + p\,\Delta y_0 + \frac{p^2-p}{2}\,\Delta^2 y_0 + \frac{p^3-3p^2+2p}{6}\,\Delta^3 y_0 + \ldots\right)dp$$

Therefore, on simplification, we have

$$I = \int_{x_0}^{x_n} y\,dx = nh\left[y_0 + \frac{n}{2}\,\Delta y_0 + \frac{n(2n-3)}{12}\,\Delta^2 y_o + \frac{n(n-2)^2}{24}\,\Delta^3 y_o + \ldots\right] \tag{5.5.3}$$

From the general formula (5.5.3), we can find different integration formulae by putting n =1, 2, 3… etc. This formula is also knows as Newton's -cotes closed quadrature formula.

5.5.1 Trapezoidal rule

Putting n=1 in (5.5.3) and taking the curve $y = f(x)$ through the points (x_0, y_0) and (x_1, y_1) as a straight line i.e. a polynomial of first degree so that differences of order higher than first become zero, we get

$$I_1 = \int_{x_0}^{x_1} y\,dx = h\left[y_0 + \frac{1}{2}\,\Delta y_0\right] = h\left[y_0 + \frac{1}{2}\,(y_1 - y_0)\right] = \frac{h}{2}\,(y_0 + y_1)$$

Similarly, $$I_2 = \int_{x_1}^{x_2} y\,dx = \frac{h}{2}\,(y_1 + y_2)$$

$$I_3 = \int_{x_2}^{x_3} y\,dx = \frac{h}{2}\,(y_2 + y_3)$$

and so on. In general, we get

$$I_n = \int_{x_{n-1}}^{x_n} y\,dx = \frac{h}{2}\,(y_{n-1} + y_n)$$

Adding all these expressions and using interval additive property of the definite integrals, we obtain.

$$I=\sum_{i=1}^{n} I_i = \int_{x_0}^{x_n} ydx = \frac{h}{2}\left[\, y_0 + 2(y_1+y_2+y_3+...+y_{n-1})+y_n\right]=\frac{h}{2}(X+2I) \quad (5.5.4)$$

where X = sum of end ordinates, I = sum of intermediate ordinates. The expression (5.5.4) is known as the trapezoidal rule. Geometrically this rule signifies that the curve $y = f(x)$ is replaced by n straight lines joining the points (x_i, y_i), $i = 0,1,2,3,\ldots n$. The area bounded by the curve $y = f(x)$ the ordinates $x = x_0$, $x = x_n$, and the *x–axis* is then approximately equivalent to the sum of the areas of the n–trapeziums so obtained.

Error estimates

Let $y = f(x)$ be a continuous and well–behaved function which possesses continuous derivatives in the interval $[x_0, x_n]$. Expanding y in a Taylor's series around $x = x_0$, we obtain

$$\int_{x_0}^{x_1} ydx = \int_{x_0}^{x_1}\left[y_0+(x-x_0)y_0'+\frac{(x-x_0)^2}{2!}y''+\cdots\right]dx$$

$$= hy_0+\frac{h^2}{2}y_0'+\frac{h^3}{6}y_0''+\frac{h^4}{24}y_0'''+\ldots \quad (5.5.5)$$

Similarly, $\frac{h}{2}(y_0+y_1)=\frac{h}{2}(y_0+y(x_0+h))=\frac{h}{2}\left[y_0+y+hy_0'+\frac{h^2}{2}y_0''+\cdots\right]$

$$= hy_0+\frac{h^2}{2}y_0'+\frac{h^3}{4}y_0''+\frac{h^4}{12}y_0'''+\cdots \quad (5.5.6)$$

From (5.5.5) and (5.5.6) the error e_1 in $[x_0, x_1]$ is obtained as

$$e_1 = \int_{x_0}^{x_1} ydx - \frac{h}{2}(y_0+y_1) = -\frac{1}{12}h^3\, y_0''+\cdots$$

Proceeding in the similar manner, we obtain

$$e_2 = \int_{x_1}^{x_2} ydx - \frac{h}{2}(y_1+y_2) = -\frac{1}{12}h^3\, y_1''+\cdots$$

$$e_3 = -\frac{1}{12}h^3\, y_2''+\cdots$$

$$e_4 = -\frac{1}{12}h^3 y_3''+\cdots$$

and so on. In general we have

$$e_n = -\frac{1}{12}h^3\, y_{n-1}''+\cdots\ .$$

Then the total error E in the interval $[x_0, x_n]$ is given by

$$E = \sum_{i=1}^{n} e_i = -\frac{h^3}{12}\left[y_0''+y_1''+y_2''+\cdots y_{n-1}''\right] \quad (5.5.7)$$

Let $y''(\bar{x})$ be the largest value of the n quantities in the braces on right hand side of (5.5.7), we obtain

$$E = -\frac{1}{12} h^3 n y''(\bar{x}) = -\frac{(b-a)}{12} h^2 y''(\bar{x}) \qquad (5.5.8)$$

since $n\,h = b - a$. Thus the total error in the evaluation of integral (5.5.2) by trapezoidal rule of order h^2.

5.5.2 Simpson's 1/3 rule

Putting $n = 2$ in (5.5.3) and taking the curve through the points (x_0, y_0), (x_1, y_1) and (x_2, y_2) as a parabola i.e. a polynomial of second degree so that differences of order higher than second vanish, we get

$$I_1 = \int_{x_0}^{x_2} y dx = 2h\left(y_0 + \Delta y_0 + \frac{1}{6}\Delta^2 y_0\right) = \frac{h}{3}(y_0 + 4y_1 + y_2)$$

Similarly, $I_2 = \int_{x_2}^{x4} y dx = \frac{h}{3}(y_2 + 4y_3 + y_4)$

$$I_3 = \int_{x_4}^{x_6} y dx = \frac{h}{3}(y_4 + 4y_5 + y_6)$$

and so on. In general, we have

$$I_n = \int_{x_{2n-2}}^{x_{2n}} y\, dx = \frac{h}{3}(y_{2n-2} + 4y_{2n-1} + y_{2n})$$

Summing up these integrals, we get

$$I = \int_{x_0}^{x_n} y dx = \frac{h}{3}\left[y_0 + 4(y_1 + y_3 + y_5 + \cdots y_{2n-1}) + 2(y_2 + y_4 + y_6 + \cdots y_{2n-2}) + y_{2n}\right]$$

$$= \frac{h}{3}(X + 4O + 2E) \qquad (5.5.9)$$

where X= sum of end ordinates, O = sum of odd ordinates and E= sum of even ordinates.

The expression (5.5.9) is known as Simpson's 1/3 -rule, or simply Simpson's rule and is most commonly used. It is observed that this rule requires the whole range i.e. the given interval must be divided into even number of equal sub-intervals, since we find the area of two strips at a time.

Error Estimate

Expanding $y = f(x)$ around $x = x_0$ by Taylor's series, we get

$$\int_{x_0}^{x_2} y dx = \int_{x_0}^{x_0+2h}\left(y_0 + (x - x_0)y_0' + \frac{(x - x_0)^2}{2!} y_0'' + \cdots\right)dx$$

$$= 2hy_0 + \frac{4h^2}{2!} y_0' + \frac{8h^3}{3!} y_0'' + \frac{16h^4}{4!} y_0''' + \frac{32h^5}{5!} y_0^{iv} + \cdots$$

$$= 2hy_0 + 2h^2 y_0' + \frac{4}{3}h^3 y_0'' + \frac{2h^4}{3} y_0''' + \frac{4}{15}h^5 + y^{iv}{}_0 + \cdots \qquad (5.510)$$

Also

$$\frac{h}{3}(y_0 + 4y_1 + y_2) = \frac{h}{3}\left[y_0 + 4\left(y_0 + hy_0' + \frac{h^2}{2!}y_0'' + \cdots\right) + \left(y_0 + 2hy_0' + \frac{4h^2}{2\,!}y_0'' + \frac{8h^3}{3\,!}y_0''' + \cdots\right)\right]$$

$$= 2hy_0 + 2h^2 y_0' + \frac{4h^3}{3}y_0'' + \frac{2h^4}{3}y_0''' + \frac{5h^5}{18}y_0{}^{iv} + \cdots \qquad (5.5.11)$$

Therefore from (5.510) and (5.5.11) the error in the subinterval (x_0, x_2) is given by

$$e_1 = \int_{x_0}^{x_2} ydx - \frac{h}{3}(y_0 + 4y_1 + y_2) = \left(\frac{4}{15} - \frac{5}{18}\right)h^5 y_0{}^{iv} + \ldots = -\frac{h^5}{90}y_o{}^{iv} + \cdots \cong -\frac{h^5}{90}y_0^{iv}$$

Similarly the errors in the subsequent intervals are given by

$$e_2 = -\frac{h^5}{90}y_2{}^{iv}, \quad e_3 = \frac{h^5}{90}y_4 \ldots,$$

and so on. Thus the total error E is given by

$$\mathrm{E} = \sum e_i = \frac{-h^5}{90}\left(y_0{}^{iv} + y_2{}^{iv} + y_4{}^{iv} + \ldots + y_{2n-2}^{iv}\right) \cong \frac{-nh^5}{90}y^{iv}{}_0(\bar{x})$$

Thus, $$\mathrm{E} = -\frac{(b-a)}{180}h^2 y^{iv}_{(\bar{x})} \qquad (5.5.12)$$

where $y^{(iv)}(\bar{x})$ is the largest value of the fourth order derivatives and $2nh = b - a$ .If the various ordinates represent equispaced cross-sectional areas, then the Simpson's rule gives the volume of the solid. As such, Simpson's rule is very useful to civil engineers for calculating the amount of earth that must be moved to fill a depression or make a dam. Similarly if the ordinates denote velocities at equal intervals of time, the Simpson's rule gives the distance traveled.

5.5.3 Simpson's 3/8 rule

Setting n=3 in (5.5.3) above and taking the curve through (x_i, y_i), i = 0, 1, 2, 3 as a polynomial of degree three so that the differences higher than the third order vanish, we get

$$I_1 = \int_{x_0}^{x_3} ydx = 3h\left(y_0 + \frac{3}{2}\Delta y_0 + \frac{3}{2}\Delta^2 y_0 + \frac{1}{8}\Delta^3 y_0\right) = \frac{3}{8}h(y_0 + 3y_1 + 3y_2 + y_3)$$

Similarly, $$I_2 = \int_{x3}^{x_6} ydx = \frac{3}{8}h(y_3 + 3y_4 + 3y_5 + y_6)$$

$$I_3 = \int_{x_6}^{x_9} ydx = \frac{3}{8}h(y_6 + 3y_7 + 3y_8 + y_9)$$

and so on. In general, we have

$$I_n = \int_{x_{3n-3}}^{x_{3n}} ydx = \frac{3}{8}h\left(y_{3n-3} + 3y_{3n-2} + 3y_{3n-1} + y_{3n}\right)$$

Summing up all these expressions, we get

$$I = \int_{x_0}^{x_{3n}} ydx = \frac{3}{8}h\left[y_0 + 3(y_1 + y_2 + y_4 + y_5 + y_7 + y_8 + \cdots + y_{3n-2} + y_{3n-1})\right.$$
$$\left. + 2(y_3 + y_6 + y_9 + \cdots + y_{3n-3}) + y_{3n}\right] \quad (5.5.13)$$

The expression (5.5.13) is known as Simpson's 3/8 rule. It is noticed that in order to apply the formula (5.5.13) the number of sub-intervals should be taken as multiples of 3. This rule is not as accurate as Simpson 1/3 rule, the dominant term in the error of this formula being $\frac{-3}{80}h^5 y^{iv}(\bar{x})$.

5.5.4 Boole's rule

Putting n=4 in (5.5.3) above and taking the curve through (x_i, y_i), i=0, 1,2,3,4 as a polynomial of degree 4, so that the difference of order higher than four are neglected, we get

$$\int_{x_0}^{x_4} ydx = 4h\left(y_0 + 2\Delta y_0 + \frac{5}{3}\Delta^2 y_0 + \frac{2}{3}\Delta^3 y_0 + \frac{7}{90}\Delta^4 y_0\right)$$
$$= \frac{2h}{45}\left(7y_0 + 32y_1 + 12y_2 + 32y_3 + 7y_4\right)$$

Similarly, $\int_{x_4}^{x_8} ydx = \frac{2h}{45}\left(7y_4 + 32y_5 + 12y_6 + 32y_7 + 7y_8\right)$ and so on.

Adding all these integrals from x_0 to x_n, where n is a multiple of 4, we get

$$I = \int_{x_0}^{x_n} ydx = \frac{2h}{45}\left[7y_o + 32\left(y_1 + y_3 + y_5 + y_7 + \cdots\right) + 12\left(y_2 + y_6 + y_{10} + \cdots\right)\right.$$
$$\left. + 14\left(y_4 + y_8 + y_{12} + \cdots\right) + 7y_n\right] \quad (5.5.14)$$

This expression (5.514) is known as Boole's rule. While applying (5.5.14) the number of sub-intervals should be taken as a multiple of 4 The leading term in the error of formula can be shown as $\frac{-8h^7}{945} y^{vi}(\bar{x})$.

5.5.5 Weddle's rule

Putting n = 6 in (5.5.3) above and taking the curve $y = f(x)$ through the points (x_i, y_i), i = 0, 1, 2, 3, 4, 5, 6 as a polynomial of degree six so that the differences of order higher than six are neglected, we obtain.

$$\int_{x_o}^{x_6} ydx = 6h\left(y_0 + 3\Delta y_0 + \frac{9}{2}\Delta^2 y_0 + 4\Delta^3 y_0 + \frac{123}{60}\Delta^4 y_0 + \frac{11}{20}\Delta^5 y_0 + \frac{41}{840}\Delta^6 y_0\right)$$

$$=\frac{3h}{10}[y_0+5y_1+y_2+6y_3+y_4+5y_5+y_6],$$

since if we replace $\frac{41}{140}\Delta^6 y_0$ *by* $\frac{3}{10}\Delta^6 y_0$, the error made will be negligible.

Similarly,

$$\int_{x_6}^{x_{12}} ydx=\frac{3h}{10}(y_6+5y_7+y_8+6y_9+y_{10}+5y_{11}+y_{12}) \text{ and so on.}$$

Adding these integrals from x_0 to x_n, where x is a multiple of 6, we get

$$\int_{x_0}^{x_n} ydx=\frac{3h}{10}[y_0+5(y_1+y_5+y_7+y_{11}+...)+(y_2+y_4+y_8+y_{10}+...)$$
$$+6(y_3+y_9+y_{15}+...)+2(y_6+y_{12}+y_{18}+...)+y_n] \quad (5.5.15)$$

The expression (5.5.15) is known as Weddle's rule. It is generally, more accurate than any of the other rules and the error in this is given by $\frac{-h^7}{140}y^{vi}(\bar{x})$. While applying (5.5.15) the number of subintervals should be taken as multiply of 6.

Example 5.5.1: Evaluate $\int_0^6 \frac{dx}{1+x^2}$ by using (*i*) Trapezoidal rule, (*ii*) Simpson's 1/3 rule, (*iii*) Simpson's rule and (*iv*) Weddle's rule and command the result the results

Solution: We divide the interval (0, 6) into six parts with $h = 1$. The values of x and $y=\frac{1}{1+x^2}$ are given by

x:	0	1	2	3	4	5	6
y:	1	0.5	0.2	0.1	0.0588	0.0385	0.027

(i) Trapezoidal rule

$$I=\int_0^6 \frac{dx}{1+x^2}=\frac{1}{2}[(1+0.027)+2(.5+.2+.1+0588+.0385)]=1.4108$$

(ii) Simpson's 1/3 rule

$$I=\int_0^6 \frac{dx}{1+x^2}=\frac{1}{3}[(1+.027)+4(0.5+0.1+0.0385)+2(0.2+0.0588)]=1.3662$$

(iii) Simpson's 3/8 rule

$$I=\int_0^6 \frac{dx}{1+x^2}=\frac{3}{8}[(1+0.027)+3(0.5+0.2+.0.0588+0.0385)+2(0.1)]=1.3571$$

(iv) Weddle's rule

$$I = \int_0^6 \frac{dx}{1+x^2} = \frac{3}{10}\left[1+5(0.5)+0.2+6(0.1)+0.0588+5(0.0385)+0.027\right]=1.3735$$

Also $I = \int_0^6 \frac{dx}{1+x^2} = \left|\tan^{-1} x\right|_0^6 = \tan^{-1} 6 = 1.4056$

This shows that the values of the integral found by Weddle's rule is the nearest to the actual value followed by Simpson's 1/3 rule.

Example 5.5.2 A solid of revolution is formed by rotating about the *x*-axis, the area between the lines *x*=0 and *x*=1 and a curve through the points with following coordinates.

x:	0.00	0.25	0.50	0.75	1.00
y:	1.0000	0.9896	0.9589	0.9089	0.8415

Estimate the volume of the solid formed by using Simpson's $\frac{1}{3}$ rule.

Solution: Here $h = 0.25$, $y_0=1$, $y_1=0.9896$ $y_2=$ .09589, $y_3=0.9089$ and $y_4=$ 0.8415. Therefore, the required volume of the solid generated, by Simpson's rule is given by

$$V = \int_0^1 \pi y^2 dx = \frac{\pi h}{3}\left[\left(y_0^2+y_4^2\right)+4\left(y_1^2+y_3^2\right)+2y_2^2\right]$$

$$= \frac{0.25\pi}{3}\left[\left\{1+(0.8415)^2\right\}+4\left\{(0.9896)^2+(0.9089)^2\right\}+2(0.9589)^2\right]$$

$$= \frac{0.25(3.1416)}{3}\left[1.7081+7.2216+1.839\right]=2.8192$$

which is the required volume.

Example 5.5.3 Apply trapezoidal and Simpson's rules to the integral $I = \int_0^1 \sqrt{1-x^2}\, dx$ successively halving the interval for better accuracy solution.

If we use 10, 20, 30, 40 and 50, subintervals successively; an electronic computer produced the following results.

No. of subintervals	Trapezoidal rule	Simpson's rule
10	0.77612958	0.78175204
20	0.78211622	0.78411177
30	0.78361079	0.78469843
40	0.78423693	0.78494384
50	0.78456713	0.78507314

The true value of the integral is $I = \int_0^1 \sqrt{1-x^2}\, dx = \frac{\pi}{4} = 0.78539816$

Thus Simpson's rule gives us more accurate result than the trapezoidal rule.

Example 5.5.4 Investigate the error in the evaluation of the integral $I = \int_1^6 \left(2 + \sin \sqrt[2]{x}\right) dx$,

dividing the interval in 10, 20, 40, and 80 subintervals, by using trapezoidal and Simpson rules.

Solution: If we use 10, 20, 40 and 80 subintervals successively, the following results are obtained.

n	h	Trapezoidal values	Error = = $0(h^2)$
10	0.5	8.19385457	–0.01037540
20	0.25	8.18604926	–0.00257006
40	0.125	8.18412019	–0.00064098
80	0.0625	8.18363936	–0.00016015
N	h	Simpson's value	Error = $0(h^2)$
5	0.5	8.18301549	0.00046371
10	0.25	8.18344750	0.00003171
20	0.125	8.18347717	0.00000204
40	0.0625	8.18347908	0.00000013
80	0.03125	8.18347920	0.00000001

Form trapezoidal rule results it is observed that when h is reduced by a factor of $1/2$ the successive errors are diminished by $\frac{1}{4}$ approximately, which congers that the order is $0(h^2)$. Also Simpson's evaluations show that when h is revived by a factor of $1/2$ the successive errors is diminished by approximately $1/16$. Thus the convergence order is $0(h^4)$.

Algorithm 5.1 Trapezoidal rule

To approximate the integral

$$I = \int_a^b y dx = \int_a^b f(x) dx \cong \frac{h}{2} [y_o + y_n] + h \sum_{k=1}^{n-1} y_k$$

by sampling $y = f(x)$ at the n+1 equally spaced points

$x_k = a + kh$ for $k = 0,1,2,\ldots n.$

Here $x_0 = a, x_n = b; y_0 = f(a), y_n = f(b), y_k = f(x_k)$

1. Input a, b, n
2. $h: = (b-a)/n$
3. Sum $= 0.0$
4. For $k = 1$ to n-1do in steps 5 to 6
5. $X = a + h\,k$
6. Sum = Sum + $f(X)$
7. Sum = h [$f(a) + f(b)$ +2*Sum]/2

8. Output *a, b, n,* Sum
9. Stop.
10. end

Algorithm 5.2 Simpson's 1/ 3 rule

To approximate the integral

$$\int_a^b y\,dx = \int_a^b f(x)dx \cong \frac{h}{3}\left[y_0 + y_n + 2\sum_{k=1}^{n-1} y_{2k} + 4\sum_{k=1}^{n} y_{2k-1}\right]$$

by sampling $y = f(x)$ at $2n$-1 equally spaced points $x_k = a + hk$ for $k = 0,1,2,...2n$.

Here $x_0 = a,\ x_{2n} = b,\ y_0 = f(a),\ y_{2n} = f(b),\ y_k = f(x_k)$.

1. Input *a, b, n.*
2. $h = (b\text{-}a)/2n$.
3. Sum Even = 0.
4. For $k = 1$ to $n - 1$ do in steps5 and 6.
5. $X = a + 2\,h\,k$.
6. Sum even = Sum even + f(X).
7. Sum odd = 0.
8. For k = 1to x^0 Do in steps 9 to 10.
9. $X = a + h\,(2k\text{–}1)$.
10. Sum odd = Sum odd + $f(X)$.
11. Sum = h [$f(a)$ +$f(b)$ +2 Sum Even +4 Sum Odd]/3.
12. Out put *a, b, n,* Sum.
13. Stop.
14. End.

5.6 Romberg integration

Romberg's method provides a simple modification to the approximate quadrature formula derived with the help if finite differences method in order to find their better approximations. As an illustration, we improve upon the value of the integral

$$I = \int_a^b y dx = \int_a^b f(x)\,dx \qquad (5.6.1)$$

by trapezoidal rule. We evaluate (5.6.1) by trapezoidal rule (5.5.4) with two different widths h_1 and h_2 to obtain the approximate values I_1 and I_2 respectively. The corresponding errors E_1 and E_2 are then given by

$$E_1 = -\frac{(b-a)h_1^2}{12}y''(\bar{x}),\ E_2 = -\frac{(b-a)h_2^2}{12}y''(\bar{\bar{x}})$$

Since $y''(\bar{\bar{x}})$ is also the largest value of $y''(x)$, so it is reasonable to assume that the quantities $y''(\bar{x})$ and $y''(\bar{\bar{x}})$ are very nearly equal Therefore we have

$$\frac{E_1}{E_2} = \frac{h_1^2}{h_2^2} \Rightarrow \frac{E_2}{E_2 - E_1} = \frac{h_2^2}{h_2^2 - h_1^2} \tag{5.6.2}$$

Now since $I = I_1 - E_1 = I_2 - E_2$, therefore

$$E_2 - E_1 = I_2 - I_1 \tag{5.6.3}$$

From (5.6.2) and (5.6.3), we have

$$E_2 = \frac{h_2^2}{h_2^2 - h_1^2}(E_2 - E_1) = \frac{h_2^2}{h_2^2 - h_1^2}(I_2 - I_1)$$

Hence

$$I_3 = I_2 - E_2 = \frac{I_1 h_2^2 - I_2 h_1^2}{h_2^2 - h_1^2} \tag{5.6.4}$$

which is a better approximation of I. In order to evaluate I systematically, we take $h_1 = h$ and $h_2 = \frac{1}{2}h$ so that (5.6.4) gives

$$I = \frac{I_1 \frac{h^2}{4} - I_2 h^2}{\frac{h^2}{4} - h^2} = \frac{4I_2 - I_1}{3}$$

i.e.

$$I\left(h, \frac{h}{2}\right) = \frac{4I\left(\frac{h}{2}\right) - I(h)}{3} \tag{5.6.5}$$

Now we use the trapezoidal rule several times successively halving h and apply(5.6.5) to each pair of values as per the following scheme.

$$\begin{array}{llll}
I(h) & & & \\
 & I\left(h, \frac{h}{2}\right) & & \\
I\left(\frac{h}{2}\right) & & I\left(h, \frac{h}{2}, \frac{h}{4}\right) & \\
 & I\left(\frac{h}{2}, \frac{h}{4}\right) & & I\left(h, \frac{h}{2}, \frac{h}{4}, \frac{h}{8}\right) \\
I\left(\frac{h}{4}\right) & & I\left(\frac{h}{2}, \frac{h}{4}, \frac{h}{8}\right) & \\
 & I\left(\frac{h}{4}, \frac{h}{8}\right) & & \\
I\left(\frac{h}{8}\right) & & &
\end{array}$$

The computation is continued till successive values are close to each other. This method, due to L.F. Richardson, is called the deferred approach to the limit and the systematic tabulation of this is called Romberg integration.

Example 5.6.1 Use Romberg's method to compute $\int_0^1 \frac{dx}{1+x^2}$, correct to 4 decimal places, by taking h=0.5,0.25 and 0.125.

Solution: We evaluate the given integral by using trapezoidal rule

(i) When h= 0.5 , we have

x:	0	0.5	1.0
$y=\frac{1}{1+x^2}$:	1	0.8	0.5

$$\therefore \quad I = \int_0^1 \frac{dx}{1+x^2} = \frac{.5}{2}[1+2(.8)+0.5] = 0.775 \textit{ by } \text{Trapezoidal rule}$$

(ii) When h = 0.25, the values of x and y are

(iii)

x:	0	0.25	0.5	0.75	1.0
$y=\frac{1}{1+x^2}$:	1	0.9412	0.8	0.64	0.5

Therefore by trapezoidal rule, we have

$$I = \int_0^1 \frac{dx}{1+x^2} = \frac{.25}{2}[1+2(0.9412+0.8+0.64)+0.5] = 0.7828$$

(iii) When h = 0.125, we find that I = 0.7848. Now using formula (5.6.5.) we obtain the table of values as

0.5	0.775		
		0.7854	
0.25	0.7828		0.7855
		0.7855	
0.125	0.7848		

Hence the value of the integral is 0.7855.

Algorithm 5.3: Romberg Integration.

To approximate the integral $I = \int_a^b ydx = \int_a^b f(x)dx$ by generating a table of approximations $R(i, k)$ for $k < i$ and using $R(j,j)$ as the final answer . The elements $R(j, 0)$ are computed by using sequential rule based on 2^j subintervals of $(a, b,)$ and then $R(j, k)$ is computed using Romberg's rule . The elements of jth row are

$$R(j, k) = R(j, k-1) + \frac{R(j,k-1) - R(j-1,k-1)}{4^k - 1}, \ 1 \le k \le j$$

The algorithm is terminated in the j*th* row when $\left| R(j,j) - R(j-1,j-1) \right| < Tol$.

1. Input a, b, n, Tol
2. Variable declaration REAL R [0,...,n; 0...n]
3. $m = 1$
4. $h = b-$a
5. close = 1
6. $j = 0$
7. $R(0,0) = h\,[f(a) + f(b)\,]/2$
8. While [close > Tol and $j < n$] or (j ,4) do steps 9 to 14
9. $j = j + 1$
10. CALL Subroutine T_γ Rule (a, h, j, m, R)
11. For $k = 1$ to j do : steps 12 to 14
12. $R(j, k) + R(j, k-1) = [R(J, k-1) - R(j-1, K-1\,]/\,4^k - 1)$
13. out put $R(j, 0)$, $R(J,1)$, $R(J, 2)$,...,R (J, J)
14. close = $|R(j-1, j-1) - R(j, j)|$
15. out put a, b, n, $R(j, j)$
16. stop
17. end

Subroutine T_γ Rule (a, h, j, m, R)

1. $H = h/2$
2. Sum = 0.0
3. For p =1 to n do step 4
4. sum = sum + $f(a + h\,(2p - 1))$
5. $R(j,0) = R(j-1,0)/2 + h^*$ sum
6. $m = 2m$
7. end
8. return.

Example 5.6.2 Use Romberg's method to compute $I = \int_0^1 \frac{dx}{1+x}$, correct to three decimal places.

Solution: We take h =0.5, 0.25 and 0.125 and use trapezoidal rule successively to obtain.

$$I(h) = I(0.5) = 0.7084,\ I\left(\frac{h}{2}\right) = I(0.25) = 0.6970$$

and
$$I\left(\frac{h}{4}\right) = I(0.125) = 0.6941$$

Now using formula (5.6.5) we obtain

$$I\left(h,\frac{h}{2}\right) = \frac{4(0.6970)-0.7084}{3} = 0.6932$$

$$I\left(h,\frac{h}{2},\frac{h}{4}\right) = \frac{4(0.6941)-0.6970}{3} = 0.6931$$

$$I\left(\frac{h}{2},\frac{h}{4}\right) = \frac{4(0.6931)-0.6932}{3} = 0.6931$$

The table of values is therefore

0.5	0.7084		
		0.6932	
0.25	0.6970		0.6931
		06931	
0.125	0.6941		

Hence the value of the given integral is 0.6931

5.6.1 Recursive formulae

Trapezoidal rule: Consider the integral

$$I = \int_a^b y dx = \int_a^b f(x)dx$$

Start with $T(0) = \frac{h}{2}[f(a)+f(b)]$. Then a sequence of trapezoidal rule $\{T(J)\}$ is generated by the recursive formula

$$T(j) = \frac{T(j-1)}{2} + h\sum_{k=1}^{2^j} f(x_{2k-1}),\ j=1,2,3...$$

where $h=(b-a)/2^j$ and $x_k = a+kh$.

Proof: For the even nodes $x_0 < x_2 <... < x_{2n-2} < x_{2n}$ we use the trapezoidal rule with step size $2h$.

$$T(j-1) = \frac{2h}{2}\left[f_0 + 2(f_2 + f_4 +...+ f_{2n-2}) + f_{2n}\right] \tag{5.6.6}$$

For all of the nodes x_0 $x_0 < x_1 < x_2 <... x_{2n-2} < x_{2n-1} < x_{2n}$ we use the trapezoidal rule with step size h.

$$T(j) = \frac{h}{2}\left[f_0 + 2(f_1 + f_2 +...+ f_{2n-1}) + f_{2n}\right] \tag{5.6.7}$$

Collecting the even and odd subscripts, the above expression leads to

$$T(j)=\frac{h}{2}\left[f_0+2(f_2+f_4+...+f_{2n-1})+f_{2n}\right]+h\sum_{k=1}^{n}f_{2k-1} \tag{5.6.8}$$

Substituting (5.6.6) in (5.6.8) we get

$$T(j)=\frac{1}{2}\mathrm{T}(j-1)+h\sum_{k=1}^{n}f_{2k-1} \tag{5.6.9}$$

This competes the proof.

Simpson's rule

Suppose that $\{T(j)\}$ is the sequence of trapezoidal rules generated by (5.6.9). If $j \geq 1$ and $S(j)$ is Simpson's rule for 2^j subintervals of $[a, b]$ then $S(j)$ and $T(j)$ obey the relational ship

$$S(j)=\frac{4T(j)-T(j-1)}{3} \; for = 1,2,3,...$$

Proof: By trapezoidal rule with step size h we have

$$\int_a^b ydx = \int_a^b f(x)dx \cong \frac{h}{2}\left[f_0+2(f_1+f_2+...+f_{2x-2})+f_{2n}\right]=T(j) \tag{5.6.10}$$

The trapezoidal rule with step size $2h$ produces

$$\int_a^b ydx = \int_a^b f(x)dx \cong \frac{2h}{2}\left[f_0+2(f_1+f_2+...+f_{2x-2})+f_{2n}\right]=T(j) \tag{5.6.11}$$

Now operating (5.6.10)-(5.6.11) we get

$$3\int_a^b f(x)dx \cong \left[f_0+4f_1+2f_2+...+2f_{2n-2}+4f_{2n-1}+f_{2x}\right]=4T(j)-T(j-1)$$

which implies that $\displaystyle\int_a^b ydx = \int_a^b f(x)dx = \frac{h}{3}\left[f_0+4f_1+2f_2+...+2f_{2n-2}+4f_{2n-1}+f_{2n}\right]$

$$=\left[4T(j)-T(j-1)\right]/3 \tag{5.6.12}$$

Thus $\quad S(j)=\left[4T(j)-T(j-1)\right]/3.$, since the middle term in (5.6.12) is the Simpson's rule.

This completes the proof.

Algorithm 5.4 Recursive trapezoidal rule

To approximate the integral $I=\int_a^b f(x)dx$, by using the trapezoidal rule successively and in creasing the number of subintervals of (a, b). The jth iteration samples $y = f(x)$ at 2n+1 equally spaced points.

1. Input a, b, n
2. Variable declaration Real T [0,1,…,n]
3. $m = 1$
4. $h = b - a$
5. $t(0) = h\,[f(a) + f(b)\,]/2$

6. For j = to n do: steps 7 to 13
7. $m = 2m$
8. $h = h/2$
9. sum = 0.0
10. For $k = 1$ to $\frac{m}{2}$ do: step 11
11. sum = sum + $f(a + h(2k-1)$
12. $T(j) = T(j-1)/2 + h$ * sum
13. output m, a, b, $T(j)$
14. stop
15. end

5.7 Gaussian integration

Consider the integral

$$I = \int_a^b y dx = \int_a^b f(x) dx \tag{5.7.1}$$

Setting $x = \frac{1}{2}u(b-a) + \frac{1}{2}(a+b)$ the integral (5.7.1) takes the form

$$I = \frac{b-a}{2} \int_{-1}^{1} f(u) du \tag{5.7.2}$$

Gauss derived a formula, which uses the same number of function values, but with different spacing in contrast to other integration formulae, which require values of the function at equally spaced points of the interval and it gives better accuracy. This formula is expressed in the form

$$\int_{-1}^{1} F(u) du = W_1 F(u_1) + W_2 F(u_2)\ W_3 F(u_3) + \ldots + W_n F(u_n) = \sum_{i=1}^{n} W_i F(u_i) \tag{5.7.3}$$

where W_i and u_i are called the weights and abscissas respectively, which are symmetrical with respect to the middle points of the interval. The weights and abscissa can be determined such that the formula is exact when $F(u)$ is a polynomial of degree not exceeding $2n-1$ as there are total $2n$ arbitrary constants. Hence we have

$$F(u) = c_0 + c_1 u + c_2 u^2 + c_3 u^3 + \ldots + c_{2n-1} u^{2n-1} \tag{5.7.4}$$

Then from (5.7.3) we have

$$\int_{-1}^{1} F(u) du = \int_{-1}^{1} \left[c_0 + c_1 u + c_2 u^2 + c_3 u^3 + \ldots + c_{2n-1} u^{2n-1}\right] du = 2c_0 + \frac{2}{3}c_2 + \frac{2}{5}c_4 + \ldots \tag{5.7.5}$$

Now (5.7.4) implies that $F(u_i) = c_0 + c_1 u_i + c_2 u_i^3 + \ldots + c_{2n-1} u_i^{2n-1}$

so that from (5.7.3) we have

$$\int_{-1}^{1} F(u)du = w_1(c_0 + c_1u_1 + c_2u_1^{\ 2} + \ldots + c_{2n-1}u_1^{\ 2n-1}) + w_2(c_0 + c_1u_2 + c_2u_2^2 + \ldots + c_{2n-1}u_1^{2n-1})$$

$$+ w_3(c_0 + c_1u_3 + c_2u_3^{\ 2} + \ldots + c_{2n-1}u_3^{\ 2n-1}) + \ldots + w_n(c_0 + c_1u_n + c_2 + \cdots + c_{2n-1}u_n^{\ 2n-1})$$

Therefore, we get

$$\int_{-1}^{1} F(u)du = c_0(w_1 + w_2 + \ldots + w_n) + c_1(w_1u_1 + w_2u_2 + \ldots w_nu_n) + c_2(w_1u_1^{\ 2} + w_2u_2^{\ 2} + \ldots + w_nu_n^{\ 2})$$

$$+ \ldots + c_{2n-1}(w_1u_1^{2n-1} + w_2u_1^{2n-1} + \ldots + w_nu_n^{2n-1}) \tag{5.7.6}$$

Now comparing expressions (5.7.5) and (5.7.6) we get

$$\left.\begin{aligned} w_1 + w_2 + w_3 + \ldots + w_n &= 2 \\ w_1u_1 + w_2u_2 + w_3u_3 + \ldots + w_nu_n &= 0 \\ w_1u_1^2 + w_2u_2^2 + w_3^2 + \ldots + w_nu_n^2 &= 2/3 \\ \ldots\ldots\ldots\ldots\ldots\ldots\ldots\ldots\ldots\ldots\ldots\ldots \\ w_1u_1^{2n-1} + w_2u_2^{2n-1} + w_3u_3^{2n-1} + \ldots + w_nu_n^{2n-1} &= 0 \end{aligned}\right\} \tag{5.7.7}$$

a system of $2n$ equations in $2n$ unknowns w_i and $u_i, (i = 1, 2, 3, \ldots, n)$.

In order to illustrate, we take n=2. Then the formula is

$$\int_{-1}^{1} F(u)du = w_1F(u_1) + w_2F(u_2) \tag{5.7.8}$$

This formula is exact when $F(u)$ a polynomial of degree not exceeding 3 is, we put successively $F(u) = 1, u, u^2$ and u^3 Then (5.7.8) provides us

$$\left.\begin{aligned} w_1 + w_2 &= 2 \\ w_1u_1 + w_2u_2 &= 0 \\ w_1u_1^2 + w_2u_2^2 &= 2/3 \\ w_1u_1^3 + w_2u_2^3 &= 0 \end{aligned}\right\} \tag{5.7.9}$$

This system of equations gives us

$$w_1 = w_2 = 1, u_2 = -u_1 = 1/\sqrt{3} \tag{5.7.10}$$

This method when applied to the general system (5.7.7) above will be extremely complicated difficult and an alternative method must be chosen to solve the non-linear system (5.7.7). It can be shown that u_i are the zeros of the $(n+1)$ Legendre polynomial $P_{n+1}(u)$, which can be generated by using the recurrence relation

$$(n+1)P_{n+1}(u) = (2n+1)uP_n(u) - nP_{n-1}(u) \tag{5.7.11}$$

where $P_0(u) = 1$ and $P_1(u) = u$ .The first five Legendre polynomials are given by

$$P_0(u) = 1, P_1(u) = u, P_2(u) = \frac{1}{2}(3u^2 - 1)$$

$$P_0(u)=1, P_1(u)=u,\ P_2(u)=\frac{1}{2}(3u^2-1),\ P_3(u)=\frac{1}{2}(5u^3-3u),$$

$$P_4(u)=\frac{1}{8}(35u^4-30u^2+3) \qquad (5.7.12)$$

It can also be shown that the corresponding weights w_i are given by

$$w_1=\int_{-1}^{1}\underset{\substack{j=0\\j\neq i}}{\overset{n}{\pi}}\left(\frac{u-u_j}{u_i-u_j}\right)du \qquad (5.7.13)$$

where u_i's are the abscissas. For example when $n=1$ we solved $P_2(u)=0$ i.e., $\frac{1}{2}(3u^2-1)=0$

which implies that $u_0=-\frac{1}{\sqrt{3}}=-\sqrt{3}/3$ *and* $u_1\ \frac{1}{\sqrt{3}}=\sqrt{3}/3$

The corresponding weights are given by

$$w_0=\int_{-1}^{1}\frac{u-u_1}{u_0-u_1}du=\frac{1}{u_0-u_1}\left[\frac{u^2}{2}-u_1u\right]_{-1}^{1}=1$$

and
$$w_1=\int_{-1}^{1}\frac{u-u_0}{u_1-u_0}du=\frac{1}{u_1-u_0}\left[\frac{u^2}{2}-u_0u\right]_{-1}^{1}=1$$

Similarly, for $n=3$ we solve $P_4(u)=0$ i.e. $\frac{1}{8}(35u^4-30u^2+3)=0$

which implies that $u_i=\pm\left[\frac{15\pm2\sqrt{30}}{35}\right]^{1/2}$ The weights w_i can then be found from (5.7.13)

The values of abscissae and weights for values of n up to $n=6$ are give in the following table

n	$\pm u_i$	w_i
2	0.57735 02692	1.0
3	0.0	0.88888 88889
	0.77459 66692	0.55555 55556
4	0.33998 10436	0.65214 51549
	0.86113 63116	0.34785 48451
5	0.0	0.5688 88889
	0.53846 93101	0.47862 86705
	0.90617 98459	0.23692 68851
6	0.23861 91861	0.46791 39346
	0.66120 93865	0.36076 15730
	0.93246 95142	0.17132 44924

Example 5.7.1 Evaluate $I = \int_0^1 x dx$, by Gauss's formula

Solution: Put $x = \frac{1}{2}(u+1)$, we get $I = \frac{1}{4}\int_{-1}^{1}(u+1)du = \frac{1}{4}\sum_{i=1}^{n} W_i F(u_i)$

where $F(u_i) = u_i + 1$. For simplicity, we take $n = 4$, we obtain

$$I = \frac{1}{4}[(-0.86114+1)(0.34785)+(-0.33998+1)(0.65214)+(0.33998+1)(0.65214)$$
$$+(0.86114+1)(0.34785)] = 0.49999...$$

where the abscissae and weights have been rounded to five decimal places.

Algorithm 5.5 Gauss-Legendre quadrature

To approximate the integral

$$\int_a^b f(x)dx = \frac{b-a}{2}\left[w_{n,1}f(t_{n,1}) + w_{n,2}f(t_{n,2}) + \ldots + w_{n,n}f(t_{n,n})\right]$$

$$= \frac{b-a}{2}\sum_{k=1}^{n} w_{n,k} f\left(\frac{a+b}{2}, \frac{b-a}{2}x_{n,k}\right), t = \frac{a+b}{2} + \frac{b-a}{2}x.$$

by sampling $y = f(x)$ at n unequally spaced points $t_{n,j}, j = 1,2,3,...n.$

Here we use $f(x)$ abscissae $x_{n,k}$ and weights $w_{n,k}$.

1. Input *a*, *b*, *n*
2. sum = 0.0
3. For $k = 1$ to n do: steps 4 and 5
4. $T = \frac{a+b}{2} + x_{n,k}\left(\frac{b-a}{2}\right)$
5. sum = sum + $w_{n,k} f(T)$
6. sum = sum + $W_{n,k} F\left(\frac{b-a}{2}\right)$
7. output *a*, *b*, *n*, sum
8. stop
9. end.

5.8 Euler-Maclaurins formula

We consider $\Delta f(x) = g(x)$ and define inverse operator Δ^{-1} as

$$f(x) = \Delta^{-1} g(x) \tag{5.8.1}$$

Now

$$f(x_1) - f(x_0) = \Delta f(x_0) = g(x_0)$$
$$f(x_2) - f(x_1) = g(x_1)$$
$$f(x_3) - f(x_2) = g(x_2)$$
$$\dots\dots\dots\dots\dots\dots\dots$$
$$f(x_n) - f(x_{n-1}) = g(x_{n-1})$$

On addition these lead to

$$f(x_n) - f(x_0) = \sum_{i=0}^{n-1} g(x_i) \tag{5.8.2}$$

where $x_i = i = 0,1,2,...n$ are $(n+1)$ equally spaced values such that $x_i = x_0 + ih.$

Now from (5.8.1), we have

$$f(x) = \Delta^{-1} g(x) = (E+1)^{-1} g(x) = \left(e^{hD} - 1\right)^{-1} g(x), \left[\because E = e^{hD}\right]$$

$$= \left[\left(1 + hD + \frac{h^2 D^2}{2!} + \frac{h^3 D^3}{3!} + \dots\right) - 1\right]^{-1} g(x) = \frac{1}{h} D^{-1}\left[1 - \frac{hD}{2} + \frac{h^2 D^2}{12} - \frac{h^4 D^4}{720} + \dots\right] g(x)$$

$$= \frac{1}{h}\int g(x)dx - \frac{1}{2} g(x)dx + \frac{h}{12} g'(x) - \frac{h^3}{720} g'''(x) + \dots \tag{5.8.3}$$

Upon putting $x = x_n$ and $x = x_0$ in (5.8.3) and then subtracting, we get

$$f(x_n) - f(x_0) = \frac{1}{h}\int_{x_0}^{x_n} g(x)dx - \frac{1}{2}[g(x_n) - g(x_0)] + \frac{h}{12}[g'(x_n) - g'(x_0)]$$

$$-\frac{h^3}{720}[g'''(x_n) - g'''(x_0)] + \dots \tag{5.8.4}$$

Now (5.8.2) and (5.8.4) provide us

$$\sum_{i=0}^{n-1} g(x_i) = \frac{1}{h}\int_{x_0}^{x_n} g(x)dx - \frac{1}{2}[g(x_n) - g(x)] + \frac{h}{12}[g'(x_n) g'(x_0)] - \frac{h^3}{720}[g'''(x_n) - g'''(x_0)] + \dots$$

which implies that

$$\frac{1}{h}\int_{x_0}^{x_n} g(x)dx = \sum_{i=0}^{n-1} g(x_i) + \frac{1}{2}[g(x_n) - g(x_0)] - \frac{h}{2}[g'(x_n) - g'(x_0)] + \frac{h^3}{720}[g'''(x_n) - g'''(x_0)] + \dots$$

$$=\frac{1}{2}\{g(x_0)+2[g(x_1)+g(x_2)+...g(x_{n-1})]+g(x_n)\}-\frac{h}{12}[g'(x_n)-g'(x_0)]+\frac{h^3}{720}[g'''(x_n)-g'''(x_0)]..$$

Hence

$$\int_{x_0}^{x_n} ydx=\frac{h}{2}[y_0+2(y_1+y_2+y_3+...+y_{n-1})+y_n]-\frac{h^2}{12}(y_n'-y_0')+\frac{h^4}{720}(y_n'''-y_0''').. \quad (5.8.5)$$

where $y=g(x)$ and $x_n=x_0\ nh.$

This is called Euler-Maclaurins formula. The first term on the right hand side of (5.8.5) represents the approximate value of the integral obtained from trapezoidal rule and other terms denote the successive corrections to this value. This formula is often used to find the sum of series of the form

$$y(x_0)+y(x_1)+y(x_2)+y(x_3)+...+y(x_n)+.. \quad \text{where } x_i=x_0+ih.$$

Example 5.8.1 Use Euler-Maclaurins formula to find the value of $\log_e^2$ from $\int_0^1\frac{dx}{1+x}$.

Solution: Here we take $y=\frac{1}{1+x}$, $x_0=0$, $n=10$, $h=0.1$ so that

$$y'=\frac{-1}{(1+x)^2}, \text{ and } y'''=\frac{-6}{(1+x)^4}.$$

Then Euler-Maclaurin's formula gives us

$$\int_0^1\frac{dx}{1+x}=\frac{0.1}{2}\left[\frac{1}{1+0}+2\left(\frac{1}{1+0.1}+\frac{1}{1+0.2}+\frac{1}{1+0.3}+...+\frac{1}{1+0.9}\right)+\frac{1}{1+1}\right]$$

$$=\frac{(0.1)^2}{2}\left[\frac{-1}{(1+1)^2}-\frac{-1}{(1+0)^2}\right]+\frac{(0.1)^4}{720}\left[\frac{-6}{(1+1)^4}-\frac{6}{(1+0)^4}\right]$$

$$=0.693773-0.000625+0.000001=0.693149$$

Also $\int_0^1\frac{dx}{1+x}=|\log(1+x)|_0^1=\log_e 2$.

Hence $\log_e 2=0.693149$, approximately.

Example 5.8.2 Use Euler-Maclaurins formula to prove that $\sum_1^n x^3=\frac{n^2(n+1)^2}{4}$

Solution: Here $y=x^3, y'=3x^2, y'''=6$ and $h=1$. From Euler-Maclaurins formula (5.8.5) is rewritten as

$$\frac{1}{2}[y_0+2(y_1+y_2+y_3+...+y_{n-1})+y_n]$$

$$=\frac{1}{h}\int_{x_0}^{x_n} ydx+\frac{h}{12}(y_n'-y_0')-\frac{h^3}{720}(y_n'''-y_0''')+\frac{h^5}{30240}(y_n^{(v)}-y_e^{(v)})-...$$

This leads to

$$\sum_1^n x^3 = \int_1^n x^3 dx + \frac{1}{2}\left(n^3+1\right) + \frac{1}{12}\left(3n^2-3\right) - \frac{1}{720}(6-6) = \frac{n^4-1}{4} + \frac{n^3+1}{2} + \frac{n^2-1}{4}$$

$$= \left(\frac{n+1}{4}\right)\left[\left(n^2+1\right)(n-1) + 2\left(n^2-n+1\right) + (n-1)\right]$$

$$= \frac{1}{4}(n+1)\left[n^3 - n^2 + n - 1 + 2n^2 - 2n + 2 + n - 1\right] = \frac{1}{4}(n+1)\left[n^3+n^2\right] = \frac{n^2(n+1)^2}{4}.$$

Example 5.8.3 Evaluate $I = \int_0^{\frac{\pi}{2}} \sin x \, dx$ by using Euler-Maclaurins formula.

Solution: Here $y = \sin x$, $x_0 = 0$ *and* $x_n = \frac{\pi}{2}$.

Then Euler-Maclaurins formula for $h = \pi/4$ provides us

$$\int_0^{\pi/2} \sin x \, dx = \frac{h}{2}\left[y_0 + 2\left(y_1 + y_2 + \ldots + y_{n-1}\right) + y_n\right]\frac{h^2}{12} + \frac{h^4}{720} + \frac{h^6}{3240} + \ldots$$

$$= \frac{\pi}{8}(0+2+0) + \frac{\pi^2}{192} + \frac{\pi^4}{184320} + \ldots = \frac{\pi}{4} + \frac{\pi^2}{192} + \frac{\pi^4}{184320}, \text{ approximately}$$

$$=0.785398+0.051404+0.000528 \quad =0.83733$$

For $h = \frac{\pi}{8}$, we obtain

$$\int_0^{\frac{\pi}{2}} \sin x \, dx = \frac{\pi}{16}[0+2(0.382683+0.707117+0.923879)+1.000000]$$

$$= 0.987119+0.012851+0.000033 = 1.000003.$$

5.9 Evaluation of singular integrals

We have so for considered the integration of the function $y = f(x)$, which can be represented either by a polynomial or can be expanded in a Taylor's series in the interval of integration [a, b]. A function $f(x)$ is said to be singular at a point if $f(x)$ or any of its derivatives is infinite at that point. In such cases the formulae discussed earlier cannot be applied, and some special methods will have to be adopted. The approach depends, in general, on the type of the problem under consideration. We describe below some methods which can be applied in certain situations.

5.9.1 Principal value integrals

Consider the integral

$$I = \int_a^b \frac{f(x)dx}{x-t} \tag{5.9.1}$$

which is singular at $t = x$. The principal value $P(I)$ of integral (5.9.1) is defined as

$$P(I) = \lim_{\epsilon \to 0}\left[\int_a^{t-\epsilon} \frac{f(x)}{x-t}dx + \int_{t+\epsilon}^b \frac{f(x)}{x-t}dx\right],\ a < t < b$$

$$= I, \text{ for } t < a \text{ or } t > b \tag{5.9.2}$$

We take $x = a + uh$ and $t = a + kh$ in (5.9.1) so that

$$P(I) = P\int_0^p \frac{f(a+uh)}{u-k}du,\ p = \frac{b-a}{h} \tag{5.9.3}$$

Upon replacing $f(a+uh)$ by Newton's forward difference formula at $x = a$ and simplifying, we have

$$I = \sum_{j=0}^{\infty} \frac{\Delta^j f(a)}{j!} c_j \tag{5.9.4}$$

where
$$c_j = P\int_0^p \frac{(u)_j}{u-k}du \tag{5.9.5}$$

Here $(u)_0 = 1,\ (u)_1 = u,\ (u)_2 = u(u-1)$, etc. Various approximate formulae can be obtained by truncating the series on the right hand side of (5.9.4). Hence we can write (5.9.4) as

$$I_n = \sum_{j=0}^{n} \frac{\Delta^j f(a)}{j!} c_j \tag{5.9.6}$$

to obtain rule of orders 1,2,3,... by setting $n = 1,2,3,\ldots$ respectively.

(a) Two point rule $(n = 1)$

$$I_1 = \sum_{j=0}^{1} \frac{\Delta^j f(a)}{j!} c_j = c_0 f(a) + c_1 \Delta f(a) = (c_0 - c_1) f(a) + c_1 f(a+h) \tag{5.9.7}$$

(b) Three point rule $(n = 2)$

$$I_2 = \sum_{j=0}^{2} \frac{\Delta^j f(a)}{j!} c_j = c_0 f(a) + c_1 \Delta f(a) + c_2 \Delta^2 f(a)$$

$$= \left(c_0 - c_1 + \frac{1}{2}c_2\right) f(a) + (c_1 - c_2) f(a+h) + \frac{1}{2}c_2 f(a+2h) \tag{5.9.8}$$

In the above relations, we have

$$c_0 = \log_e\left|\frac{p-k}{k}\right|, \; c_1 = p + c_0 k, \; c_2 = \frac{p^2}{2} + p(k-1) + c_0 k(k-1) \tag{5.9.9}$$

5.9.2 Generalized quadrature

Consider the integral

$$I(s) = \int_a^b f(t)\phi(t-s)dt \tag{5.9.10}$$

where $f(t)$ is continuous but $\phi(u)$ may have an integrable singularity by adopting the forms of the type $|s-t|^{\alpha}$, $\alpha > -1$ or $\log|s-t|$ etc. For the numerical evaluation, we divide the range (a, b) such that $t_j = a + jh$, $j = 0,1,2,\ldots n$ with $nh = b - a$.

Then (5.9.10) becomes

$$I(s) = \sum_{j=0}^{n-1} \int_{t_j}^{t_{j+1}} f(t)\phi(t-s)dt \tag{5.9.11}$$

Now we approximate $f(t)$ by a linear interpolating function $f_n(t)$ as

$$f_n(t) = \frac{1}{h}\left[(t_{j+1}-t)f(t_j) + (t-t_j)f(t_{j+1})\right] \tag{5.9.12}$$

so that (5.9.11) takes the form

$$I(s) = \frac{1}{h}\sum_{j=0}^{n-1}\int_{t_j}^{t_{j+1}}\left[(t_{j+1}-t)f(t_j) + (t-t_j)f(t_{j+1})\right]\phi(t-s)\,dt$$

$$= h\sum_{j=0}^{n-1}\int_0^1\left[(1-p)f(t_j) + pf(t_{j+1})\right]\phi(t_j + ph - s)\,dp$$

where $t = t_j + ph$. This can be rewritten as

$$I(s) = \sum_{j=0}^{n-1}\left[\alpha_j f(t_j) + \beta_j f(t_{j+1})\right] \tag{5.9.13}$$

where $$\alpha_j = h\int_0^1 (1-p)\phi\left(t_j + ph - s\right)dp$$

and $$\beta_j = h\int_0^1 p\phi\left(t_j + ph - s\right)dp \tag{5.9.14}$$

Clearly if $\phi(u) = 1$, then $\alpha_j = \beta_j = \dfrac{h}{2}$ and hence (5.9.13) gives us

$$I(s) = \frac{h}{2}\left[f(t_0) + 2\{f(t_1) + f(t_2) + \ldots + f(t_{n-1})\} + f(t_n)\right] \tag{5.9.15}$$

which is the trapezoidal rule. Therefore the rule defined by (5.9.13) and (5.9.14) is called generalized trapezoidal rule due to K. E. Atkinson. For $\phi(u)=\log|u|$, this rule finds important applications in the numerical solution of certain singular integral equations. In general, the computation of the weights α_j and β_j may be difficult but they can be evaluated once and for all, for a given $\phi(u)$. In a similar fashion we can also deduce the generalized Simpson's rule, analogous to the ordinary Simpson's rule, by approximating $f(t)$ by means of a quadratic interpolating function in the interval (t_j, t_{j+1}). The error in generalized quadrature can also be estimated by the method used in case of ordinary quadrature formulae. For example it can be shown that the error in the generalized trapezoidal rule is of order h^2, assuming that f'' is continuous in $[a,b]$.

5.10 Double integration

The double integral of the type

$$I = \int_c^d \int_a^b f(x, y)\, dx dy \tag{5.10.1}$$

is evaluated numerically by two successive integration in x and y direction considering one variable at a time by repeated applications of trapezoidal or Simpson's rule.

5.10.1 Trapezoidal rule

We divide intervals (a,b) and (c,d) into n and m equal subintervals each of length h and k respectively, so that we have

$$x_i = x_0 + i\,h,\ \ x_0 = a,\ \ x_n = b, i = 0,1,2,\dots n$$
$$y_j = y_0 + jk,\ \ y_0 = c,\ \ y_m = d,\ \ j = 0,1,2,\dots m$$

Using trapezoidal rule in both direction, we get

$$I = \int_{y_0}^{y_m}\int_{x_0}^{x_n} f(x,y)dxdy = \frac{h}{2}\int_{y_0}^{y_m}\{f(x_0,y)+f(x_n,y)+2[f(x_1,y)+f(x_2,y)+\dots+f(x_{n-1},y)]\}dy$$

$$= \frac{hk}{4}[(f_{00}+f_{om})+2(f_{01}+f_{02}+\dots+f_{0,m-1})+(f_{n0}+f_{nm})+2(f_{n1}+f_{n2}+\dots+f_{n,m-1})$$

$$+2\sum_{i=1}^{n-1}\{(f_{i0}+f_{im})+2(f_{i1}+f_{i2}+\dots+f_{i,m-1})\}], \qquad f_{ij} = f(x_i, y_j) \tag{5.10.2}$$

5.10.2 Simpson's rule

In this case we divide the interval (a,b) in $2n$ equal subintervals each of width h and the interval (c,d) into $2m$ equal subintervals each of width k so that we have

$$x_i = x_0 + ih,\ \ y_j = y_0 + jk,\ x_0 = a, x_{2n} = b, y_0 = c, y_{2m} = d$$

Then by Simpson's rule in both directions, we have

$$\int_{y_{j-1}}^{y_{j+1}} \int_{x_{i-1}}^{x_{i+1}} f(x,y)\,dxdy = \frac{h}{3}\int_{y_{j-1}}^{y_j} [f(x_{i-1},y)+4f(x_i,y)+f(x_{i+1},y)]dy$$

$$= \frac{hk}{9}[(f_{i-1,j-1}+4f_{i-1,j}+f_{i-1,j+1})+4(f_{i,j-1}+f_{i,j}+f_{i,j+1})+(f_{i+1,j-1}+4f_{i+1,j}+f_{i=1,j+1})]$$

$$= \frac{hk}{9}[f_{i-1,j-1}+f_{i-1,j+1}+f_{i+1.j-1}+f_{i+1,j+1}+4(f_{i-1,j}+f_{i,j-1}++f_{i,j+1}+f_{i+1,j})+16f_{i,j}]$$

Adding all such intervals, we obtain the value of

$$I = \int_{y_0}^{y_{2m}} \int_{x_0}^{x_{2n}} f(x,y)\,dx\mathrm{d}y$$

Solution: We take $h = 0.2,\ k = 0.3$ so that $m = n = 2$. Then the values of the integrand $f(x,y) = \dfrac{1}{xy}$ are given as:

y \ x	4	4.2	4.4
2	0.185	0.1190476	0.1136363
2.3	0.186956	0.1035196	0.0988142
2.6	0.0961538	0.091575	0.0874125

(i) Trapezoidal rule

$$= \frac{hk}{4}[(f(4,2)+f(4,2.6))+2(f(4,2.3))$$

$$+(f(4.4,2)+f(4.4,2.6))+2(f(4.4,2.3)+2(f(4.2,2)+2f(4.2,2.3)+f(4.2,2.6))]$$

$$= \frac{0.06}{4}[0.4222026+2(0.4181324)+4(0.1035196)] = \frac{0.06}{4}[1.6725458] = 0.0250881$$

(ii) Simpson's rule

$$I = \frac{hk}{9}[f(4,2)+4\{f(4,2.3)+f(4.2,2)+4f(4.2,2.3)+f(4.4,2.3)$$

$$+f(4.2,2.6)\}+f(4,4,2.6)+f(4.6,2.0)+f(4.4,2.6)]$$

$$= \frac{0.06}{9}[0.4222026+4(0.4181324)+16(0.1035196)]$$

$$= \frac{0.06}{9}[0.4222026+1.6725296+1.6563136] = 0.0250069$$

Example 5.10.2 Evaluate the integral $I = \int_0^1 \int_0^1 e^{x+y}\,dxdy$

by using (i) trapezoidal rule and (ii) Simpson's rule

Solution: We take $h = k = 0.5$ and $f(x, y) = e^{x+y}$. Then the values of $f(x, y)$ are give as

y \ x	0	0.5	1.0
0	1	1.6487	2.7183
0.5	1.6487	2.7183	4.4817
1.0	2.7183	4.4817	7.3891

(i) Trapezoidal rule

$$I = \frac{0.25}{4}\left[1.0 + 4(1.6487) + 6(2.7183) + 4(4.4817) + 7.3891\right] = \frac{12.3050}{4} = 3.07625$$

(ii) **Simpson's rule**

$$I = \frac{0.25}{9}\left[1.0 + 2.7183 + 7.3891 + 2.7183 + 4(1.6487 + 4.4817 + 4.4817 + 1.6487) + 16(2.7183)\right]$$

$$= \frac{26.59042}{9} = 2.9545$$

The exact value of the double integral is 2.9525 and thus the result obtained by Simpson's rule is sixty time more accurate than that given by trapezoidal rule.

Exercises

1. Find first and second derivatives of $f(x)$ at $x = 1.5$ from the data

x:	1.5	2.0	2.5	3.0	3.5	4.0
$f(x)$:	3.375	7.000	13.625	24.000	38.875	59.000

2. Find $y'(0)$ and $y''(0)$ from the following table:

x:	0	1	2	3	4	5
y:	4	8	15	7	6	2

3. Calculate the angular velocity and angular acceleration at the instant $t = 0.06$ from the table below

θ:	0.052	0.105	0.168	0.242	0.327	0.408	0.489
t:	0	0.02	0.04	0.06	0.08	0.010	0.12

4. Find the first and second derivatives of the function tabulated below at $x = 1.1$, 1.45 and 1.9

x:	1.0	1.2	1.4	1.6	1.8	2.0
$f(x)$:	0	0.128	0.544	1.296	2.432	4.00

5. A rod is rotating in a plane. Find its angular velocity at $t = 0.6$ from the following data values given below

x:	0	0.2	0.4	0.6	0.8	1.0	1.2
y:	0	0.122	0.493	1.123	2.022	3.200	4.666

6. Find $\frac{dy}{dx}$ and $\frac{d^2y}{dx^2}$ at $x = 1.00$, 1.25 and 1.15 from the table below

x:	1.00	1.05	1.10	1.15	1.20	1.25	1.30
y:	1.0000	1.0247	1.0488	1.0723	1.0954	1.1180	1.1401

7. Find $\frac{dy}{dx}$ and $\frac{d^2y}{dx^2}$ at (i) $x = 1$, (ii) $x = 3$ and $x = 6$ from the data

x:	0	1	2	3	4	5	6
y:	6.9897	7.4036	7.7815	8.1291	8.4510	8.7506	9.0339

8. Calculate the values of $\frac{dy}{dx}$ and $\frac{d^2y}{dx^2}$ at $x = 2.01$ from the table below

x:	1.96	1.98	2.00	2.02	2.24
y:	0.7825	0.7739	0.7651	0.7563	0.7473

9. Find the value of cos 1.74 from the data

x:	1.70	1.74	1.78	1.82	1.86
$\sin x$:	0.9916	0.9857	0.9781	0.9691	0.9584

10. Calculate $\frac{dy}{dx}$ at x = 1 from the following data by constructing a central difference table

x:	0.7	0.8	0.9	1.0	1.1	1.2	1.3
y:	0.644218	0.717356	0.783327	0.841471	0.891207	0.932039	0.963558

11. Find the value of x for which y is maximum from the table of values:

x:	1.2	1.3	1.4	1.5	1.6
y:	0.9320	0.9636	0.9855	0.9975	0.9996

Also find corresponding maximum value of y.

12. Use trapezoidal rule to evaluate $\int_0^1 x^3 dx$, corresponding five subintervals.

13. Evaluate $\int_1^3 \frac{dx}{x}$ by Simpson's rule with 8 strips. Hence obtain approximation value of $\log_e 3$.

14. Evaluate (i) $\int_0^\pi t \sin t \, dt$ and (ii) $\int_{-2}^{2} \frac{t\,dt}{5+2t}$, using trapezoidal rule.

15. Find the value of $\int_3^7 x^2 \log x \, dx$ by taking 4 strips.

16. Evaluate $\int_0^1 \frac{dx}{1+x^2}$ by using (i) Trapezoidal rule taking h = 0.25, (ii) Simpson's $\frac{1}{3}$ rule taking h = 0.25, (iii) Simpson's $\frac{3}{8}$ rule taking $h = \frac{1}{6}$, (iv) Weddle's rule taking $h = \frac{1}{6}$.

17. Evaluate $\int_0^1 e^x dx$ by Simpson's rule given that $e = 2.72$, $e^2 = 7.39$, $e^3 = 20.09$, $e^4 = 54.6$ and compare it with the actual value

18. Calculate the value of $\int_0^{\frac{\pi}{2}} \sin x \, dx$, by Simpson's $\frac{1}{3}$ rule using 11 ordinates.

19. Integrate numerically $\int_0^{\frac{\pi}{2}} \sqrt{\cos\theta}\, d\theta$.

20. A reservoir discharging through sluices at depth h below the water surface has a surface area A for various values of h as given below:

$h(ft)$:	10	11	12	13	14
$A(sq.ft)$	950	1070	1200	1350	1350

21. Find the distance moved by the particle in 12 seconds and the acceleration at t = 2 secs, by using the data

t (seconds):	0	2	4	6	8	10	12
$v(m/\text{sec})$:	4	6	16	34	60	94	136

22. Use Euler-Maclaurins formula to find the value of π from the formula $\frac{\pi}{4} = \int_0^1 \frac{dx}{1+x^2}$.

23. Find the sum of the series $\frac{1}{51^2}+\frac{1}{53^2}+\frac{1}{55^2}+...+\frac{1}{99^2}$ by using Euler-Maclaurins summation formula.

24. Drive Gauss Ian integration formula when $n-2$ and apply this to evaluate the integral $\int_{-1}^{1}\frac{dx}{1+x^2}$.

25. Use there point Gauss-Legendre formula to evaluate the integral $\int_0^{\frac{\pi}{2}} \sin x\, dx$

Compare this result with that obtained by Simpson's rule using seven points.

26. Use Romberg's method to compute $\int_0^1 \frac{dx}{1+x}$ with h = 0.5, 0.25 and 0.125. Hence evaluate $\log_e^2$ correct to four decimal places.

27. Apply Romberg's method to evaluate given that

x:	4.0	4.2	4.4	4.6	4.8	5.0	5.2
$\log_e 2$:	1.3863	1.4351	1.4816	1.526	1.5686	1.6094	1.6486

28. Use Euler-Maclaurins formulae to prove that $\sum x^2 = \frac{n(n+1)(2n+1)}{6}$.

29. Evaluate $\int_0^1 \int_0^1 xe^y\, dx\, dy$ using Trapezoidal rule $(h = k = 0.5)$.

30. Apply Trapezoidal rule to evaluate $\int_1^5 \int_1^5 \frac{dxdy}{\sqrt{x^2+y^2}}$, taking two subintervals.

31. Evaluate $\int_1^{2.6} \int_2^{3.2} \frac{dxdy}{x+y}$, using Simpson's rule.

CHAPTER 6

ORDINARY DIFFERENTIAL EQUATIONS

6.1 Introduction

Most of the problems in engineering and science can be formulated in terms of ordinary differential equations satisfying certain given conditions. The analytic methods of solving differential equations are applicable only to limited class of equations and oftenly differential equations appearing in physical problems do not belong to any of these familiar types and one is obliged to resort to numerical methods. These methods are of greater importance nowadays in this age of computation and information technology. In this chapter, we shall describe methods for solving initial and boundary value problems involving ordinary differential equations.

6.2 Solution of ordinary differential equations

The solution of an ordinary differential equation means to find an explicit expression for the dependent variable y in terms of a finite number of elementary functions of x. Such a solution of differential equation is called closed or finite form of the solution. In most of the numerical methods, we replace the differential equation by a difference equation and then solve it. The methods developed and applied to solve ordinary differential equations will yield the solution in one of the following two forms:

(i) A power series in x for y, from which the values of y can be obtained by direct substitution.

(ii) A set of tabulated values of x and y.

In single step methods such as Taylor's series and Picard's approximations method, the information about the curve represented by a differential equation at one point is utilized and the solution is not iterated. The methods of Euler, Runge Kutta, Miline, Adams-Moulton etc. belong to step by step methods or marching methods and in these methods the next point on the curve is evaluated in short steps ahead, for equal intervals of width h of the independent variable, by performing iterations till the desired level of accuracy is achieved.

Definition 6.2.1 An equation involving independent variable, dependent variable and derivatives of dependent variable with respect to independent variable, of the form

$$f\left(x, y, \frac{dy}{dx}, \frac{d^2y}{dx^2}, \ldots \frac{d^ny}{dx^n}\right) = 0, \qquad (6.2.1)$$

is called an ordinary differential equation of order n.

Definition 6.2.2 A relation of the type $\phi(x, y, c_1, c_2, ..., c_n) = 0$, satisfying the differential equation (6.2.1), involving independent and dependent variables along with arbitrary constants whose number equals to the order of the equation, is called its general solution.

Definition 6.2.3 To obtain its particular solution, we need to determine n-constants $c_1, c_2, c_3, ..., c_n$ for which n-conditions are required to be given. If these conditions are prescribed at one point x_0 (say) only, then the differential equation together with the conditions is known as initial value problem of nth order. If the conditions are prescribed at two or more points, the problem is called boundary value problem.

Now, we outline and discuss the developments and verifications of different single-step and marching methods to solve various types of initial and boundary value problems involving ordinary differential equations.

6.3 Single step methods

6.3.1 Taylor's series method

Consider the differential equation

$$\frac{dy}{dx} = f(x, y),\ y(x_0) = y_0 \tag{6.3.1}$$

Let $y = y(x)$ be a continuously differentiable function satisfying the equation (6.3.1). Expanding y in terms of Taylor's series around the point $x = x_0$, we get

$$y = y_0 + (x - x_0)y_0' + \frac{(x - x_0)^2}{2!}y_0'' + \frac{(x - x_0)^3}{3!}y_0''' + \cdots \tag{6.3.2}$$

the value of the differential coefficients $y_0', y_0'', y_0''', ...$ *at* $x = x_0$ can be computed from equation (6.3.1) as follows:

Given that

$$y' = f(x, y)$$

Differentiating, we get

$$y'' = f_x + f_y \frac{dy}{dx} = f_x + f\, f_y$$

Differentiating again, we get

$$y''' = f_{xx} + f\, f_{xy} + f\, f_{xy} + f^2 f_{yy} + f_x f_y + f\, f_y^2$$

$$= f_{xx} + 2f\, f_{xy} + f_x f_y + f\, f_y^2 + f^2 f_{yy}$$

and so on differentiating successively, we get y^{iv}, y^{v} etc. These leads to the values of $y_0', y_0'', y_0''', ...$ at (x_0, y_0) and when used in equation (6.3.2) gives us the desired solution. This is single step method and works well so long as the successive derivatives can be calculated easily.

Example 6.3.1 Use Taylor's series method to find $y(0.1)$ and $y(0.2)$ from the equation

$$y' = y^2 + x, y(0) = 1$$

Solution: The given differential equation is

$$y' = y^2 + x\ y(0) = 1$$

Here $$f(x, y) = y^2 + x,\ x_0 = 0,\ y_0 = 1$$

Therefore, we have

$$y'_0 = 1^2 + 0 = 1 + 0 = 1.0,\ y'' = 2y\,y' + 1 \Rightarrow y'_0 = 2.1.1 + 1 = 3$$

$$y''' = 2(y')^2 + 2y\,y'' \Rightarrow y'''_0 = 2(1)^2 + 2.1.3 = 8$$

$$y^{iv} = 4y'y'' + 2y\,y' = 6y'y'' + 2y\,y''' \Rightarrow y_0^{iv} = 6.1.3 + 2.1.8 = 34$$

$$4y^{v} = 2y\,y^{(iv)} + 8y'y''' + 6y''^2 \Rightarrow y_0^{(v)} = 186$$

Thus the Taylor's series solution is

$$y = y_0 + \left(x\,y'_0 + \frac{x^2}{2!}y''_0 + \frac{x^3}{3!}y'''_0 + \frac{x^4}{4!}y + \cdots\right)$$

$$= 1 + x + \frac{3x^2}{2} + \frac{4}{3}x^3 + \frac{17}{12}x^4 + \frac{31}{20}x^5 + \cdots$$

Here $y(0.1) = 1.11647$ and $y(0.2) = 1.27296$

Example: 6.3.2 Find by Taylor's series method, the value of y at $x = 0.1$ and $x = 0.2$ to five decimal places from $\frac{dy}{dx} = x^2 y - 1, y(0) = 1$.

Solution: Here $y_0 = 1$ and $y' = x^2 y - 1$

Differentiating successively and substituting, we get

$$y' = x^2 y - 1 \Rightarrow y'_0 = -1$$

$$y'' = 2x\,y + x^2 y' \Rightarrow y''_0 = 0$$

$$y''' = 2y + 4x\,y' + x^2 y'' \Rightarrow y'''_0 = 2$$

$$y^{iv} = 6y' + 6x\,y'' + x^2 y''' \Rightarrow y_0^{iv} = -6$$

and so on. Thus by Taylor's series method, we have

$$y = 1 + x(-1) + \frac{x^2}{2!}(0) + \frac{x^3}{3!}(2) + \frac{x^4}{4!}(-6) + \cdots$$

$$= 1 - x + \frac{x^3}{3} - \frac{x^4}{4} + \cdots$$

Hence $y(0.1) = 0.90033$ and $y(0.2) = 0.80227$.

6.3.2 Picard's method of approximations

Consider the differential equation (6.3.1)

$$\frac{dy}{dx} = f(x, y),\ y(x_0) = y_0$$

Integrating this equation, we get

$$y_1 = y_0 + \int_{x_0}^{x} f(x, y)dx \tag{6.3.3}$$

This is an integral equation equivalent to (6.3.1) and hence the problem of solving differential equation (6.3.1) has been reduced to solve the integral equation (6.3.3). For the first approximation y_1 to the solution, we put $y = y_0$ in $f(x, y)$ and obtain

$$y_1 = y_0 + \int_{x_0}^{x} f(x, y_0)dx$$

The second approximation is given by

$$y_2 = y_0 + \int_{x_0}^{x} f(x, y_1)dx$$

Similarly,

$$y_3 = y_0 + \int_{x_0}^{x} f(x, y_2)dx$$

$$y_4 = y_0 + \int_{x_0}^{x} f(x, y_3)dx$$

and so on.

Continuing this process, we obtain $y_5, y_6, \ldots, y_{n+1}$

where

$$y_{n+1} = y_0 + \int_{x_0}^{x} f(x, y_n)dx \tag{6.3.4}$$

Hence this method gives us a sequence of approximations $y_1, y_2, y_3, \ldots, y_n$. This sequence $\{y_n\}$ $n = 1,2,3,\ldots$ converges to the exact solution provided that the function $f(x,y)$ is bounded in some region in the neighborhood of (x_0, y_0) and it satisfies the Lipschits condition *viz.* there exists a constant K such that $|f(x,y) - f(x,\bar{y})| \le K|y - \bar{y}|$, for all x.

Example 6.3.3 Find $y(0.1)$ by Picard's method from the equation $\frac{dy}{dx} = \frac{y-x}{y+x}$, $y(0) = 1$

Solution: The given equation is

$$\frac{dy}{dx} = \frac{y-x}{y+x}, y(0) = 1$$

$$\Rightarrow \qquad y = 1 + \int_0^x \frac{y-x}{y+x} dx$$

Therefore, by Picard's method we get

$$y_1 = 1 + \int_0^x \frac{1-x}{1+x} dx = 1 + \int_0^x \left(-1 + \frac{2}{1+x}\right) dx = 1 - x + 2\log(1+x)$$

$$y_2 = 1 + \int_0^x \frac{1 - x + 2\log(1+x) - x}{1 - x + 2\log(1+x) + x} dx = 1 + \int_0^x \left(1 - \frac{2x}{1 + 2\log(1+x)}\right) dx$$

which is quite difficult to integrate. Hence using y_1 and taking $n = 0.1$ we get $y(0.1) = 1 - (.1) + 2\log(1.1) = 0.9828.$

Example 6.3.4 Solve the equation

$$\frac{dy}{dx} = x + y^2, \ y(0) = 1.0$$

Solution: The given equation is

$$\frac{dy}{dx} = x + y^2, \ y(0) = 1.0$$

We take $y^{(0)} = 1$ and obtain

$$y^{(1)} = 1 + \int_0^x (x+1) dx = 1 + x + \frac{1}{2}x^2$$

Similarly

$$y^{(2)} = 1 + \int_0^x [x + \left(1 + x + \frac{1}{2}x^2\right)^2 dx$$

$$= 1 + x + \frac{3}{2}x^2 + \frac{2}{3}x^3 + \frac{1}{4}x^4 + \frac{1}{20}x^5 \cdots$$

and so on. It is clear that the integration on right hand side become more and more difficult as we proceed to higher approximations.

6.4 Marching methods

6.4.1 Euler's method

Consider the differential equation

$$\frac{dy}{dx} = f(x,y), \quad y(x_0) = y_0 \tag{6.4.1}$$

Integrating (6.4.1), we obtain

$$y = y_0 + \int_{x_0}^x f(x,y) dx \tag{6.4.2}$$

Suppose we want to find the solution of the equation (6.4.1) for values of y at $x = x_r = x_0 + rh, \mathrm{r} = 1,2,3...$

Then from equation (6.4.2), we have

$$y_1 = y_0 + \int_{x_0}^{x_1} f(x, y)\,dx \tag{6.4.3}$$

Assuming that $f(x, y) = f(x_0, y_0)$ in $x_0 \le x \le x_1$, the equation (6.4.3) leads to

$$y_1 = y_0 + h\,f(x_0, y_0) \tag{6.4.4}$$

This is called Euler's formula.

Similarly, for the range $x_1 \le x \le x_2$, we have

$$y_2 = y_1 + \int_{x_1}^{x_2} f(x, y)\,dx = y_1 + h\,f(x_1, y_1)$$

and for the range $x_2 \le x \le x_3$, we have

$$y_3 = y_2 + h\,f(x_2, y_2)$$

and so on.

Proceeding in this way, we obtain the general formula

$$y_{n+1} = y_n + h\,f(x_n, y_n), n = 0,1,2,3... \tag{6.4.5}$$

Error Estimate

Let $y_n = y(x_n)$ be the actual solution of differential equation (6.4.1) at $x = x_n$ and the approximate solution is given by the relation (6.4.5).

Then

$$y_{n+1} = y_n + h\,f(x_n, y_n) = y_n + h\,y'_n \tag{6.4.6}$$

Assuming the existence of the higher derivative, $y(x_{n+1})$ can be expanded by Taylor series about $x = x_n$, we obtain

$$y(x_{n+1}) = y(x_n) + h\,y'(x_n) + \frac{h^2}{2!}y''(\zeta_n),\ x_n \le \zeta_n \le x_{n+1} \tag{6.4.7}$$

Therefore, truncation error T_{n+1} is given by

$$T_{n+1} = y(x_{n+1}) - y(x_n) = \frac{1}{2}h^2 y''(\zeta_n)$$

Thus the truncation error is of $O(h^2)$. In addition to the truncation error, there will be another error due to rounding off in computing y_{n+1} using (6.4.5). Suppose R_{n+1} be the round off error, so that

$$\bar{y}_{n+1} = \bar{y}_n + h\,f(x_n, \bar{y}_n) - R_{n+1}$$

where $\bar{y}_n$ is a new approximation obtained from (6.4.5) by allowing rounding off error. Then the total error is defined by

$$E_{n+1} = y(x_{n+1}) - \bar{y}_{n+1}$$

$$= y(x_n) + h\,y'(x_n) + \frac{h^2}{2!} y''(\zeta_n) - \{\bar{y}_n + h\,f(x_n, y_n) - R_{n+1}\}$$

$$= [y(x_n) - \bar{y}_n] + h[y'(x_n) - f(x_n, y_n)] + T_{n+1} + R_{n+1} \qquad (6.4.8)$$

We assume that $\frac{\partial f}{\partial y}$ is continuous and on using Mean-value theorem, we have

$$f(x_n, y(x_n)) - f(x_n, y_n) = [y(x_n) - y_n] f_y(x_n, \zeta_n), \quad y(x_n) \le \zeta_n \le y_n$$

Therefore, (6.4.8) takes the form

$$E_{n+1} = E_n\{1 + h\,f_y(x_n, \zeta_n)\} + T_{n+1} + R_{n+1} \qquad (6.4.9)$$

which is the recurrence formula for finding the total error. The first term on the right hand side is the inherited error i.e. the propagation of the error from the previous step y_n to y_{n+1}.

Here ignoring the round off error $R_{n+1,}$ we get

$$E_{n+1} = E_n[1 + h\,f_y(x_n, \xi_n)] + T_{n+1} \qquad (6.4.9a)$$

We now assume that over the interval of interest

$$|f_y(x, y)| < L, \ |y''(x)| < Y$$

where L and Y are fixed positive constants. Taking absolute value in (6.4.9), we obtain

$$|E_{n+1}| \le |E_n| + h\,L|E_n| + \frac{h^2}{2} Y = (1 + h\,L)|E_n| + \frac{h^2}{2} Y \qquad (6.4.9b)$$

It can be shown by the principal of mathematical induction that the solution of the difference equation

$$\xi_{n+1} = (1 + hL)\xi_n + \frac{h^2}{2} Y$$

with $\xi_0 = 0$ dominants the solution of (6.4.9a)*i.e.* we can show that

$$\xi_n \ge |E_n|, n = 0,1,2\ldots$$

Thus the solution ξ_n of above now-homogeneous difference equation provides an upper bound for the discretisation error e_n. The solution of (6.4.9b) is $\xi_n = c(1 + hL)^n - B$

where c is an arbitrary constant and $B = \frac{hY}{2L}$

Using the condition $\xi_0 = 0$, we see that $c = B$, so that we get

$$\xi_n = B(1 + hL)^n - B$$

$$\le B(e^{nhL} - 1), \text{ since } e^x \ge 1 + x, \forall x$$

$$= \frac{hy}{L}\left(e^{(x_n - x_0)L} - 1\right)$$

where $x_n = x_0 + nh$. Since $|E_n| \le \xi_n$, we have proved the following theorem.

Theorem 6.4.1: Let y_n be the approximate solution (6.4.5) of $y' = f(x, y)$, $y(x_0) = y_0$ generated by Euler's method. If the exact solution $y(x)$ has a continuous second order derivative on the interval $[x_0, b]$ and if on this interval,

$$|f_y(x, y)| \le L, \; |y''(x)| \le Y$$

are satisfied for fixed constants L and Y, the error $E_n = y(x_n) - y_n$ at $x_n = x_0 + nh$ is bounded by

$$|E_n| \le \frac{hY}{L}\left(e^{(x_n - x_0)L} - 1\right)$$

This shows that the error is of $0(h)$ i.e. the error tends to zero as $h \to 0$, for $x = x_n$ fixed

Example 6.4.1 Use Euler's method to find $y(1)$ from the differential equation $\frac{dy}{dx} = x + y, y(0) = 1$.

Solution: We take $n = 10$ and $h = 0.1$. Then, we have

Iteration n	x	y	$x + y = dy/dx$	$y_{n+1} = y + hf(x_n, y_n)$
0	0.0	1.00	1.00	1.10
1	0.1	1.10	1.20	1.22
2	0.2	1.22	1.42	1.36
3	0.3	1.36	1.66	1.53
4	0.4	1.53	1.93	1.72
5	0.5	1.82	2.22	1.94
6	0.6	1.94	2.54	2.19
7	0.7	2.19	2.89	2.48
8	0.8	2.48	3.29	2.81
9	0.9	2.81	3.71	3.18
10	1.0	3.18		

Thus the required value of $y = 3.18$

6.4.2 Modified Euler's method

The process used in Euler's method is very slow and obtain reasonable accuracy we need to take a smaller value of h. Because of this restriction, the method is unsuitable for practical use and hence a modification of it, known as the modified Euler method, which gives more accurate results, is described below. Instead of approximating $f(x,y)$ by $f(x_0,y_0)$ in equation 6.4.1), we now approximate the integral in (6.4.2) by means of the trapezoidal rule to obtain

$$y_1^{(1)} = y_0 + \frac{h}{2}\left[f(x_0, y_0) + f\left(x_1, y_1^{(0)}\right)\right] \qquad (6.4.10)$$

where $y_1^{(0)} = y_0 + hf(x_0, y_0)$ is computed from Euler's formula.

Similarly,

$$y_1^{(2)} = y_0 + \frac{h}{2}\left[f(x_0,y_0) + f\left(x_1,y_1^{(1)}\right)\right]$$

$$y_1^{(3)} = y_0 + \frac{h}{2}\left[f(x_0,y_0) + f\left(x_1,y_1^{(2)}\right)\right]$$

$$y_1^{(4)} = y_0 + \frac{h}{2}\left[f(x_0,y_0) + f\left(x_1,y_1^{(3)}\right)\right]$$

and so on. In general, we have

$$y_1^{(n+1)} = y_0 + \frac{h}{2}[f(x_0, y_0) + f\left(x_1, y_1^{(n)}\right)] \quad ,n = 0,1,2,3,... \qquad (6.4.11)$$

where $y_1^{(n)}$ is the nth approximation to y_1. The iteration formula (6.4.11) can be started by choosing $y_1^{(0)}$ from the Euler's formula and can be terminated at each step if the condition $\left| y_n^{(i)} - y_n^{(i-1)} \right| < \epsilon$, ϵ being arbitrarily small number to be selected at random depending on the desired level of accuracy to be achieved, is satisfied. Suppose this happen for $i = k$, then we take $y_n = y_n^{(k)}$ and proceed to compute the value of y at the next point by repeating the above procedure.

Example 6.4.2: Use modified Euler's method to find an approximate value of y when $x = 0.3$ given that

$$\frac{dy}{dx} = x + y, \quad y(0) = 1.$$

Solution: Here $f(x,y) = x + y, x_0 = 0, y_0 = 1$ and we take $h = 0.1$. Then, we have $x_1 = 0.1, y_1 = y_0 + hf(x_0, y_0) = 1 + 0.1(0+1) = 1.10$

Iteration (n)	x	$y' = x + y$	$y_1^{n+1} = y_0 + \frac{h}{2}\left[f(x_0,y_0) + f(x_1,y_1^{(n)})\right]$
1	0.0	0+1=1	1.0+.1 (1.0)=1.1
2	0.1	0.1+1.1=1.2	1.11
3	0.1	.1+1.11=1.21	1.1105
4	0.1	.1+1.1105=1.2105	1.1105

Hence $y(0.1) = 1.1105$

1	0.1	1.2105	1.1105+0.1(1.2105)=1.2316
2	0.2	.2 + 1.2426 = 1.316	1.2426
3	0.2	0.2 + 1.2426 =1.4426	1.2432
4	0.2	0.2 + 1.2432 =1.4432	1.2432

Hence $y(0.2) = .2432$

1	0.2	1.4432	1.2432+0.1(1.4432)=1.3875
2	0.3	0.3+1.3875=1.6875	1.3997
3	0.3	0.3+1.3997=1.6997	1.4003
4	0.3	0.3+1.4003=1.1.7003	1.4004
5	0.3	0.3+1.4004=1.7004	1.4004

Hence y (0.3) = 1.2432

Example 6.4.3 Determine the value of y at $x = 0.1$, given that $y(0) = 1$ and $y' = x^2 + y$

Solution: Here $\frac{dy}{dx} = x^2 + y$, therefore $f(x, y) = x^2 + y, x = 0, y = 1$. We take $h = 0.05$.

Then we have

Iteration (n)	x	$y' = x^2 + y$	$y^{(n+1)} = y_0 + \frac{h}{2}\left[f(x_0, y_0) + f(x_1, y_1^n)\right]$
0	0	0+1=1	1.0 +. 05 (1.0) = 1.05
1	0.55	1.0525	1.0513
2	0.55	1.0538	1.0513

Hence $y(0.05) = 1.0513$

0	0.05	1.0538	1.1040
1	0.1	1.1140	1.1055
2	0.1	1.1155	1.1055

Hence $y(0.1) = 1.1055$

6.4.3 Runge – Kutta methods

In addition to overcome the inefficiency and unsuitability of Euler's method due to the requirement of smallness of h for attaining reasonable accuracy, the Runge – Kutta methods are designed to give greater accuracy and they possess the advantage of requiring only the function values at some selected points on the subinterval. Consider the differential equation

$$\frac{dy}{dx} = f(x, y), \quad y(x_0) = y_0$$

Integrating, we get

$$y = y_0 + \int_{x_0}^{x} f(x, y)\, dx$$

For the interval $x_0 \leq x \leq x_1$, this gives us

$$y_1 = y_0 + \int_{x_0}^{x_1} f(x, y)\, dx$$

Upon using modified Euler's method, we get

$$y_1 = y_0 + \frac{h}{2}[f(x_0, y_0) + f(x_1, y_1)]$$

For $x_1 = x_0 + h$, the Euler's method provides us $y_1 = y_0 + h\, f(x_0, y_0)$ and we have

$$y_1 = y_0 + \frac{h}{2}[f(x_0, y_0) + f(x_0 + h, y_0 + h\, f(x_0, y_0))]$$

$$= y_0 + \frac{h}{2}[f_0 + f(x_0 + h, y_0 + h\, f_0)], \quad f_0 = f(x_0, y_0)$$

$$= y_0 + \frac{1}{2}[h\, f_0 + h\, f(x_0 + h, \ y_0 + h\, f_0)]$$

Thus

$$y_1 = y_0 + \frac{1}{2}(k_1 + k_2) \tag{6.4.12}$$

where

$$k_1 = h\, f_0, \quad f_0 = f(x_0, y_0)$$

$$k_2 = h\, f(x_0 + h, \ y_0 + k_1) \tag{6.4.13}$$

The equation (6.4.12) is known as the second order Runge-Kutta formula.

More generally, if we set

$$k_1 = h\, f_0$$

$$k_2 = h\, f(x_0 + \alpha_0\, h, \ y_0 + \beta_0\, k_1) \tag{6.4.14}$$

and

$$y_1 = y_0 + w_1\, k_1 + w_2\, k_2 \tag{6.4.15}$$

Then on applying Taylor's series to both sides of equation (6.4.15)we get

$$y_0 + h\, f_0 + \frac{h^2}{2}(f_x + f_0\, f_y) + O(h^3)$$

$$= y_0 + (w_1 + w_2)\, h\, f_0 + w_2\, h^2 (\alpha_0\, f_x + \beta_0\, f_0\, f_y) + O(h^3)$$

$$\Rightarrow \qquad w_1 + w_2 = 1, w_2\, \alpha_0 = \frac{1}{2} = w_2\, \beta_0 \qquad (6.4.16)$$

Clearly, $\alpha_0 = \beta_0$ and if α_0 is assigned any value arbitrarily, then the remaining parameters can be determined uniquely. If we set, for example $\alpha_0 = \beta_0 = 1$, then we obtain $w_1 = w_2 = 1/2$, which gives us the formula (6.4.12). Therefore, there are several second order Runge-Kutta methods and that the formula (6.4.12) is just one of several such formulae.

The error in Runge-Kutta second order formula (6.4.12) can be shown to be of order h^3. Expanding both sides of (6.4.12) in Taylor's series, we get

$$l.h.s = y_0 + h\, y_0' + \frac{h^2}{2} y_0'' + \frac{h^3}{6} y_0''' + \ldots \qquad (6.4.17)$$

$$k_2 = h\, f(x_0 + h, y_0 + h\, f_0) = h\left[f_0 + h\, f_x + h\, f_0 f_y + 0(h^2)\right] \text{ at } (x_0, y_0)$$

$$= h\, f_0 + h^2 f_0' + o(h^3)$$

and hence right hand side of (6.4.12)gives us

$$y_0 + \frac{1}{2}\left[h\, f_0 + h\, f_0 + h^2 f_0' + 0(h^3)\right]$$

$$= y_0 + h\, f_0 + \frac{1}{2} h^2 f_0' + 0(h^3)$$

$$= y_0 + h\, y_0' + \frac{1}{2} h^2\, y_0'' + O(h^3) \qquad (6.4.18)$$

On comparing (6.4.17) and (6.618) it follows that the Taylor's expansions of both sides of (6.4.12) agree up to the terms of order h^2, which means that the error in this formula is of order h^3.

Higher order Runge-Kutta formulae

The Runge-Kutta formulae of order higher than second also exist, of which we mention only the fourth order formulae defined by

$$y_1 = y_0 + w_1\, k_1 + w_2\, k_2 + w_3\, k_3 + w_4\, k_4 \qquad (6.4.19)$$

where

$$\left.\begin{aligned} k_1 &= h\, f(x_0, y_0) \\ k_2 &= h\, f(x_0 + \alpha_0 h,\ y_0 + \beta_0 k_1) \\ k_3 &= h\, f(x_0 + \alpha_1\, h,\ y_0 + \beta_1\, k_1 + v_1\, k_2) \\ \text{and} \qquad k_4 &= h\, f(x_0 + \alpha_2\, h,\ y_0 + \beta_2\, k_1 + v_2\, k_2 + \delta_1\, k_3) \end{aligned}\right\} \qquad (6.4.20)$$

Here the parameters are to be determined by expanding both sides of equation (6.4.19) by Taylor's series and securing agreement of terms up to and including those containing h^4. The choice of parameters in this method is also arbitrary and hence we have several fourth order Runge-Kutta formulae. If, we set

$$\alpha_0 = \beta_0 = 1/2 = \alpha_1 \quad \alpha_2 = 1, \; \beta_1 = \frac{\sqrt{2}-1}{2}, \quad \beta_2 = 0$$

$$v_1 = 1 - \frac{1}{\sqrt{2}}, \quad v_2 = -\frac{1}{\sqrt{2}}, \quad \delta_1 = 1 + \frac{1}{\sqrt{2}}, \tag{6.4.21}$$

$$w_1 = w_4 = 1/6, \quad w_2 = \frac{1}{3}\left(1 - \frac{1}{\sqrt{2}}\right), \quad w_3 = \frac{1}{3}\left(1 + \frac{1}{\sqrt{2}}\right)$$

then we obtain the method of Gill. The choice of parameters

$$\alpha_0 = \alpha_1 = 1/2, \quad \beta_0 = v_1 = 1/2, \quad \beta_1 = \beta_2 = v_2 = 0, \quad \alpha_2 = \delta_1 = 1,$$

$$w_1 = w_4 = 1/6, \quad w_2 = w_3 = 2/6 \tag{6.4.22}$$

leads to the fourth order Runge-Kutta formula, which is most commonly used one in practice. In this case the equations (6.4.19) and (6.4.20) respectively reduce to

$$y_1 = y_0 + \frac{1}{6}(k_1 + 2k_2 + 2k_3 + k_4) \tag{6.4.23}$$

where

$$k_1 = hf(x_0, y_0)$$

$$k_2 = hf\left(x_0 + \frac{1}{2}h, y_0 + \frac{1}{2}k_1\right) \tag{6.4.24}$$

$$k_3 = hf\left(x_0 + \frac{1}{2}h, \; y_0 + \frac{1}{2}k_2\right)$$

$$k_4 = hf(x_0 + h, y_0 + k_3)$$

The error in this formula is of order h^5. The one of the advantages of these methods is that the operation is identical whether the differential equation is linear or non-linear. The working rule for finding the increment k of y corresponding to an increment h of x by Runge-Kulta fourth order method from

$$\frac{dy}{dx} = f(x, y), \qquad y(x_0) = y_0$$

is as follows:

Calculate successively $\quad k_1 = hf(x_0, y_0)$.

$$k_2 = hf\left(x_0 + \frac{1}{2}h, y_0 + \frac{1}{2}k_1\right)$$

$$k_3 = hf\left(x_0 + \frac{1}{2}h, \; y_0 + \frac{1}{2}k_2\right)$$

$$k_4 = h\; f(x_0 + h, \; y_0 + k_3)$$

Finally, compute $\quad k = \frac{1}{6}(k_1 + 2k_2 + 2k_3 + k_4),$

This gives us the required approximate value as $y_1 = y_0 + k$.

Algorithm 6.4.1 Runge-Kutta second order method

For the equation

$$y' = f(x, y) \qquad y(x_0) = y$$

generate approximations y_n to $y(x_0 + nh)$, for h fixed and $n = 0,1,2,3,\ldots$, using the recursion formula

$$y_{n+1} = y_n + \frac{1}{2}(k_1 + k_2) \tag{6.4.25}$$

with $$k_1 = h\, f(x_n, y_n),\; k_2 = h\, f(x_n + h, y_n + k_1)$$

The local error in (6.4.25) is of the form

$$y(x_{n+1}) - y_{n+1} = \frac{h^3}{12}\left(f_{xx} + 2f\, f_{xy} + f^2 f_{yy} - 2f_x f_y - 2f\, f_y^2\right) + o(h^4)$$

Algorithm 6.4.2 Runge-Kutta fourth order method for the equation $y' = f(x, y)$, $y(x_0) = y_0$, generate approximations y_n to $y(x_0 + nh)$ for h fixed and for $n = 0,1,2,3,\ldots$, using the recursion formula

$$y_{n+1} = y_n + \frac{1}{6}(k_1 + 2k_2 + 2k_3 + k_4) \tag{6.4.26}$$

where $k_1 = h\, f(x_n, y_n)$, $k_2 = h\, f\left(x_n + \frac{h}{2}, y_n + \frac{1}{2}k_1\right)$, $k_3 = h\, f\left(x_n + \frac{h}{2}, y_n + \frac{1}{2}k_2\right)$,

$$k_4 = hf(x_n + h, y_n + k_3)$$

The local discretisation error in this algorithm is $o(h^5)$. The formula (6.4.26) is widely used in practice with considerable success. It has the important advantage that it is self-starting i.e., it requires only the value of y at a point $x = x_n$ to find by and y' at $x = x_{n+1}$.

Example 6.4.4 Find $y(0.1)$ and $y(0.2)$ correct to four decimal places, given that

$$\frac{dy}{dx} = y - x,\; y(0) = 2 \tag{6.4.27}$$

Solution: We take $h = 0.1$ and use second order Runge-Kutta method.

Here $$f(x, y) = y - x,\; x_0 = 0, y_0 = 2$$

$$k_1 = h\, f(x_0, y_0) = 0.1(2 - 0) = .2$$

$$k_2 = h\, f(x_0 + h, y_0 + k_1) = 0.1(2.2 - 1) = 0.21$$

$\therefore$ $$y(0.1) = y_1 = y_0 + \frac{1}{2}(k_1 + k_2) = 2 + \frac{1}{2}(.2 + .21) = 2.2050$$

$\therefore$ $$y(0.1) = 2.2050$$

To find $y(0.2)$, we take $x_0 = 0.1, y_0 = 2.2050$ so that $k_1 = 0.1(2.105) = 0.2105$

And $k_2 = 0.1(2.4155 - 0.2) = 0.22155$

Then, we have
$$y_2 = 2.2050 + \frac{1}{2}(0.2105 + 0.22155) = 2.4210$$

$$\therefore \qquad y(0.2) = 2.4210$$

Proceeding in a similar way, we obtain

$$y_3 = y(0.3) = 2.6492$$
$$y_4 = y(0.4) = 2.8909$$

The analytical solution of equation (6.4.27) is given by $y = x + 1 + e^x$ and the exact values of $y(0.2)$ and $y(0.4)$ are respectively 2.4214 and 2.8918. Thus, we have

x	Computed y	Exact y	Difference	Ratio
0.2	$h = 0.1$, 2.4210 $h = 0.2$, 2.4200	2.4214	0.0004	3.5
0.4	$h = 0.1$, 2.8909 $h = 0.2$, 2.8880	2.8918	0.0009 0.0038	4.2

This table shows that the method has an h^3 order of convergence.

Example 6.4.5 Solve $\frac{dy}{dx} = \frac{y^2 - x^2}{y^2 + x^2}$ with $y(0) = 1$ at $x = 0.2, 0.4$ by using Runge-Kulta fourth order method

Solution: We have

$$f(x, y) = \frac{y^2 - x^2}{y^2 + x^2},$$

To find $y(0.2)$, we take $x_0 = 0 \quad y_0 = 1 \ , h = 0.2$.

$$k_1 = h\,f(x_0, y_0) = 0.2\frac{1^2 - 0^2}{1^2 + 0^2} = 0.2000$$

Then
$$k_2 = h\,f\left(x_0 + \frac{1}{2}h, y_0 + \frac{1}{2}k_1\right) = 0.19672$$

$$k_3 = h\,f\left(x_0 + \frac{1}{2}h, y_0 + \frac{1}{2}k_2\right) = 0.1967$$

$$\therefore \qquad y(0.2) = y_0 + \frac{1}{6}(k_1 + 2k_2 + 2k_3 + k_4)$$

$$= 1 + \frac{1}{6}0.2 + 2((0.2 + 2(0.19672 + 0.1967) + 0.1891)) = 1.196$$

To find $y(0.4)$ we take $x_0 = 0.2$, $y_0 = 1.196$, $h = 0.2$ and repeat the above procedure to obtain

$$k_1 = 0.1891 \ , \ k_2 = 0.1795 \ , \ k_3 = 0.1793 \ , \ k_4 = 0.1688$$

so that $y(0.4) = y_0 + \frac{1}{6}(k_1 + 2k_2 + 2k_3 + k_4)$

$$= 1.196 + \frac{1}{6}(\ 0.1891 + 2(0.1795 + 0.1793) + 0.1688) = 1.3752$$

Example 6.4.6 Find $y(0.2)$, $y(0.4)$ and $y(0.6)$ by using Runge-Kulta fourth order method given that

$$\frac{dy}{dx} = 1 + y^2 \ , \ y(0) = 0 \ .$$

Solution: We take $x_0 = 0 = y_0$, $h = 0.2$ and $f(x, y) = 1 + y^2$.

Then, we have

$$k_1 = hf(x_0, y_0) = 0.2$$

$$k_2 = hf\left(x_0 + \frac{h}{2}, y_0 + \frac{1}{2}k_1\right) = 0.202$$

$$k_3 = hf\left(x_0 + \frac{h}{2}h, y_0 + \frac{1}{2}k_2\right) = 0.20204$$

$$k_4 = hf(x_0 + h, y_0 + k_3) = 0.20816$$

So that

$$y(0.2) = y_0 + \frac{1}{6}(k_1 + 2k_2 + 2k_3 + k_4) = 0 + \frac{1}{6}(0.2 + 2(0.202 + 0.20204) + 0.20816) = 0.2027$$

correct to four decimal places.

To find $y(0.4)$, we take $x_0 = 0.2,\ y_0 = 0.2027$ and $h = 0.2$.

Then, we obtain $\qquad k_1 = 0.2027,\ k_2 = 0.2188,\ k_3 = 0.2195,\ k_4 = 0.2356,$

$$y(0.4) = 0.2027 + \frac{1}{6}(0.2082 + 2(0.2188 + 2195) + 0.2356) = 0.4228,$$

correct to four decimal places.

Finally, we take $x_0 = 0.4,\ y_0 = 0.4228$ and $h = 0.2$, and proceed as above to obtain $y(0.6) = 0.6841$.

6.5 Predictor corrector methods

In the methods so far described to solve a differential equation over an interval, only the value of y at the beginning of the interval was required. In the predictor-corrector methods, four prior values are needed for finding the value of y at x_i. These methods though slightly complex, have the advantage of giving an estimate of error from successive approximations to y_i. If x_i and x_{i+1} be two connective mesh points, such that $x_{i+1} = x_i + h$, then in Euler's method we have

$$y_{i+1} = y_i + h\,f(x_0 + ih, y_i),\quad i = 0,1,2,3,... \tag{6.5.1}$$

The modified Euler's method gives us

$$y_{i+1} = y_i + \frac{h}{2}[f(x_i, y_i) + f(x_{i+1}, y_{i+1})] \tag{6.5.2}$$

The value of y_{i+1} is first estimated by (6.5.1) and then used in right hand side of (6.5.2) giving a better approximation of y_{i+1}. This value of y_{i+1} is again substituted in (6.5.2) to find a still better approximation of y_{i+1}. This procedure is repeated till two consecutive iterated values of y_{i+1} agree. This technique of refining an initially crude estimate of y_{i+1} by means of a more accurate formula is known as predictor-corrector method. The equation (6.5.1) is therefore called the predictor while (6.5.2) serves as a corrector of y_{i+1}. In this section we describe two such methods, namely, Adams-Moulton method and Milne's method.

6.5.1 Adams-Moulton method

Consider the differential equation

$$\frac{dy}{dx} = f(x, y),\; y(x_0) = y_0$$

Integrating, we get

$$y = y_0 + \int_{x_0}^{x} f(x, y)dx$$

In the range $x_0 \le x \le x_1$, this implies that

$$y_1 = y_0 + \int_{x_0}^{x_1} f(x, y)dx \tag{6.5.3}$$

By Newton's backward difference formula, we have

$$f(x, y) = f_0 + n\nabla f_0 + \frac{n(n+1)}{2}\nabla^2 f_0 + \frac{n(n+1)(n+2)}{6}\nabla^3 f_0 + ... \tag{6.5.4}$$

where $$n = \frac{x - x_0}{h} \text{ and } f_0 = f(x_0, y_0)$$

Using (6.5.4) in right hand side of (6.5.3) we get

$$y_1 = y_0 + \int_{x_0}^{x_1}\left[f_0 + n\nabla f_0 + \frac{n(n+1)}{2}\nabla^2 f_0 + \ldots \right] dx$$

$$y_1 = y_0 + h\int_0^1\left[f_0 + n\nabla f_0 + \frac{n(n+1)}{2}\nabla^2 f_0 + \ldots \right] dn$$

$$= y_0 + h\left[1 + \frac{1}{2}\nabla + \frac{5}{12}\nabla^2 + \frac{3}{8}\nabla^3 + \frac{251}{720}\nabla^4 + \ldots \right] f_0$$

Thus it is seen that the right hand side of the above relation depends only on $y_0, y_{-1}, y_{-2}, \ldots$ all of which are known and hence this formula can be used to compute y_1. Therefore, we write

$$y_1^p = y_0 + h\left[1 + \frac{1}{2}\nabla + \frac{5}{12}\nabla^2 + \frac{3}{8}\nabla^3 + \frac{251}{720}\nabla^4 + \ldots \right] f_0 \qquad (6.5.5)$$

This is called Adams-Bashforth formula and is used as a predictor formula. A corrector formula can be derived in a similar manner by using Newton's backward difference formula at f_1 viz.,

$$f(x, y) = f_1 + n\nabla f_1 + \frac{n(n+1)}{2}\nabla^2 f_1 + \frac{n(n+1)(n+2)}{6}\nabla^3 f_1 + \ldots \qquad (6.5.6)$$

Upon using (6.5.6) in (6.5.3) we get

$$y_1 = y_0 + \int_{x_0}^{x_1}\left[f_1 + n\nabla f_1 + \frac{n(n+1)}{2}\nabla^2 f_1 + \ldots \right] dx$$

$$= y_0 + h\int_{-1}^{0}\left[f_1 + n\nabla f_1 + \frac{n(n+1)}{2}\nabla^2 f_1 + \right] dx$$

$$= y_0 + h\left[1 - \frac{1}{2}\nabla - \frac{1}{12}\nabla^2 - \frac{1}{24}\nabla^3 - \frac{19}{720}\nabla^4 + \right] f_1 \qquad (6.5.7)$$

The right hand side of (6.5.7) depends on $y_1, y_0, y_{-1}, y_{-2}, \ldots$, where y_1^p is used for y_1. The new value of y_1 is given by

$$y_1^c = y_0 + h\left[1 - \frac{1}{2}\nabla - \frac{1}{12}\nabla^2 - \frac{1}{24}\nabla^3 - \frac{19}{720}\nabla^4 \right] f_1^p \qquad (6.5.8)$$

where $f_1^p = f_1\left(x_1, y_1^p\right)$

The relation (6.5.8) is called Adams-Moulton corrector formula. Neglecting fourth and higher order differences and expressing the remaining difference operators in their function values, the formula (6.5.5) and (6.5.8) become

$$y_1^{(p)} = y_0 + \frac{h}{24}(55 f_0 - 59 f_{-1} + 37 f_{-2} - 9 f_3) \tag{6.5.9}$$

and

$$y_1^{(C)} = y_0 + \frac{h}{24}\left(9 f_1^P + 19 f_0 - 5 f_{-1} + f_{-2}\right) \tag{6.5.10}$$

respectively. The formula (6.5.9) and (6.5.19) are quite convenient to use in practical applications. The corrector formula (6.5.10) is repeatedly applied by calculating an improved value of f_1 at each stage, to find a better value of y_1 unless it becomes stable and remains unchanged and then we proceed to calculate y_2. The error in formulae (6.5.9) and (6.5.10) are approximately $\frac{251}{720}h^5 f_0^{(4)}$ and $-\frac{19}{720}h^5 f_0^{(4)}$ respectively. Such formulae expressed in ordinate form, are often called explicit predictor-corrector formula. To apply Adams-Moulton method, we require four starting values of y which are calculated by means of Picard's method or Taylor's series method or Euler's method or Runge-Kutta methods. In practice Adams-Moulton formula have been found to be most useful.

In general the Adams-Bashforth and Adams-Moulton formulae are given by

$$y_{n+1}^{(p)} = y_n + \frac{h}{24}\left(55 f_n - 59 f_{n-1} + 37 f_{n-2} - 9 f_{n-3}\right) \tag{6.5.11}$$

and

$$y_{n+1}^{(c)} = y_n + \frac{h}{24}\left(9 f_{n+1} + 19 f_n - 5 f_{n-1} + f_{n-2}\right) \tag{6.5.12}$$

respectively, with local errors estimates

$$\left.\begin{aligned} E_{AB} &= \frac{251}{720}h^5 y^V(\xi_1) \\ E_{AM} &= -\frac{19}{720}h^5 y^V(\xi_2) \end{aligned}\right\} \tag{6.5.13}$$

Algorithm 6.5.1

For the differential equation $y' = f(x, y)$, $y(x_0) = y_0$ with h fixed and $x_n = x_0 + nh$ and with $(y_0, f_0), (y_1, f_1), (y_2, f_2), (y_3, f_3)$ given for each fixed $n = 3,4,5,...$

1. Compute $y^{(P)}{}_{n+1}$ using the formula (6.5.11)
2. Compute $y^{(P)}{}_{n+1} = f\left(x_{n+1}, y_{n+1}^{(p)}\right)$
3. Compute $y_{n+1}^{(k)} = y_n + \frac{h}{24}\left[f\left(x_{n+1}, y_{n+1}^{(k-1)}\right) + 19 f_n - 5 f_{n-1} + f_{n-2}\right]$, for $k = 1,2,3,...$ by using the formula $(6.5.12)$
4. Iterate on k until $\left| y_{n+1}^{(k)} - y_{n+1}^{(k-1)} \right| < \in$, for $\in$ prescribed

Let $y_{n+1}^{(0)}$ represents the value of y_{n+1} obtained from $(6.5.11)$ and $y^{(1)}{}_{n+1}$ the result obtained with one application of Algorithm 6.5.1. If $y(x_{n+1})$ represents the exact value of y at x_{n+1} and the values of f are assumed to be exact at all points including x_n, then from (6.5.13) we obtain the order estimates

$$y(x_{n+1}) - y_{n+1}^{(0)} = \frac{251}{720} h^5 y^v(\zeta_1)$$

$$y(x_{n+1}) - y_{n+1}^{(1)} = \frac{-19}{720} h^5 y^v(\zeta_2)$$

which leads to the estimate of $y^{(V)}$, if we assume that over the interval of interest $y^V(x)$ is approximately constant, as $h^5 y^V = \frac{720}{270}\left[y_{n+1}^{(1)} - y_{n+1}^{(0)}\right]$

Using in the second relation above, we get

$$y(x_{n+1}) - y_{n+1}^{(1)} = \frac{-19}{270}\left(y_{n+1}^{(1)} - y_{n+1}^{(0)}\right) \approx -\frac{1}{14}\left(y_{n+1}^{(1)} - y_{n+1}^{(0)}\right) = D_{n+1} \qquad (6.5.14)$$

Thus the error of the corrected value is approximately $-1/14$ of the difference between the corrected and the predicted values .It is advisable to use the corrector only once. If the accuracy as determined by (6.5.14.) is insufficient, it is better to reduce the step size than to correct more than once. In a general purpose routine for solving differential equations, the error estimate is obtained in the following manner. Let us assume that we wish to keep the local error per unit step bounded as in Runge-Kutta method so that

$$E_1 \le \frac{|D_n + 1|}{h} \le E_2$$

and that starting values have been provided. We proceed as follows.

1. Use (6.5.11) to obtain $y_{n+1}^{(0)}$. Compute $f_{n+1}^{(0)}$.
2. Use (6.5.12) to obtain $y_{n+1}^{(1)}$. Compute $f_{n+1}^{(1)}$.
3. Compute $|D_{n+1}|$ from (6.5.14).
4. If $E_1 \le |D_{n+1}|/h \le E_2$, proceed to the next integration step, using some value of h.
5. If $|D_{n+1}|/h > E_2$, the step size h is too large and shall be reduced. Usually replace h by $\frac{h}{2}$, re-compute four starting values and then return to step 1.
6. If $|D_{n+1}|/h \le E_2$, more accuracy is being obtained than necessity. Hence we can save computer time by replacing h by $2h$. Recomputed four new starting values at interval of length $2h$ and return to step 1.

Example 6.5.1 Use Adams-Moulton method to find $y(1.4)$ from the differential equation $\frac{dy}{dx} = x^2(1+y)$, given that $y(1) = 1$, $y(1.1) = 1.233$, $y(1.2) = 1.543$, $y(1.3) = 1.979$.

Solution: Here $f(x, y) = x^2(1+y)$ and $h = 0.1$

Then we have

x	y	$f = x^2(1+y)$	
1.0	1.000	2.000	f_{-3}
1.1	1.233	2.702	f_{-2}
1.2	1.543	3.669	f_{-1}
1.3	1.979	5.035	f_{-0}
1.4	2.573	7.004	

by using predictor formula (6.5.9) .

Using the corrector formula (6.5.10), we obtain

$$y_1^c = y_0 + \frac{h}{24}[9 f_1 + 19 f_0 - 5 f_1 + f_2]$$

$$= 1.979 + \frac{0.1}{24}\left(9(7.004) + 19(5.035) - 5(3.669) + 2.702\right) = 2.575$$

Hence $y(1.4) = 2.575$.

Example 6.5.2 Use Adams-Moulton method to find $y(0.8)$ and $y(1.0)$ given that $\frac{dy}{dx} = 1 + y^2, y(0) = 0$.

Solution: Given differential equation is

$$\frac{dy}{dx} = 1 + y^2, y(0) = 0$$

We take $h = 0.2$ with $x_0 = 0, y_0 = 0$ and use Runge-Kutta fourth order method to obtain

$$k_1 = 0.2$$

$$k_2 = 0.2(1.01) = 0.202$$

$$k_3 = 0.2(1 + 0.010201) = 0.20204$$

$$k_4 = 0.2(1 + 0.040820) = 0.20816$$

correct up to four decimal places.

To compute $y(0.4)$, we take

$$x_0 = 0.2, y_0 = 0.2027 \text{ and } h = (0.4)$$

we get $k_1 = 0.2082, \quad k_2 = 0.2188, \quad k_3 = 0.2195, \quad k_4 = 0.2356$

and $y0.4 = 0.2027 + 0.2201 = 0.4228$, correct to four decimal places.

Now taking $x_0 = 0.4 \quad y_0 = 0.4228$ and $h = 0.2$ using the above procedure we obtain $y(0.6) =$ 0.6841.

Thus, we have

x	y	$f(x,y)=1+y^2$
0	0	1.0
0.2	0.2027	1.041099
0.4	0.4228	1.17876
0.6	0.6841	1.46799

By Adams-Bashforth predictor method, we obtain

$$y^P(0.8) = 0.6841 + \frac{0.2}{24}[55(1.46799) - 59(1.17876) + 37(1.04109) - 9] = 1.02337$$

Using Adams-Moulton corrector method, we obtain

$$y^c(0.8) = 0.6841 + \frac{0.2}{24}[9(2.04729) + 19(1.46799) - 5(1.7876) + (1.04109)] = 1.0296$$

correct to four decimal places.

6.5.2 Miline's method

Consider the differential equation

$$\frac{dy}{dx} = f(x,y), \quad y(0) = 0$$

Integrating, we get

$$y = y_0 + \int_{x_0}^{x} f(x,y)\,dx$$

This in the range $x_0 \le x \le x_4$ implies that

$$y = y_0 + \int_{x_0}^{x_4} f(x,y)\,dx \qquad (6.5.15)$$

Newton's forward difference formula provides us

$$f(x,y) = f_0 + n\Delta f_0 + \frac{n(n-1)}{2}\Delta^2 f_0 + \frac{n(n-1)(n-2)}{6}\Delta^3 f_0 + \ldots \qquad (6.5.16)$$

Upon using (6.5.16) in right hand side of (6.5.15), we get

$$y_4 = y_0 + \int_{x_0}^{x_4}\left[f_0 + n\Delta f_0 + \frac{n(n-1)}{2}\Delta^2 f_0 + \cdots\right]dx$$

$$= y_0 + h\int_0^4\left[f_0 + n\Delta f_0 + \frac{n(n-1)}{2}\Delta^2 f_0 + \cdots\right]dn$$

$$= y_0 + h\left[4f_0 + 8\Delta f_0 + \frac{20}{3}\Delta^2 f_0 + \frac{8}{3}\Delta^3 f_0 + \cdots\right]$$

$$= y_0 + \frac{4}{3}h\left[2f_1 - f_2 + 2f_3\right] \tag{6.5.17}$$

after neglecting fourth and higher order differences and expressing the differences $\Delta f_0, \Delta^2 f_0$ and $\Delta^3 f_0$ in terms of the function values. This formula can be used to predict the value of y_4 when those of y_0, y_1, y_2 and y_3 are known. Having determined y_4, we obtain a first approximation to

$$f_4 = f(x_0 + 4h, y_4).$$

Then a better value of y_4 is found by Simpson's rule as

$$y_4 = y_2 + \frac{h}{2}(f_2 + 4f_3 + f_4) \tag{6.5.18}$$

which is called a corrector. The formula (6.5.18) can also be obtained by using Newton's forward difference formula (6.5.16) in the interval $x_0 \le x \le x_2$. Then an improved value of f_4 is computed and again the corrector is applied to find a still better value of y_4. We repeat the step until y_4 remains unchanged. Once y_4 and f_4 are obtain to desired degree of accuracy, $y_5 = (x_0 + 5h)$ is determined from the predictor as $y_5 = y_1 + \frac{4}{3}h(2f_2 - f_3 + 2f_4)$ and $f_5 = f(x_0 + 5h, y_5)$ is calculated. Then a better approximation to the value of y_5 is obtained from the corrector as

$$y_5 = y_3 + \frac{h}{3}(f_3 + 4f_4 + f_5).$$

We repeat this step till y_5 becomes stable and then proceed to calculate y_6 as before. This is Miline's predictor-corrector method. To ensure greater accuracy, we must first improve the accuracy of the starting values and then sub-divide the intervals.

In general, the predictor and corrector formulae are given by

$$y_{n+1}^{(p)} = y_{n-3} + \frac{4h}{3}(2f_n - f_{n-1} + 2f_{n-2}) \tag{6.5.19}$$

and

$$y_{n+1}^{(c)} = y_{n-1} + \frac{h}{3}\left(f_{n+1}^{(p)} + 4f_n + f_{n-1}\right) \tag{6.5.20}$$

with respective error estimates as

$$\left.\begin{aligned} E^{(p)} &= \frac{28}{29}h^5 y^v(\xi_1) \\ E_M^{(P)} &= -\frac{1}{90}h^5 y^v(\xi_2) \end{aligned}\right\} \tag{6.5.21}$$

Proceeding as in Adams -Moulton method, we can show that a local error estimate is provided by

$$D_{n+1} = \frac{-1}{29}\left(y_{n+1}^{(1)} - y_{n+1}^{(0)}\right) \tag{6.5.22}$$

The error estimate for Milne method appears to be somewhat more favorable than for Adams-Moulton method, but we shall see, (6.5.20) is subject to numerical instability in some cases.

Example 6.5.3 Using Runge-Kutta method of order 4, find *y* for *x = 0.1, 0.2, 0.3* given that $\frac{dy}{dx} = xy + y^2, y(0) = 1$. Continue the solution at $x = 0.4$ by using Miline's method.

Solution: The given differential equation is $\frac{dy}{dx} = xy + y^2, y(0) = 1$

Here $f(x, y) = xy + y^2$, $x_0 = 0, y_0 = 0, h = 0.1$

$\therefore\ k_1 = hf(x_0, y_0) = 0.1000$, $k_2 = hf\left(x_0 + \frac{1}{2}h, y_0 + \frac{1}{2}k_1\right) = 0.1155$

$k_3 = hf\left(x_0 + \frac{1}{2}h, y_0 + \frac{1}{2}k_2\right) = 0.1172$, $k_4 = hf\left(x_0 + h, y_0 + k_3\right) = 0.13598$

$$y(0.1) = y_0 + \frac{1}{6}(k_1 + 2k_2 + 2k_3 + k_4) = 1.1169$$

Again taking $x_0 = 0.1$, $y_0 = 1.1169$ and h = 0.1 and using the above method we obtain $y(0.2) = 1.2773$.

Similarly, with $x_0 = 0.2$, $y_0 = 1.2773$, $h = 0.1$, we get $y(0.3) = 1.504 = 1.504$

Now the starting values for Miline's methods are:

x	y	$f = f(x, y)$
0.0	1.0000	1.0000
0.1	1.01169	1.3591
0.2	1.2773	1.8869
0.3	1.5049	2.7132

Using predictor, $y_4 = y_0 + \frac{4}{3}h\left(2f_1 - f_2 + 2f_3\right)$, we get

$$x_4 = 0.4\ ,\ y_4 = 1.8344\ ,f_4 = 4.0988$$

Now using corrector, $y_4 = y_2 + \frac{h}{3}\left(f_2 + 4f_3 + f_4\right)$, we obtain

$$y_4 = 1.8386\ ,\quad f_4 = 4.1159$$

Again using the corrector we get

$$y_4 = 1.2773 + \frac{0.1}{3}(1.8869 + 4(2.7132) + 4.1182) = 1.8392$$

and $f_4 = 4.1182$

Again using the corrector, we get

$$y_4 = 1.2773 + \frac{0.1}{3}(1.8869 + 4(2.7132) + 44.1182) = 1.8392$$

Hence $y(0.4) = 1.8392$

Example 6.5.4 Use Milne's method to find $y(0.8)$ and $y(1.0)$ from $\frac{dy}{dx} = 1 + y^2$, $y(0) = 0$, given that $y(0.2) = 0.2027$, $y(0.4) = 0.4228$ and $y(0.6) = 0.6841$

Solution: The given differential equation is

$$\frac{dy}{dx} = 1 + y^2 \ , \ y(0) = 0$$

with $h = 0.2$, the starting values for Miline's method are

x	y	$y' = 1 + y^2$
0.0	0.0	1.0
0.2	0.2027	1.0411
0.4	0.4228	1.1787
0.6	0.6841	1.4681

Using Miline's predictor formula, we get

$$y(0.8) = 0 + \frac{0.8}{3}[2(1.0411) - 1.1787 + 2(1.4681)] = 1.0239$$

This gives us $y' = (0.8) = 2.0480$

Now the Miline's corrector formula (6.5.14) provides us

$$y(0.8) = 0.4228 + \frac{0.2}{3}[1.1787 + 4(1.4681) + 2.0480] = 1.0294$$

Proceeding on similar lines, we obtain $y(1.0) = 1.5549$. The accuracy in the values of $y(0.8)$ and $y(1.0)$ can be improved with repeated use of formula (6.5.14).

6.6 Simultaneous and higher order equations

6.6.1 Simultaneous differential equations

The simultaneous differential equations of the type

$$\frac{dy}{dx} = f(x, y, z) \tag{6.6.1}$$

and $$\frac{dz}{dx} = \phi(x, y, z) \tag{6.6.2}$$

with initial conditions $y(x_0) = y_0$ and $z(x_0) = z_0$ can be solved by the methods discussed in the previous sections, especially Picard's or Runge-Kutta methods.

(i) Picard's method

This method gives us

$$y_1 = y_0 + \int f(x, y_0, z_0)\,dx, \quad z_1 = z_0 + \int \phi(x, y_0, z_0)\,dx$$

$$y_2 = y_0 + \int f(x, y_1, z_1)\,dx, \quad z_2 = z_0 + \int \phi(x, y_1, z_1)\,dx$$

$$y_3 = y_0 + \int f(x, y_2, z_2)\,dx, \quad z_3 = z_0 + \int \phi(x, y_2, z_2)\,dx$$

and so on.

(ii) Taylor's series method

This method is used as follows.

If h be the step-size $y_1 = y(x_0 + h)$ and $z_1 = z(x_0 + h)$. Then Taylor's algorithm for (6.6.1) and (6.6.2) gives

$$y_1 = y_0 + h\,y_0' + \frac{h^2}{2!} y_0'' + \frac{h^3}{3!} y_0''' + \cdots \tag{6.6.3}$$

$$z_1 = z_0 + h\,z_0' + \frac{h^2}{2!} z_0'' + \frac{h^3}{3!} z_0''' + \cdots \tag{6.6.4}$$

Differentiating (6.6.1) and (6.6.2) successively, we get y'', z'' etc. so that

the values $y_0', y_0'', y_0''' \cdots$ and $z_0', z_0'', z_0''' \cdots$ are known. Substituting these values in (6.6.3) and (6.6.4), we obtain y_1, z_1 for the next step. Similarly, we have

$$y_2 = y_1 + h\,y_1' + \frac{h^2}{2!} y_1'' + \frac{h^3}{3!} y_1''' + \cdots \tag{6.6.5}$$

$$z_2 = z_1 + h\,z_1' + \frac{h^2}{2!} z_1'' + \frac{h^3}{3!} z_1''' + \cdots \tag{6.6.6}$$

Since y_1 and z_1 are known, we can calculate $y_1', y_1'', y_1''' \cdots$ and $z_1', z_1'', z_1''' \cdots$. Substituting these in (6.6.4) and (6.6.5), we get y_2 and z_2. Proceeding in this way, we can find other values of y and z step by step.

(iii) Runge-Kutta method

This method is applied by starting at (x_0, y_0, z_0) and taking the step-sizes for x, y, z to be h, k, *and* l respectively. We have

$$k_1 = h\,f(x_0, y_0, z_0)$$

$$l_1 = h\phi(x_0, y_0, z_0)$$

$$k_2 = hf\left(x_0 + \frac{1}{2}h, y_0 + \frac{1}{2}k_1, z_0 + \frac{1}{2}l_1\right)$$

$$l_2 = h\phi\left(x_0 + \frac{1}{2}h, y_0 + \frac{1}{2}k_1, z_0 + \frac{1}{2}l_1\right) \tag{6.6.6}$$

$$k_3 = hf\left(x_0 + \frac{1}{2}h, y_0 + \frac{1}{2}k_2, z_0 + \frac{1}{2}l_2\right)$$

$$l_3 = h\phi\left(x_0 + \frac{1}{2}h, y_0 + \frac{1}{2}k_2, z_0 + \frac{1}{2}l_2\right)$$

$$k_4 = hf(x_0 + h, y_0 + k_3, z_0 + l_3)$$

$$l_4 = h\phi(x_0 + h, y_0 + k_3, z_0 + l_3)$$

Hence

$$y_1 = y_0 + \frac{1}{6}(k_1 + 2k_2 + 2k_3 + k_4) \tag{6.6.7}$$

and

$$z_1 = z_0 + \frac{1}{6}(l_1 + 2l_2 + 2l_3 + l_4) \tag{6.6.8}$$

To compute y_2 and z_2 we simply replace x_0, y_0, z_0 by x_1, y_1, z_1 respectively in the above formulae. In a similar manner, one can extend the Runge-Kutta method to a system of *n* equations. We now consider the second order differential equation

$$y'' = f(x, y, y') \tag{6.6.9}$$

with initial conditions

$$y(x_0) = y_0 \text{ and } y'(x_0) = y_0' \tag{6.6.10}$$

By setting $y' = z$ the problem (6.6.9) and (6.6.10) can be reduced to the problem of solving the system

$$y' = z \quad z' = F(x, y, z) \tag{6.6.11}$$

with the conditions

$$y(x_0) = y_0 \text{ and } z(x) = z_0 \tag{6.6.12}$$

This can be solved by the method described above. Similarly, any higher order differential equation in which we can solve the highest order derivatives may be reduced to a system of first order differential equations.

6.6.2 Second order differential equations

Consider the second order differential equation

$$\frac{d^2y}{dx^2} = f\left(x, y, \frac{dy}{dx}\right) \tag{6.6.13}$$

We write $\frac{dy}{dx} = z$ in order to reduce it to two first order simultaneous differential equations

$$\frac{dy}{dx} = z, \frac{dz}{dx} = f(x, y, z) \tag{6.6.14}$$

These equations can be solved by the methods outlined above. The procedure to solve such equations is explained in the following example.

Example 6.1 Use Runge-Kutta method to solve $y'' = xy'^2 - y^2$ for x= 0.2, correct to four decimal places, subject to the initial conditions $x = 0, y = 1, y' = 0$.

Solution: The given equation is

$$y'' = x\,y'^2 - y^2 \tag{6.6.15}$$

Let
$$\frac{dy}{dx} = z = f(x,y,z)$$

so that
$$\frac{dz}{dx} = x\,z^2 - y^2 = \phi(x, y, z)$$

we have $x_0 = 0,\ y_0 = 1,\ z_0 = 0, h = 0.2$

Therefore, Runge-Kutta formulae become

$k_1 = h\,f(x_0, y_0, z_0) = 0.2(0) = 0,\ l_1 = h\phi\,(x_0, y_0, z_0) = 0.2(-1) = -0.2$

$$k_2 = h\,f\left(x_0 + \frac{1}{2}h, y_0 + \frac{1}{2}k_1, z_0 + \frac{1}{2}l_1\right) = -0.02,\ l_2 = h\phi\left(x_0 + \frac{1}{2}h, y_0 + \frac{1}{2}k_1, z_0 + \frac{1}{2}l_1\right) = -0.1998$$

$$k_3 = h\,f\left(x_0 + \frac{1}{2}h, y_0 + \frac{1}{2}k_2, z_0 + \frac{1}{2}l_2\right) = -0.02,\ l_3 = h\phi\left(x_0 + \frac{1}{2}h, y_0 + \frac{1}{2}k_2, z_0 + \frac{1}{2}l_2\right) = -0.1958$$

$$k_4 = h\,f\,(x_0 + h, y_0 + k_3, z_0 + l_3) = -0.0392,\quad l_4 = h\phi\,(x_0 + h, y_0 + k_3, z_0 + l_3) = -0.1905$$

$$k = \frac{1}{6}\,(k_1 + 2k_2 + 2k_3 + k_4) = -0.0199,\ l = \frac{1}{6}\,(l_1 + 2l_2 + 2l_3 + l_4) = -0.1970$$

Hence at x=0.2, we have

$$y = y_0 + k = 1 - 0.0199 = 0.9801$$

$$y' = z = z_0 + l = 0 - 0.1970 = -0.1970$$

This is the required solution.

Example 6.2 Use Picard's method to obtain the second approximation to the solution of

$$\frac{d^2y}{dx^2} = x^3\,\frac{dy}{dx} + x^3 y \text{ so that } y(0) = 1.0, y'(0) = \frac{1}{2}.$$

Solution: The given equation is

$$\frac{d^2y}{dx^2} = x^3\frac{dy}{dx} + x^3 y$$

Let $\frac{dy}{dx} = z = f(x, y, z)$ so that $\frac{dz}{dx} = x^3 z + x^3 y = x^3(y+z) = \phi(x, y, z)$

We have $x_0 = 0\ y_0 = 1, z_0 = \frac{1}{2}$ then

$$y = y_0 + \int_{x_0}^{x} f(x, y, z)\,dx$$

and

$$z = z_0 + \int_{x_0}^{x} \phi(x, y, z)\,dx$$

First approximations

$$y_1 = y_0 + \int_{x_0}^{x} f(x, y_0, z_0)\,dx = 1 + \int_0^1 (1/2)\,dx = 1 + x/2$$

$$z_1 = z_0 + \int_{x_0}^{x} \phi(x, y_0, z_0)\,dx = \frac{1}{2} + \int_0^1 x^3\left(1 + \frac{1}{2}\right)dx = \frac{1}{2} + \frac{3}{2}\frac{x^4}{4} = \frac{1}{2} + \frac{3}{8}x^4$$

Second approximations

$$y_2 = y_0 + \int_{x_0}^{x} f(x, y_1, z_1)\,dx = 1 + \int_0^x \left(\frac{1}{2} + \frac{3}{8}x^4\right)dx = 1 + \left|\frac{x}{2} + \frac{3}{40}x^5\right|_0^x = 1 + \frac{x}{2} + \frac{3}{40}x^5$$

$$z_2 = z_0 + \int_{x_0}^{x} \phi(x, y_1, z_1)\,dx = \frac{1}{2} + \int_0^x x^3\left(1 + \frac{x}{2} + \frac{1}{2} + \frac{3}{8}x^4\right)dx$$

$$= \frac{1}{2} + \int_0^x \left(\frac{3}{2}x^3 + \frac{x^4}{2} + \frac{3}{8}x^7\right)dx$$

$$= \frac{1}{2} + \frac{3}{2}\frac{x^4}{4} + \frac{x^5}{10} + \frac{3}{8}.\frac{x^8}{8} = \frac{1}{2} + \frac{3}{8}x^4 + \frac{x^5}{10} + \frac{3}{64}x^8$$

Hence the required second approximation to the solution of given equation is

$$y_2 = 1 + \frac{x}{2} + \frac{3}{40}x^5.$$

6.7 Boundary value problems

The boundary value problems require solution of a differential equation in a region R subject to the various conditions on the boundary of R. Practical applications give rise to many boundary value problems. Some simple examples of two point linear boundary value problems are:

(a) $y''(x) + f(x)y'(x) + g(x)y(x) = r(x)$ (6.7.1)

with the boundary conditions

$$y(x_0)=a,\ y(x_n)=b,\ y'(x)+p(x)y(x)=q(x) \tag{6.7.2}$$

Also with the conditions $y(x_0)=y'(x_0)=a,\ \ y(x_n)=y'(x_n)=b$

There exists many numerical methods of solving such boundary value problems, the method of finite-difference is a popular one and will be described here.

6.7.1 Finite difference method

The Finite-difference method for the solution of a two point boundary value problem consists in replacing the derivatives occurring in the differential equation and the boundary conditions by means of their finite-difference approximations and solving the resulting linear system of equations by a standard procedure. To obtain the appropriate finite-difference approximations to the derivatives, we proceed as follows: If *y(x)* and its derivatives are single-valued continuous functions of *x* then Taylor's series expansion leads to

$$y(x+h)=y(x)+hy'(x)+\frac{h^2}{2}y''(x)+\frac{h^3}{6}y'''(x)+\ldots \tag{6.7.3}$$

which implies that

$$y'(x)=\frac{y(x+h)-y(x)}{h}=\frac{h}{2}y''(x)+\frac{h^2}{6}y'''(x)+\ldots$$

Thus we have

$$y'(x)=\frac{1}{h}[y(x+h)-y(x)]+o(h) \tag{6.7.4}$$

which is the forward difference approximation of $y'(x)$ with an error of order *h*.

Similarly, expansion of $y(x-h)$ in Taylor's series gives

$$y(x-h)=y(x)-hy'(x)+\frac{h^2}{2}y''(x)-\frac{h^3}{6}y'''(x)+\ldots \tag{6.7.5}$$

from which we obtain

$$y'(x)=\frac{1}{h}[y(x)-y(x-h)]+o(h) \tag{6.7.6}$$

which is the backward difference approximation $y'(x)$ with an error of the order *h*.

Subtracting (6.7.3) and (6.7.5), we obtain

$$y'(x)=\frac{1}{2h}[y(x+h)-y(x-h)]+o(h^2) \tag{6.7.7}$$

which is the central difference approximation of $y'(x)$ with an error of the order h^2. Clearly this central difference approximation to $y'(x)$ is better than the forward or backward difference approximations and hence should be preferred. Again adding (6.7.3) and (6.7.5), we get

$$y''(x) = \frac{1}{h^2}\left[y(x+h) - 2y(x) + y(x-h)\right] + o\left(h^2\right) \tag{6.7.8}$$

which is the central difference approximation to $y''(x)$. In a similar manner, it is possible to derive finite-difference approximations to higher order derivatives. To solve the boundary value problems defined by (6.7.1) and (6.7.2), we divide the range $\left[x_0, x_n\right]$ into n equal subintervals of width h so that

$$x_i = x_0 + ih, i = 1,2,3 \ldots n.$$

The corresponding values of y at these points are denoted by

$$y(x_i) = y_i = y(x_0 + ih), i = 1,2,3 \ldots \tag{6.7.9}$$

Hence the working expressions for the central difference approximations to the first four derivatives of y_i are as under:

$$y_i' = \frac{1}{2h}\left(y_{i+1} - y_{i-1}\right)$$

$$y_i'' = \frac{1}{h^2}\left(y_{i+1} - 2y_i + y_{i-1}\right) \tag{6.7.10}$$

$$y_i''' = \frac{1}{2h^3}\left(y_{i+2} - 2y_{i+1} - y_{i-2}\right)$$

$$y_i^{iv} = \frac{1}{h^4}\left(y_{i+2} - 4y_{i+1} + 6y_i - 4y_{i-1} + y_{i-2}\right)$$

The accuracy of this method depends on the size of the subinterval viz h and also on the order of approximations. As we reduce h, the accuracy improves although the number of equations to be solved increases.

Example 6.7.1 To solve the equation $y'' + y + 1 = 0$ with the boundary conditions $y(0) = 0$ and $y(1) = 0$.

Solution: The given equation is

$$y'' + y + 1 = 0$$

we divide the interval (0,1) into four sub intervals so that $h = 0.25$ and the pivot points are $x_0 = 0, x_1 = 0.25, x_2 = 0.5, x_3 = 0.75$ and $x_4 = 1.0$. Then the differential equation is approximated as

$$\frac{1}{h^2}\left[y_{i+1} - 2y_i + y_{i-1}\right] = -1 - y_i$$

or

$$16y_{i+1} - 31y_i + 16y_{i-1}, = -1, i = 1,2,3$$

Using $y_0 = y_4 = 0$, we get the system of equations

$$16y_2 - 31y_1 = -1$$

$$16y_3 - 31y_2 + 16y_1 = -1$$
$$-31y_3 + 16y_2 = -1$$

Their solutions are

$$y_1 = 0.1047, y_2 = 0.1403, y_3 = 0.1047$$

up to four decimal places. The exact solution of the given differential equation is

$$y(x) = \frac{1 - e^{-x}}{1 - e^{-1}} - x$$

The error at each nodal point is given in the table below:

X	Computed value of $y(x)$	Exact value of $y(x)$	Error
0.25	0.1047	0.099932	0.004768
0.5	0.1403	0.1224593	0.0178407
0.75	0.1047	0.0847038	0.0199961

Example 6.7.2 The deflection of a beam is governed by the equation $\frac{d^4y}{dx^4} + 81y = \phi(x)$, where $\phi(x)$ is given by

x	$\frac{1}{3}$	$\frac{2}{3}$	1
$\phi(x)$	81	162	243

and boundary conditions $y(0) = y'(0) = y''(1) = y'''(1) = 0$. Evaluate the deflection at the pivot points of the beam using three subintervals.

Solution: Here $h = 1/3$ and the pivot points are $x_0 = 0, x_1 = 1/3, x_2 = 2/3$ and $x_3 = 1$.

The corresponding values of y are $y_0 = 0, y_1, y_2, y_3$. The given differential equation can be approximated to

$$\frac{1}{h^4}[y_{i+2} - 4y_{i+1} + 6y_i - 4y_i + y_{i-2}] + 81\,y_i = \phi(x_i), i = 1.2.3$$

This leads to system of equations

$$\begin{aligned} y_3 - 4y_2 + 7y_1 - 4y_0 + y_{-1} &= 1 \\ y_4 - 4y_3 + 7y_2 - 4y_1 + y_0 &= 2 \\ y_5 - 4y_4 + 7y_3 - 4y_2 + y_1 &= 3 \end{aligned} \qquad (6.7.11)$$

where $y_0 = 0$. Since $y_i' = \frac{1}{2h}(y_{i+1} - y_{i-1})$, therefore for *i= 0,* we have

$$0 = y_0' = \frac{1}{2h}(y_1 - y_{-1}) \Rightarrow y_{-1} = y_1 \tag{6.7.12}$$

Also $y_i'' = \frac{1}{h^2}(y_{i+1} - 2y_i + y_{i-1})$, for $i = 3$ gives us

$$0 = y_3'' = \frac{1}{h^2}(y_4 - 2y_3 + y_2) \Rightarrow y_4 = 2y_3 - y_2 \tag{6.7.13}$$

Again $y_i''' = \frac{1}{2h^3}(y_{i+2} - 2y_{i+1} + 2y_{i-1} - y_{i-2})$, for $i = 3$ leads to

$$0 = y_3''' = \frac{1}{2h^3}(y_5 - 2y_4 + 2y_2 - y_1) \Rightarrow y_5 = 2y_4 - 2y_2 + y_1 \tag{6.7.14}$$

Using equations (6.7.12) to (6.7.14) in equation (6.7.11) we obtain

$$\begin{aligned} y_3 - 4y_2 + 8y_1 &= 1 \\ -y_3 + 3y_2 - 2y_1 &= 1 \\ 3y_3 - 4y_2 + 2y_1 &= 3 \end{aligned} \tag{6.7.15}$$

Solving the system of equations (6.7.15), we get

$$y_1 = 8/13,\ y_2 = 22/13,\ y_3 = 37/13$$

Hence the required solution is

$$y(1/3) = 0.6154,\ y(2/3) = 1.6923,\ y(1) = 2.8462$$

correct to four decimal places.

6.7.2 Shooting method

This method can be applied to both linear and non-linear problems and required good initial guesser for the slope. It is easy and convenient to apply. The men steps involve in this method are:

(i) transformation of the boundary value problem into an initial value problem

(ii) solution of the initial value problem by Taylor' series method ,or Runge-Kutta method etc. and

(iii) solution of the given boundary value problem

Consider the second order boundary value problem

$$y''(x) = f(x), y(0) = 0, y(1) = 1 \tag{6.7.16}$$

Let us assume that the true value of $y'(0)$ be m. We take two initial guesses for m as m_0 and m_1. Let $y(m_0,1)$ and $y(m_1,1)$ be the corresponding values of $y(1)$ obtained by initial value method. The by using liner interpolation, we option a better approximation m_2 for m, given by

$$\frac{m_2 - m_0}{y(1) - y(m_0,1)} = \frac{m_1 - m_0}{y(m_1,1) - y(m_0,1)} \tag{6.7.17}$$

which implies that

$$m_2 = m_0 + (m_1 - m_0)\frac{y(1) - y(m_0,1)}{y(m_1,1) - y(m_0,1)} \quad (6.7.17)$$

We now solve that initial problem

$y''(x) = f(x), y(0) = 0$ *and* $y'(0) = m_2$ and obtain $y(m_2 1)$.we again apply interpolation with $(m_1, y(m_1,1))$and$(m_1, y(m_2,1))$ to obtain a better approximation m_3 for *m* and so on. The process is repeated until the convergence and desisted level of accuracy is obtained i.e., until the value of $y(m_1,1)$ agree with $y(1)$ of to the desired level of accuracy. The spaced of convergence depends, on how good the initial guesses are chosen. However, the method will be difficult to apply to higher order boundary value problem and in the case of now linear problems.

Example 6.7.2 Solve the boundary value problem $y''(x) = y(x), y(0) = 0$ and $y(1) = 1.1752$ by shooting method, taking $m_0 = 0$ and $y(1) = 0.9$.

Solution: Appling Taylor's series method, we obtain

$$y(x) = y'(0)\cdot\left[x + \frac{x^3}{6} + \frac{x^5}{120} + \frac{x^7}{5040} + \frac{x^9}{362800} + \cdots\right]$$

Hence $$y(1) = y'(0)\left[1 + \frac{1}{6} + \frac{1}{120} + \frac{1}{5040} + \frac{1}{362800} + \cdots\right] = 1.1752\, y'(0)$$

With $m_0 = 0.8$ and $m_1 = 0.9$, we get

$$y(m_0,1) = 0.9402 \text{ and } y(m_1,1) = 1.578$$

Using liner interpolation formula (6.7.18), we get

$$m_2 = 0.8 + (0.1)\frac{1.1752 - 0.9402}{1.0578 - 0.9404} = 0.8 + 0.1998 = 0.9998$$

which is closer to the exact value of $y'(0) = 1$ We now solve the initial value problem $y''(x) = y(x), y(0) = 0, m_2$ .Again using Taylor's series solution, we get $y(m_2,1) = 1.174$ which is also closer to the exact value $y(1) = 1.175$. This problem can also be solve by using Runge-Kutta fourth order method.

Exercises

1. Use Taylor's series to solve the differential equations

   (a) $\frac{dy}{dx} = 1 + xy, y(0) = 1$ at x = 0.1 correct to three decimal places

   (b) $\frac{dy}{dx} = \frac{1}{x^2 + y}$, $y(4) = 4$. Obtain $y(4.1)$ and $y(4.2)$

2. Using Picard's method to tabulate the value of $y(0.1), y(0.2), \cdots y(1.0)$ form the solution of $\frac{dy}{dx} = x(1 + x^3 y), y(0) = 3$

3. Solve $\frac{dy}{dx} = 1 + xy, y(0)$ to find the solution by Picard's method

4. Use Euler's method

   (a) $\frac{dy}{dx} + 2y = 0, y(0) = 1$, take h = 0.1 and obtain $y(0.1)$ and $y(0.2)$

   (b) $\frac{dy}{dx} = 1 + y^2, y(0) = 0$, take h=0.1 and obtain $y(0.1)$ and $y(0.2)$

   (c) $\frac{dy}{dx} = x + y, y(0) = 0$, take h=0.2 and compute $y(0.4)$ and $y(0.6)$.

5. Apply Euler's modified to find $y(0.02)$, $y(0.04)$ and $y(0.06)$ form the differential $\frac{dy}{dx} = x^2 + y, y(0) = 1$.

6. Given that $\frac{dy}{dx} = 2 + \sqrt{xy}$, $y(1) = 1$, find the value of $y(2)$ in steps of 0.1 by using the Euler's modified method

7. Use the Runge-Kutta fourth order method to find the value of y when $x = 1$ given that $\frac{dy}{dx} = \frac{y - x}{y + x}$, $y(0) = 1$

8. Use the Runge-Kutta method to solve $10\frac{dy}{dx} = x^2 + y^2$, $y(0) = 1$ for the interval $0 < x \leq 0.4$ with h=0.1

9. Use predictor-corrector method for tabulating a solution of $10\frac{dy}{dx} = x^2 + y^2, y(0) = 1$ for the range $0.5 \leq x \leq 1.0$

10. Tabulate the solution of $\frac{dy}{dx} = x + y, y(0) = 0$ for $0.4 < x \leq 1.0$ with h=0.1, using predictor corrector formulae

11. Solve the system of differential equations $\frac{dx}{dt} = y - t$, $\frac{dy}{dt} = x + t$ with $x = 1$, $y = 1$ when $t = 0$ taking $\Delta t = h = 0.1$

12. Compute $y(1.1)$ and $z(1.1)$ given that $\frac{dy}{dx} = xyz$, $\frac{dz}{dx} = \frac{xy}{z}$ and $y(1) = \frac{1}{3}$, $z(1) = 1$.

13. Solve the equation $\frac{d^2y}{dx^2} + y = 0$ with the conditions $y(0)=1$ and $y'(0) = 0$. Compute $y(0.2)$ and $y(0.4)$

14. Solve the boundary value problem $y'' - 64y + 10 = 0$ with $y(0) = y(1) = 0$, by finite difference method. Compute y (0.5) and compute it with the true value.

15. Solve the boundary value problems

    (a) $y''(x) = y(x)$, $y(0) = 0$, $y(1) = 1$

    (b) $y'' - 64y + 10 = 0$, $y(0) = y(1) = 0$

    by shooting method.

CHAPTER 7

PARTIAL DIFFERENTIAL EQUATIONS

7.1 Introduction

Partial differential equations occur in many branches of applied mathematics for example, in hydrodynamics, elasticity, quantum mechanics, electromagnetic theory, heat transfer, seismology and boundary layer flow. The analytical treatment of these equations is too involved a process and requires application of advanced mathematical methods. In most of the cases it is easier to develop approximate solutions by numerical methods. Of all the numerical methods available for the solution of partial differential equations, the method of finite differences is most commonly used. In this method, the derivations appearing in the equation and the boundary conditions are replaced by their finite difference approximations and hence the given equation is changed to a system of linear equations, which are solved by iterative procedures. This process is although slow but produces good results in many boundary value problems. Moreover, this method is well suited for computations carried out by electronic computers. In this chapter we describe methods for solving boundary value problems involving partial differential equations.

7.2 Classifications of partial differential equations

The general second order linear partial differential equation is of the from

$$A\frac{\partial^2 u}{\partial x^2}+B\frac{\partial^2 u}{\partial x\partial y}+C\frac{\partial^2 u}{\partial y^2}+D\frac{\partial u}{\partial x}+E\frac{\partial u}{\partial y}+Fu=G$$

which can be rewritten as

$$Au_{xx}+Bu_{xy}+Cu_{yy}+\phi(x,y,u,u_x,u_y)=G \qquad (7.2.1)$$

where $A, B, C, \ldots G$ are all functions of x and y. Such a partial differential equation is said to be

$$\left.\begin{array}{l}\text{(i) elliptic if } B^2-4AC<0\\ \text{(ii) parabolic if } B^2-4AC=0\\ \text{(iii) hyperbolic if } B^2-4AC>0\end{array}\right\} \qquad (7.2.2)$$

In the following analysis we will restrict ourselves to three simple and important types of equations (7.2.1), namely

$$u_{xx}+u_{yy}=0 \text{ (Laplace equation)} \qquad (7.2.3)$$

$$u_{xx}=\frac{1}{c^2}u_{tt} \text{ (Wave equation)} \qquad (7.2.4)$$

$$u_{xx} = \frac{1}{K} u_t \quad \text{(Diffusion equation)} \tag{7.2.5}$$

where (x, y) are space coordinates and t is the time coordinate. It can be easily seen that Lapace equation is elliptic, the wave equation is hyperbolic and that the diffusion equation is of parabolic type. In the study of partial differential equations, usually three types of problems arise:

(i) Dirichlet's problem

Given a continuous function f on the boundary C of a region R, to find a function u such that

$$u_{xx} + u_{yy} = 0, \text{ in } R \tag{7.2.6}$$

and

$$u = f, \text{ on } C$$

(ii) Cauchy's problem

(a) To find u such that

$$u_{tt} = u_{xx}, \text{ for } t > 0$$

$$u(x,0) = f(x) \tag{7.2.7}$$

and

$$u_t(x,0) = g(x),$$

$f(x)$ and $g(x)$ being arbitrary.

(b) To find u such that

$$u_t - u_{xx} = 0 \text{ for } t > 0 \tag{7.28}$$

and

$$u(x,0) = f(x)$$

These problems are all well defined or (well posed) and it is proved in text books of partial differential equations that these possess unique solutions. In Contrast with ordinary differential equations, the form of a partial differential equation is always connected with particular type of associated conditions. Thus the problem of Laplace equation with Cauchy boundary conditions, viz,

$$u_{xx} + u_{yy} = 0$$

$$u(x,0) = f(x)$$

and

$$u_y(x,0) = g(x) \tag{7.2.9}$$

is an ill-posed problem.

7.3 Integral transforms method

The integral transforms are essentially a mathematical tool, which can be used to solve several problems in science and engineering. Linear integral transforms of function $f(t)$, defined on a finite or infinite interval $a < t < b$ are particularly useful in dealing with problems in linear differential equations. A general linear integral transformation $f(t)$ is represented by the equation.

$$\bar{f}(s) = T\{f(t)\} = \int_a^b k(s,t) f(t) dt \tag{7.3.1}$$

where $k(s,t)$ is called its kernel. It represents a function $\bar{f}(s)$ the image or transform of the function $f(t)$. The kernels and limits of integration of some of the integral transforms are given in table7.1

Table 7.1: Integral transforms and their kernels

Name of transform	$k(s,t)$	**a**	**b**
Laplace	$\bar{e}^{st}$	0	∞
Fourier transform	$\frac{1}{\sqrt{2\pi}} e^{ist}$	$-\infty$	∞
Fourier sine transform	$\sqrt{\frac{2}{\pi}} \sin st$	0	∞
Fourier cosine transform	$\sqrt{\frac{2}{\pi}} \cos st$	0	∞
Hankel transform	$t\,J_n(st)$	0	∞
Mellin transform	t^{s-1}	0	∞

The integral transforms defined above are applicable either for semi-infinite or infinite minute domains. Similarly, finite integral transforms can be defined on finite domains.

Theorem 7.1 Prove that the application of an integral transform to a partial differential equation reduces the independent variable by one.

Proof: Suppose we want to solve the partial differential equation of the type.

$$A(x)\frac{\partial^2 f}{\partial x^2} + B(x)\frac{\partial f}{\partial x} + C(x) f + Lf = \phi \tag{7.3.2}$$

where f and ϕ are functions of the independent variable $x, y, z...$ and L is a linear differential operator in the variables $y, z...$ but not in x. Let $k(x,s)$ be the kernel of the integral transform. Now

$$kA\frac{\partial^2 f}{\partial x^2} - +\frac{\partial^2 (Ak)}{\partial x^2} = \frac{\partial}{\partial x}\left(Ak\frac{\partial f}{\partial x}\right) - \frac{\partial}{\partial x}\left(f\left(\frac{\partial (Ak)}{\partial x}\right)\right) = \frac{\partial}{\partial x}\left[Ak\frac{\partial f}{\partial x} - f\frac{\partial (Ak)}{\partial x}\right]$$

$$KB\frac{\partial f}{\partial x} + f\frac{\partial (kB)}{\partial x} = \frac{\partial}{\partial x}(fkB)$$

$$\hbar Cf - F\hbar C = 0$$

Therefore $$\int_a^b \left(A\frac{\partial^2 f}{\partial x^2} + B\frac{\partial f}{\partial x} + Cf\right) kdx - \int_a^b f\left(\frac{\partial^2 (Ak)}{\partial x^2} - \frac{\partial (kB)}{\partial x} + kC\right) dx$$

$$= \left(Ak\frac{\partial^2 f}{\partial x^2} + f\frac{\partial (Ak)}{\partial x} + fkB \right)_a^b = g(s, y, z, ...) \text{ (say)} \tag{7.3.3}$$

We choose the function k in such a way that

$$\frac{\partial^2 (Ak)}{\partial x^2} - \frac{\partial}{\partial x}(kB) + kC = \lambda k \tag{7.3.4}$$

Then equation (7.3.3) becomes

$$\int_a^b \left(A\frac{\partial^2 f}{\partial x^2} + B\frac{\partial f}{\partial x} + Cf \right) k\, dx = \int_a^b \lambda\, fk\, dx + g(s, y, z, ...) = \lambda \bar{f}(s) + g(s, y, z, \cdots) \tag{7.3.5}$$

Multiplying (7.3.3) throughout by k and integrating between the limits a to b and choosing k as given in (7.3.4) and upon using (7.3.5), we obtain.

$$L(\bar{f}) + \lambda \bar{f} + g(s, y, z, ...) = \bar{\phi}$$

Therefore $$(L + \lambda)\bar{f} = \bar{\phi} - g(s, y, z, ...) = \bar{\Phi} \text{ (say)} \tag{7.3.6}$$

The equation (7.3.6) is a partial differential equation for $\bar{f}$ in the independent variables $y, z, ...$ but not in x. Thus we conclude that the application of an integral transform to a partial differential equation reduces the number of independent variables by one. Hence proof of the theorem.

Laplace Transforms

This transform was first introduced by Laplace (1749-1827), a French mathematician, in the year 1790 in his work on probability theorems. This technique became popular when Heaviside applied it to the solutions of the ordinary differential equations representing problems in electrical engineering.

Definition 7.3.1: Let $f(t)$ be a piecewise continuous function. Then $f(t)$ is said to be of exponential order if there exists a real and finite positive number M such that

$$\lim_{t \to \infty} |f(t)| e^{-\alpha t} \le M$$

and we write $$|f(t)| = O(e^{\alpha t})$$

Definition 7.3.2: Let $f(t)$ be a continuous and single valued function of the real variable t defined for all t, $0 < t < \infty$, and is of exponential order. Then the Laplace transform of $f(t)$ is defined as.

$$L\{f(t)\} = \bar{f}(s) = \int_0^\infty e^{-st} f(t) dt \tag{7.3.7}$$

provided the integral on right hand exists. Here s is a parameter, real or complex and clearly $\bar{f}(s)$ is a function of s.

Inverse Laplace Transform Integral

We have $L\{f(t)\} = \bar{f}(s)$, which implies that $f(t) = L^{-1}\{\bar{f}(s)\}$, where L and L^{-1} respectively, represent Laplace and inverse Laplace transform operations. The transformed solutions of many problems yield results not contained in tables of Lap lace transforms. What is needed, therefore is a so called 'first –principle' method of inverting a transformed solution that can be applied regardless of the availability of the tabulated values. To develop the needed inversion process, it is essential to extend our theory of the Laplace transform by letting s represent a complex variable and then an extension of the Cauchy integral formula provides the desired result.

From complex variables Cauchy's integral formula states if $f(z)$ is an analytic function inside a closed curve C and z_0 is a point within C, then

$$f(z_0)=\frac{1}{2\pi i}\int_C \frac{f(z)dz}{z-z_0} \tag{7.3.8}$$

An extension of this formula is stated below.

Theorem 7.2 Let $f(z)$ be analytic for $\text{Re}(z)\geq\gamma$, where γ is a real constant greater than zero. Then, for $\text{Re}(z_0)>\gamma$,

$$f(z_0)=-\frac{1}{2\pi i}\lim_{\beta\to\infty}\int_{\gamma-i\beta}^{\gamma+i\beta}\frac{f(z)\,dz}{z-z_0} \tag{7.3.9}$$

Proof: This theorem can be proved by choosing a closed rectangular contour $C(C_1+C_2+C_3+C_4)$ enclosing the point z_0 and having sides parallel to the real and imaginary axes as shown in Figure 7.3.1.

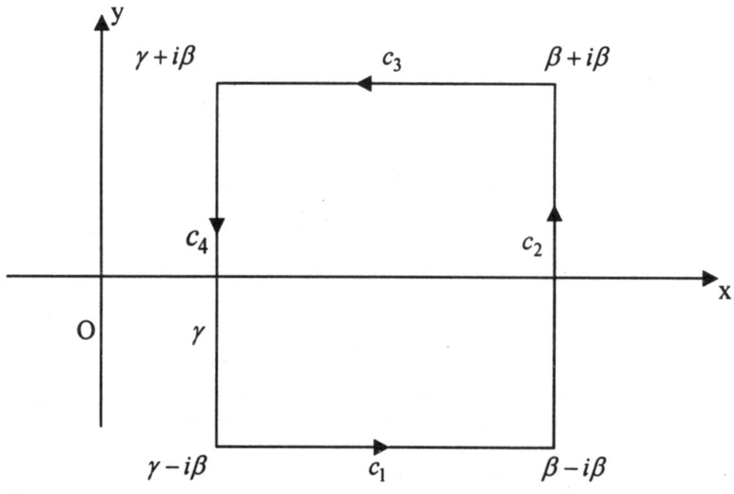

Figure 7.3.1 Contour of integration

Then we have

$$f(z_0)=\frac{1}{2\pi i}\left[\int_{C_1}+\int_{C_2}+\int_{C_3}+\int_{C_4}\right]\frac{f(z)}{z-z_0}dz \tag{7.3.10}$$

By some limiting arguments and inequalities, it can be shown that the contribution from the paths C_1, C_2 and C_3 vanishes on letting $\beta \to \infty$ and the expression (7.3.10) becomes.

$$f(s) = \frac{1}{2\pi i} \lim_{\beta \to \infty} \int_{\gamma - i\beta}^{\gamma + i\beta} \frac{f(z)}{s - z} dz$$

where z_0 has been replaced by $s = x + iy$ and $f(s)$ is assumed to be analytic in the half-plane $\text{Re}(s) > \gamma$. Applying the inverse Laplace transform on the sides of this equation , we get.

$$f(t) = L^{-1}\{f(s)\} = \frac{1}{2\pi i} L^{-1} \left\{ \lim_{\beta \to \infty} \int_{\gamma - i\beta}^{\gamma + i\beta} \frac{f(z)}{s - z} dz \right\}$$

$$= \frac{1}{2\pi i} \lim_{\beta \to \infty} \int_{\gamma - i\beta}^{\gamma + i\beta} f(z) L^{-1}\left(\frac{1}{s - z}\right) dz = \frac{1}{2\pi i} \lim_{\beta \to \infty} \int_{\gamma - i\beta}^{\gamma + i\beta} e^{st} f(z) ds \quad \left[\because L^{-1}\left(\frac{L}{s - z}\right) = e^{zt}\right]$$

where the dummy variable z has been represented by s. Thus, in general, Laplace transform inverse is given by

$$f(t) = \frac{1}{2\pi i} \int_{\gamma - i\infty}^{\gamma + i\infty} \bar{f}(s) e^{st} ds \qquad (7.3.11)$$

for $t > 0$ and γ being positive constant.

The path of integration in this inversion integral of Laplace transform, is often called the 'Bromwich contour', since Bromwich first devised this method of handling certain integrals that arose in operation mathematics. The basic concept that will underline the actual evaluation of inverse Laplace transforms by use of inversion integral (7.3.11) is the application of contour integration in the complex plane. The details of the calculations will depend on the nature of transformed function $\bar{f}(s)$.

7.3.1 Numerical inversion of Laplace transform

Now we shall discuss a numerical technique to evaluate the integral on the right hand side of equation (7.3.11). The inversion formula (7.3.11) for Laplace transform is given as

$$f(t) = \frac{1}{2\pi i} \int_{\gamma - i\infty}^{\gamma + i\infty} \bar{f}(s) e^{st} ds$$

where γ is an arbitrary real number greater than all the real parts of the singularities of $\bar{f}(s)$.

We take $s = \gamma + iy$, so that

$$f(t) = \frac{e^{\gamma t}}{2\pi} \int_{-\infty}^{\infty} \bar{f}(\gamma + iy) e^{ity} dy \qquad (7.3.12)$$

Expanding the function $h(t)=f(t)e^{\gamma t}$ in a Fourier series in the interval $[0,2l]$, we obtain the approximation formula

$$f(t)=f_{\infty}(t)+E_D \tag{7.3.13}$$

where

$$f_{\infty}(t)=\frac{C_0}{2}+\sum_{k=1}^{\infty}C_k, 0\le t\le 2l \tag{7.3.14}$$

$$C_k=\exp(\gamma t/l)\mathrm{Re}\left[\exp(ik\pi t/l)\bar{f}(\gamma+ik\pi/l)\right] \tag{7.3.15}$$

and E_D the discretisation error, can be made arbitrarily small by choosing γ large enough.

Because the infinite series in equation (7.3.14) can only be summed up to a finite number N of terms, the approximate value of $f(t)$ becomes

$$f_N(t)=\frac{C_0}{2}+\sum_{k=1}^{N}C_K, \textit{for} \le t\le 2l \tag{7.3.16}$$

Using this formula to evaluate $f(t)$, we introduce a truncation error E_T that must be added to the discretisation error to produce the total approximate error. The "Korrector" method is used to reduce the discretisation error and the ϵ-algorithm is used to reduce the truncation error and hence to accelerate the convergence. The Korrector method formula to evaluate the function $f(t)$ is

$$f(t)=f_{\infty}(t)-\exp(-2\gamma t)f_{\infty}(2l+t)+E'_D \tag{7.3.17}$$

where $|E'_D|\langle|E_D|$. Thus the approximate value of $f(t)$ becomes

$$f_{N_k}(t)=f_N(t)-\exp(-2\gamma t)f_{N'}(2l+t) \tag{7.3.18}$$

where N' is an integer such that $N'\angle N$. We shall also describe the ϵ – algorithm that is used to accelerate the convergence of the series in equation (7.3.16). Let N be an odd natural number and let $s_m=\sum_{K=1}^{m}C_k$, be the sequence of partial sums of the equation (7.3.16). We define the ϵ-sequence by

$$\epsilon_{0,m}=0,\ \epsilon_{1,m}=s_m \text{ and } \epsilon_{n+1,m}=\epsilon_{n-1,m+1}+\frac{1}{\epsilon_{n,m+1}-\epsilon_{n,m}}, n,m=1,2,3\cdots$$

It can be shown that the sequence $\epsilon_{1,1},\epsilon_{3,1},\epsilon_{5,1},\ldots,\epsilon_{n,1}$ converges to $f(t)+E_D-C_{0/2}$ faster than the sequence of partial sums s_m, m = 1,2,3,...The actual procedure used to invert the Laplace transform consists of using equation (7.318) together with the ϵ – algorithm.

7.3.2 Solution of partial differential equations

A large number of problems in science and engineering involve the solution of linear partial differential equations. A function of two or more variables may also have a Laplace transform. Suppose that $u(x,t)$ is a function of two independent real variables x and t. When Laplace transform is applied with respect to the variable t, the partial differential equation is reduced to an ordinary differential equation of the t-transform $\bar{u}(x,s)$. The general solution of ordinary differential equation is then subjected to the boundary conditions of the original problem. Finally, the solution $u(x,t)$ is obtained by using the complex inversion formula (7.3.11) or numerically by using the relation (7.3.18). Thus the Laplace transform is specially suited to solve initial boundary value problems (IBVP), when conditions are prescribed at $t = 0$.

Example: 7.3.1 Use Laplace transform method to solve the initial value problem

$$K\frac{\partial u}{\partial t} = \frac{\partial^2 u}{\partial x^2},\ 0 \angle x \angle l,\ 0 \angle t \angle \infty$$

subject to the conditions

$$u(0,t) = 0,\ u(l,t) = g(t),\ \ 0 \angle t \angle \infty$$

and
$$u(x,0) = 0,\ \ 0 \angle x \angle l$$

Solution: Applying Laplace transform *w .r. t.* 't' to the given partial differential equation , we get

$$\mathrm{K}\ L\left(\frac{\partial u}{\partial t}\right) = L\left(\frac{\partial^2 u}{\partial x^2}\right)$$

$$\Rightarrow \qquad K[s\bar{u}(x,s) - u(x,0)] = \frac{d^2\bar{u}}{\partial x^2}$$

or
$$\frac{d^2\bar{u}}{dx^2} - Ks\bar{u} = 0,\ \text{ as } \bar{u}(x,0) = 0$$

Therefore,

$$\bar{u}(x,s) = A\ \cosh\left(\sqrt{ksx}\right) + B\ \sinh\left(\sqrt{ksx}\right) \qquad (7.3.19)$$

Taking Laplace transform of boundary conditions, we get

$$\bar{u}(0,s) = 0,\ \ \bar{u}(l,s) = g(s) \qquad (7.3.20)$$

Using equation (7.3.20) in (7.3.19),we obtain

$$A = 0$$

and

$$B = \frac{\bar{g}(s)}{\sinh\left(\sqrt{ks}\ 1\right)}$$

Taking the inverse Laplace transform, we get

$$u(x,t) = \frac{1}{2\pi i}\int_{\gamma - i\infty}^{\gamma + i\infty} \bar{g}(s)e^{st}\frac{\sinh\left(\sqrt{ks}x\right)}{\sinh\left(\sqrt{ksl}\right)}ds \qquad (7.3.21)$$

To evaluate the integral on the right hand side of equation (7.3.21), we use the method of residues. The poles of the integrand are given by

$$\sinh\left(\sqrt{\mathrm{ks}l}\right) = 0 \Rightarrow e^{\sqrt{ksl}} - e^{-\sqrt{ksl}} = 0$$

$$\Rightarrow \qquad e^{2\sqrt{ksl}} = 1 = e^{2n\pi i}$$

or $$s_n = -\frac{n^2\pi^2}{k^2l^2}, n = 0, \pm 1, \pm 2, \pm 3, \ldots$$

Now by Cauchy residue theorem, we have

$$u(x,t) = \frac{1}{2\pi i}\int_{\gamma - i\infty}^{\gamma + i\infty} \bar{g}(s)e^{st}\frac{\sinh\left(\sqrt{ks}x\right)}{\sinh\left(\sqrt{ks}l\right)}ds = \sum R_i$$

where R_1 = residue at the *ith* pole. Here residue at $s = 0$ is zero. The residue at $s = s_n$ is

$$\lim_{s\to s_n} \bar{g}(s)e^{st}\frac{\sinh\sqrt{ks}x}{\frac{d}{ds}\left[\sinh\left(\sqrt{ks}l\right)\right]} = \lim_{s\to s_n}\frac{2\sqrt{s}\ \bar{g}(s)}{l\sqrt{k}[\cosh(\sqrt{k}sl)]}e^{st}\sinh\left(\sqrt{ks}x\right)$$

$$\frac{2\sqrt{-\frac{n^2\pi^2}{kl^2}}\,\bar{g}\left(-\frac{n^2\pi^2}{kl^2}\right)}{l\sqrt{k}\ \cosh\left(-\sqrt{-\frac{n^2\pi^2}{l^2}}\right)}\exp\left(\frac{-x^2\pi^2}{kl^2}\right)\text{sine}\left[\left(\sqrt{\frac{-n^2\pi^2}{l^2}}\right)x\right] \text{n} = 1,2,3,\cdots$$

$$= \frac{2\,in\,\pi\bar{g}\left(\frac{-n^2\pi^2}{kl^2}\right)}{kl^2\cos\left(in\pi\right)}\sinh\left(\frac{in\pi}{l}x\right)e^{-\frac{n^2\pi^2}{kl^2}}, n = 1,2,3$$

Therefore

$$u(x,t) = \frac{2n\pi(-1)^n}{kl^2}\sin\left(\frac{n\pi}{l}x\right)\bar{g}\left(\frac{-n^2\pi^2}{kl^2}\right)\bar{e}^{\frac{n^2\pi^2}{kl^2}}, n = 1,2,3,\ldots$$

as $\cosh(in\pi) = \cos n\pi$ and $\sinh\left(\frac{in\pi}{l}x\right) = i\sin\left(\frac{n\pi}{l}x\right)$

Thus the required solution is

$$u(x,t) = \frac{2\pi}{kl^2}\sum_{n=1}^{\infty}(-1)^n n\bar{g}\left(\frac{n^2\pi^2}{kl^2}\right)\sin\left(\frac{n\pi}{l}x\right)e^{\frac{-n^2\pi^2}{kl^2}t}$$

Example 7.3.2 A string is stretched and fixed between two points (0, 0) and $(l, 0)$. Motion is initiated by displacing the string in the form $u = \sin\left(\frac{\pi x}{l}\right)$ and released from rest at time $t = 0$. Find the displacement of any point on the string at any time *t*.

Solution: The displacement $u(x,t)$ of the string is governed by

$$\frac{\partial^2 u}{\partial t^2} = c^2\frac{\partial^2 u}{\partial x^2}, 0 \angle x \angle l, t > 0$$

subject to the conditions

$$u(x,0)=\sin(\pi x/l),\ \frac{\partial u}{\partial t}(x,0)=0,\ u(0,t)=u(l,t)=0$$

Taking the Laplace transform of the given partial differential equation, we get

$$s^2\bar{u}(x,s)-su(x,0)-\frac{\partial u}{\partial t}(x,0)=c^2\frac{d^2\bar{u}}{dx^2}$$

$$\frac{\partial^2\bar{u}}{dx^2}-\frac{s^2}{c^2}\bar{u}=-\frac{s^2}{c^2}\sin\frac{\pi x}{l}$$

Its general solution is given by,

$$\bar{u}(x,s)=Ae^{sx/c}+Be^{-sx/c}+\frac{s\ \sin(\pi x/l)}{s^2+\pi^2c^2/l^2} \tag{7.3.22}$$

The Laplace transform of the boundary conditions is given by

$$\bar{u}(0,s)=0,\quad \bar{u}(l,s,)=0$$

Applying these conditions in equation on (7.3.22), we get

$$A+B=0$$

$$A\,e^{sl/c}+B\bar{e}^{sl\,c}=0$$

Solving these equations, we get

$$A=B=0$$

Therefore, $$\bar{u}(x,s)=\frac{s\sin(\pi x/l)}{s^2+\pi^2c^2/l^2}$$

Taking inverse Laplace transform, we get

$$u(x,t,)=\cos\left(\frac{\pi c}{l}t\right)\sin\frac{\pi x}{l}$$

which in the required solution.

7.3.3 Fourier and other integral transforms with their application to partial differential equations

Joseph Fourier, a French mathematician, had invented a method called Fourier transform in 1801, to explain the flow of heat around an anchor ring. Since, then it has become a powerful tool in diverse fields of science and engineering. It can provide a means of solving unwieldy equations that describe dynamic responses to electricity, heat or light and it can also identify the regular contributions to a fluctuating signal, there by helping to make sense of observations in astronomy, medicine and chemistry. Fourier transforms has become indispensable in the numerical calculations needed to design electric circuits, to analyze mechanical vibrations, and to study wave propagation. Fourier transform techniques have been widely used to solve problems involving semi-infinite or totally infinite range of the variables or unbounded regions. In this section, we shall deal with some applications of Fourier, Hankel and Meline transforms to diffusion, wave and Laplace equations.

Example 7.3.3: Solve the heat conduction equation given by

$$k\frac{\partial^2 u}{\partial t^2}=\frac{\partial u}{\partial t},\ -\infty\angle x\angle\infty,\ t>0$$

subject to the initial and boundary conditions

$$u(x,t) \text{ and } \frac{\partial u}{\partial k}(x,t)\to 0, \text{ as } |x|\to\infty$$

$$u(x,0)=f(x),\ -\infty<x<\infty$$

Solution: Taking Fourier transform of the partial differential equation, we get,

$$\frac{\partial \bar{u}}{\partial t}(\alpha,t)+k\alpha^2\bar{u}(\alpha,t)=0$$

since $\bar{u}(\alpha,t,)$ and $\frac{d}{dx}\bar{u}(\alpha,t)\to 0$

Its general solution is

$$\bar{u}(\alpha,t)=A\ \bar{e}^{k\alpha^2 t}$$

The Fourier transform of initial condition gives us

$$\bar{u}(\alpha,o)=\bar{f}(\alpha),\ -\infty<x<\infty$$

Using this condition in the solution, we get

$$\bar{u}(\alpha,t)=\bar{f}(\alpha)\bar{e}^{k\alpha^2 t}$$

Taking inverse Fourier transforms, we get

$$u(x,t)=\frac{1}{\sqrt{2\pi}}\int_{-\infty}^{\infty}\bar{f}(\alpha)\bar{e}^{-i\alpha x-k\alpha^2 t}\,d\alpha \tag{7.3.23}$$

We know that

$$\int_{-\infty}^{\infty}\exp\left(-ax^2-2bx\right)dx=\sqrt{\frac{\pi}{a}}\exp\left(b^2/a\right)$$

Therefore

$$g(x)=\frac{1}{\sqrt{2\pi}}\int_{-\infty}^{\infty}\bar{e}^{k\alpha^2 t}\bar{e}^{i\alpha x}\,d\alpha=\frac{1}{\sqrt{2\pi}}\int_{-\infty}^{\infty}\bar{e}^{k\alpha^2 t-i\alpha x}\,d\alpha$$

$$=\frac{1}{\sqrt{2\pi}}\frac{\sqrt{\pi}}{\sqrt{kt}}\exp\left(\frac{-x^2}{4kt}\right)=\frac{1}{\sqrt{2kt}}\exp\left(\frac{-x^2}{4kt}\right)$$

since here $a=kt$ and $b=\frac{ix}{2}$ and the Fourier transform of $g(x)$ is $\bar{e}^{k\alpha^2 t}$. Thus using convolution theorem to (7.3.23), we have

$$u(x,t)=\frac{1}{\sqrt{2\pi}}\int_{-\infty}^{\infty}f(\alpha)g(x-\alpha)d\alpha=\frac{1}{\sqrt{2\pi}}\int_{-\infty}^{\infty}f(\alpha)\frac{1}{\sqrt{2kt}}\exp\left[\frac{(x-\alpha)^2}{4kt}\right]d\alpha$$

$$= \frac{1}{\sqrt{4\pi kt}} \int_{-\infty}^{\infty} f(\alpha) \exp\left[\frac{-(x-\alpha)^2}{4kt}\right] d\alpha$$

$$\therefore \qquad u(x,t) = \frac{1}{\sqrt{\pi}} \int_{-\infty}^{\infty} f\left(x + \sqrt{4kt}\, z\right) \bar{e}^{z^2}\, dz,$$

where $z = (\alpha - x)/\sqrt{4kt}$. This is the required solution.

Example 7.3.4: A uniform string of length l is stretched tightly between two fixed points at $x = 0$ and $x = l$. If it is displaced a small distance d at a point $x = b, 0 < b < l$ and released from rest at time $t = 0$, find an expression for the displacement at subsequent times.

Solution: If $u(x,t)$ denote the displacement by of the string, the initial boundary value problem is described by

$$\frac{\partial^2 u}{\partial t^2} = c^2 \frac{\partial^2 u}{\partial x^2},\ 0 \angle \mathrm{x} \angle l,\ \mathrm{t} > 0$$

subjected to the initial and boundary conditions given by

$$u(x,0) = \begin{cases} \dfrac{dx}{db}, & 0 < \mathrm{x} < b \\ \dfrac{d(x-l)}{b-l}, & b < \mathrm{x} < l \end{cases}$$

$$\frac{\partial u}{\partial t}(x,0) = 0 \quad \text{and} \quad u(0,t) = u(l,t) = 0,\ t \geq 0.$$

Taking finite Fourier sin transform of the partial differential equation, we get.

$$\int_0^l \frac{\partial^2 u}{\partial t^2} \sin\frac{n\pi x}{l}\, dx = c^2 \int_0^l \frac{\partial^2 u}{\partial x^2} \sin\frac{n\pi x}{l}\, dx$$

which on using boundary condition leads to

$$\frac{d^2\bar{u}}{dt^2} + \frac{n^2\pi^2c^2}{l^2}\bar{u} = 0$$

Its general solution is found to be

$$\bar{u}(n,t) = A\cos\frac{n\pi ct}{l} + B\sin\frac{n\pi ct}{l} \qquad (7.3.24)$$

Now taking finite Fourier sine transform of initial conditions, we obtain

$$\bar{u}(n,0) = \int_0^b \frac{dx}{b} \sin\frac{n\pi x}{l}\, dx + \int_b^l \frac{d(x-l)}{b-l} \sin\frac{n\pi x}{l}\, dx = \frac{dl^3}{n^2\pi^2 b(l-b)} \sin\frac{n\pi b}{l}$$

From equation (7.3.24) for $t = 0$, we have

$$\bar{u}(n,0) = A = \frac{dl^3}{n^2\pi^2 b(l-b)} \sin\frac{n\pi b}{l}$$

Also taking finite Fourier sine transform of the second initial condition, we get

$$\frac{d\overline{u}}{dt}=0$$

This when used in equation (7.3.24) leads to $B=0$. Thus, we get

$$\overline{u}(n,t)=\frac{dl^3}{n^2\pi^2 b(l-b)}\sin\frac{n\pi b}{l}\cos\frac{n\pi ct}{l}$$

Inverting the Fourier transform, we obtain

$$u(x,t)=\frac{2dl^2}{n^2\pi^2 b(l-b)}\sum_{n=1}^{\infty}\frac{1}{n^2}\sin\frac{n\pi b}{l}\sin\frac{n\pi x}{l}\cos\frac{n\pi ct}{l}$$

which in the required solution.

Example 7.3.5 Solve the boundary value problem in the half- plane $y>0$ described by

$$u_{xx}+u_{yy}=0,\ -\infty<x<\infty,\ y>$$

subject to the boundary conditions,

u is bounded as

$$y\to\infty,\ u \text{ and } \frac{\partial u}{\partial x} \text{ both vanish, as } |x|\to\infty$$

Solution: Since x has an infinite range of values, we take the Fourier exponential transform of partial differential equation w. r. t. the variable x and use the conditions that u and $\frac{\partial u}{\partial x}$ both vanish, $as|x|\to\infty$, to get

$$-\alpha^2\overline{u}(\alpha,y)+\frac{1}{\sqrt{2\pi}}\int_{-\infty}^{\infty}u_{yy}e^{-i\alpha x}dx=0$$

$$\Rightarrow \qquad \frac{d^2\overline{u}(\alpha,y)}{dy^2}+\alpha^2\overline{u}(\alpha,y)=0.$$

Its general solution is given by

$$\overline{u}(\alpha,y)=\mathrm{A}e^{\alpha y}+Be^{-\alpha y} \tag{7.3.25}$$

Because, u must be bounded as $y\to\infty$, $\overline{u}$ is also bounded as $y\to\infty$, implying that $A=0$, for $\alpha>0$ and $B=0$, for $\alpha<0$. Thus for any α, $\overline{u}(\alpha,y)=C\ e^{-|\alpha|y}$, C being constant.

Now the Fourier transforms of the boundary condition yields

$$\overline{u}(\alpha,0)=\bar{f}(\alpha)$$

which leads us to

$$\bar{f}(\alpha)=c$$

Hence $$\bar{u}(\alpha, y) = \bar{f}(\alpha)\bar{e}^{|\alpha|y} = \frac{1}{\sqrt{2\pi}}\int_{-\infty}^{\infty} f(x)\bar{e}^{|\alpha|y}\,dx$$

Taking inverse transforms, we obtain

$$u(x, y) = \frac{1}{\sqrt{2\pi}}\int_{-\infty}^{\infty}\left[\frac{1}{\sqrt{2\pi}}\int_{-\infty}^{\infty} f(\xi)\bar{e}^{|\alpha|y}e^{i\alpha\xi}\,d\xi\right]\bar{e}^{i\alpha x}\,d\alpha$$

$$= \frac{1}{\sqrt{2\pi}}\int_{-\infty}^{\infty} f(\xi)\,d\xi\int_{-\infty}^{\infty}\exp\left[\{\alpha\,[i(\xi - x)] - |\alpha|y\}\right]d\alpha$$

$$= \frac{y}{\pi}\int_{-\infty}^{\infty}\frac{f(\xi)\,d\xi}{(\xi - x)^2 + y^2}$$

since $$\frac{1}{2\pi}\int_{-\infty}^{\infty}\exp\{\alpha[i(\xi - x)] - |\alpha|y\}\,d\xi = \frac{1}{\pi}\frac{y}{(\xi - x)^2 + y^2}$$

Thus $$u(x, y) = \frac{y}{\pi}\int_{-\infty}^{\infty}\frac{f(\xi)\,d\xi}{(\xi - x)^2 + y^2}$$

This is a well known Poisson integral formula and is valid for $y > 0$, when $f(x)$ is bounded and piecewise continuous for all real x.

Example 7.3.6. Use Hankel transform to derive the solution of the equation

$$\frac{\partial^2 v}{dr^2} + \frac{1}{r}\frac{\partial v}{\partial r} + \frac{\partial^2 v}{\partial z^2} = 0, \; r \geq 0, z \geq 0$$

satisfying the conditions

(i) $v \to 0$, *as* $z \to \infty$ and $r \to \infty$, (ii) $v = f(r)$, on $z = 0$, $r \geq 0$.

Solution: Applying Hankel transforms of order zero to the partial differential equation, we get

$$\frac{d^2\bar{v}}{dz^2} - \xi^2\bar{v} = 0 \qquad (7.3.26)$$

where $$\int_{-\infty}^{\infty}\left(\frac{\partial^2 v}{\partial r^2} + \frac{1}{2}\frac{\partial v}{\partial r}\right) rJ_0(\xi r)\,dr = -\xi^2\bar{v},\; \bar{v} = \int_{-\infty}^{\infty} vrJ_0(\xi r)\,dr$$

after using condition (i) and since $J_0(\xi r)$ is a solution of

$$\frac{d^2 f}{dr^2} + \frac{1}{r}\frac{df}{dr} + \xi^2 f = 0$$

The general solution of equation (7.3.26) is

$$\bar{v}(\xi, z) = A\,e^{\xi z} + B\bar{e}^{\xi z}$$

Now $\bar{v} \to 0$, *as* $z \to \infty$ and $\bar{v} \to \bar{f}(\xi)$ *on* $z = 0$, we have

$$B = 0 \text{ and } B = \bar{f}(\xi).$$

Thus $$\bar{v}(\xi, z) = \bar{f}(\xi)\bar{e}^{-\xi z}.$$

Taking inverse of the Hanker transforms , we get

$$V(r,z) = \int_0^\infty \xi\, \bar{v}(\xi, z) J_0(\xi r) d\xi$$

$$= \int_0^\infty \xi\, \bar{v}(\xi) \bar{e}^{-\xi z} J_0(\xi r) d\xi$$

which is the required solution. The integral can be evaluated numerically by standard methods depending upon the nature of the function $f(s)$.

Example 7.3.7 The function $v(r, \theta)$ satisfies the differential equation.

$$\frac{\partial^2 v}{\partial r^2} + \frac{1}{r}\frac{\partial v}{\partial r} + \frac{1}{r^2}\frac{\partial^2 v}{\partial \theta^2} = 0$$

in the wedge-shaped region $|\theta| \le \alpha$ and the boundary condition $V = f(r)$ when $\theta = \pm\alpha$. Shows with help of Mellin transform that it can be expressed in the form

$$v(r,\theta) = \frac{1}{2\pi i}\int_{r-i\infty}^{r+i\infty} \frac{\cos(\xi\theta)}{\cos(\xi\alpha)} \bar{f}(\xi) \bar{r}^{-\xi} d\xi,\ \bar{f}(\xi) = \int_0^\infty f(r) r^{\xi-1} dr$$

Solution: Applying the Mellin transform with respect to *r*, we get,

$$\int_0^\infty \left(\frac{\partial^2 v}{\partial r^2} + \frac{1}{r}\frac{\partial v}{\partial r} \frac{1}{r^2}\frac{\partial^2 v}{\partial \theta^2} \right) r^{s+1} dr = 0 \Rightarrow$$

$$s^2\bar{v} + \frac{d^2\bar{v}}{d\theta^2} = 0, \quad \bar{v} = \int_0^\infty v\, r^{s-1} dr$$

and in all physical problems $v, \frac{\partial v}{\partial r} \to 0$ as $r \to \infty$.

The general solution in this case is given by

$$\bar{v}(s,\theta) = A\ \cos(s\theta) + B\sin(s\theta)$$

where *A* and *B* are arbitrary constants. Now $v = f(r)$, for $\theta = \pm\alpha$

$$\Rightarrow \qquad \bar{v} = \bar{f}(s), \quad \text{for } \theta = \pm\alpha, \bar{f}(s) = \int_0^\infty f(r) r^{s-i} dr.$$

New *v* must be on even function of *s*, and hence we have

$$B = 0 \ \text{ and A} = \frac{\bar{f}(s)}{\cos(s\alpha)}$$

Therefore, $$\bar{v}(s,\theta) = \frac{\bar{f}(s)}{\cos(s\alpha)}\cos(s\theta)$$

Inverting the transform, we get,

$$v(r,\theta)=\frac{1}{2\pi i}\int_{r-i\infty}^{r+i\infty}\bar{v}(s,\theta)\bar{r}^{s}\,ds=\frac{1}{2\pi i}\int_{\gamma-i\infty}^{r+i\infty}\frac{\bar{f}(s)\bar{r}^{s}\cos(s\theta)}{\cos(s\alpha)}\,ds$$

which is the required result.

7.3.4 Numerical inversion of a combination of Fourier and Laplace transforms

Let $f(x,z,t)$ be a sufficiently well behaved function of x, z and t in its domain of definition. This function $f(x,z,t)$ assumed to satisfy initial and regularity conditions in addition to boundary conditions of the physical problem. We apply Laplace transform with respect to 't' defined by

$$\bar{f}(x,z,p)=\int_0^{\infty} f(x,z,t)\bar{e}^{pt}\,dt \qquad (7.3.27)$$

and then Fourier transform with respect to x defined by

$$\hat{f}(q,z,p)=\frac{1}{\sqrt{2\pi}}\int_{-\infty}^{\infty}\bar{f}(x,z,p)\bar{e}^{iqx}\,dx \qquad (7.3.28)$$

In order obtain the solution of various problems in the physical domain once it is determined in the transform domain, we have to invert the transforms in equation (7.3.27) and (7.3.28). The details of the process are outlined in [39]. First we invert Fourier transform, which gives the Laplace transform expression of $\bar{f}(x,z,p)$ of the function $f(x,z,t)$ as

$$\bar{f}(x,z,p)=\frac{1}{\sqrt{2\pi}}\int_{-\infty}^{\infty}\hat{f}(q,z,p,)\exp(iqx)dq=\sqrt{\frac{2}{\pi}}\int_0^{\infty}\left\{\cos(qx)\hat{f}_e+i\sin(qx)\hat{f}_o\right\}dq \qquad (7.3.29)$$

where $\hat{f}_e$ and $\hat{f}_o$ denote the even and odd parts of the function $\hat{f}(q,z,p)$ respectively, with respect to q. For fixed values of q, x and z, the function inside braces in the last integral (7.3.29) can be considered as a Laplace transform $\bar{g}(p)$ of some function $g(t)$. The inversion formula for Laplace transform is

$$g(t)=\frac{1}{2\pi i}\int_{\gamma-i\infty}^{\gamma+i\infty}\bar{g}(p)\exp(pt)dp \qquad (7.3.30)$$

where γ is an arbitrary real number greater than all the real parts of the singularities of $\bar{g}(p)$. The integral (7.3.20) can be numerically evaluated by setting $p=\gamma+iy$ and expanding the function $g(t)\bar{e}^{\gamma t}$ in a Fourier series in the interval $[0,2l]$ by adopting the procedure outlined in section 7.3.1 in order to obtain approximation formula (7.3.18). Here again the discretisation and truncation error are controlled and minimized by using the "Korrector method" and $\in$ – algorithm respectively as discussed in section 7.3.1. The values of γ and l are to be chosen according to the criteria outlined by Honig and Hirdes [20].

The last step in the inversion process is to evaluate the integral (7.3.29). This can be done by using Romberg integration with adaptive step size. This method uses the results from successive refinements of the extended trapezoidal rule followed by extrapolation of the results to the limit when the step size tends to zero. The details of which can be found in press et al. [32].

7.3.5 Numerical inversion of a combination of Hankel and Laplace transforms

Suppose $f(r,z,t)$ be a sufficiently well behaved function of r, z and t in its domain of definition. It is assumed to satisfy initial, regularity and boundary conditions of the physical problem. We apply Laplace transform with respect to 't' defined by

$$\bar{f}(r,z,p) = \int_0^{\infty} f(r,z,t)e^{-pt}\,dt \tag{7.3.31}$$

and then we apply Hankel transform w. r .t. 'r' defined by

$$\hat{\bar{f}}(q,z,p) = \int_0^{\infty} \bar{f}(r,z,p) r J_n(qr) dr \tag{7.3.32}$$

where $J_n(x)$ is a Bessel function of order n. The value of n can be selected depending upon the nature of the function $f(r,z,p)$. In order to obtain the solution of various problems in the physical domain once it is found in the transformed domain, we have to invert the transforms in equations (7.3.31) and (7.3.32) which is quite difficult task analytically most of the time. But it can be conveniently managed through numerical evaluations of the inversion integrals [40, 42]. First, we invert the Hankel transform, which gives the Laplace transform expression $\bar{f}(r,z,p)$ of the function $f(r,z,t)$ as

$$\bar{f}(r,z,p) = \int_0^{\infty} q\,\hat{\bar{f}}(q,z,p) J_n(qr) dq \tag{7.3.33}$$

For fixed values of q, r and z the function under integral in (7.3.33) can be considered as a Lap lace transform $\bar{g}(p)$ of some function $g(t)$. The integral (7.3.33) can be approximated and evaluated by adopting the procedure and technique outlined above in case of a combination of Fourier and Laplace transforms and in section 7.3.1. Some suitable approximations or equivalent forms of Bessel function $J_n(x)$ can also be used during the inversion process, if needed. The errors can again be controlled and minimized by using the above-suggested techniques.

7.4 Finite difference approximations to partial derivatives

We consider a rectangular region R in the $x-y$ plane. We divide this region into a rectangular network sides $\Delta x = h$ and $\Delta y = k$ as shown in figure 7.4.1. The points of intersection of the dividing lines are called mesh points, nodal points or grid points. Then, we have

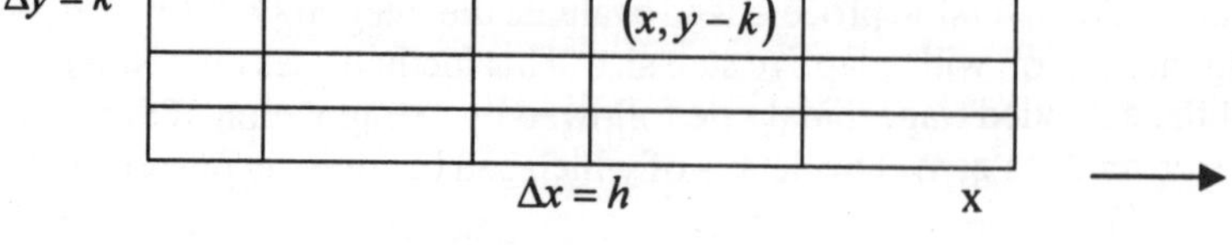

Figure 7.4.1

$$\frac{\partial u}{\partial x}=\frac{u(x+h,y)-u(x,y)}{h}+o(h)=\frac{u(x,y)-u(x-h,y)}{h}+o(h)$$

$$=\frac{u(x+h,y)-u(x-h,y)}{2h}+o\left(h^2\right)$$

and

$$\frac{\partial^2 u}{\partial x^2}=\frac{u(x-h,y)-2u(x,y)+u(x+h,y)}{h^2}+o\left(h^2\right)$$

If we write $u(x,y)=u(ih, jk,)$ as $u_{i,j}$, the above approximations become

$$u_x=\frac{u_{i+1,\,j}-u_{i,\,j}}{h}+O(h)=\frac{u_{i,j}-u_{i-1,j}}{h}+O(h)\quad=\frac{u_{i+1,j}-u_{i-1,j}}{2h}+O\left(h^2\right) \tag{7.4.1}$$

and

$$u_{xx}=\frac{u_{i-1,j}-2u_{i,j}+u_{i+1,j}}{h^2}+O\left(h^2\right) \tag{7.4.2}$$

Similarly, we have

$$u_y=\frac{u_{i,j+1}-u_{i,j}}{k}+O(k)\quad=\frac{u_{i,j}-u_{i,j-1}}{k}+O(k)\quad=\frac{u_{i,j}-u_{i,j-1}}{h}+O\left(k^2\right)$$

$$=\frac{u_{i,j+1}-u_{i,j-1}}{2k}+O\left(k^2\right) \tag{7.4.3}$$

$$u_{yy}=\frac{u_{i,j+1}-2u_{i,j}+u_{i,j+1}}{k^2}+O\left(k^2\right) \tag{7.4.4}$$

The derivatives in any partial differential equation can be replaced by their corresponding difference approximations (7.4.1) to (7.4.4), we obtain the finite-difference analogues of the given equation.

7.5 Elliptic equations

In this section we shall study various techniques for solving Laplace and Poisson equation which are elliptic in nature. Various physical phenomenon, are governed by these well-known equations. A few of them, frequently encountered in physical and engineering applications are steady heat condition, seepage through porous media, rotational flow of an ideal fluid, distributional potential, steady viscous flow, equilibrium stresses in elastic structures etc

7.5.1 Solution of Laplace and Poisson equations

We consider the Laplace equation in two dimensions

$$\frac{\partial^2 u}{\partial x^2}+\frac{\partial^2 u}{\partial y^2}=0 \tag{7.5.1}$$

We take a rectangular region R for which $u(x, y)$ is known at the boundary. Assuming that an exact sub-division of R is possible, we divide this region into a network of square mesh of side h as shown in figure 7.5.3

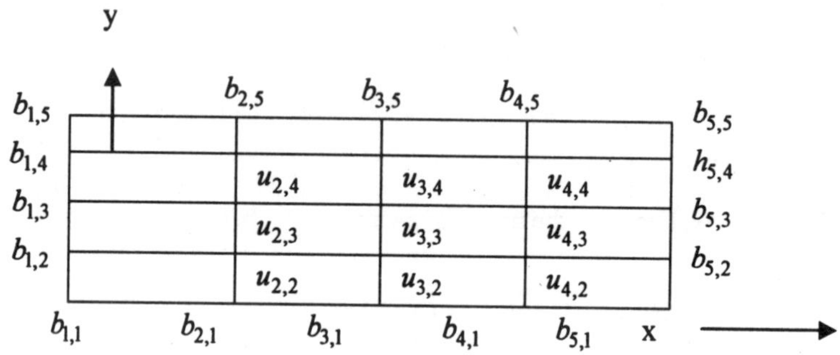

Figure 7.5.1

Replacing the derivatives in (7.5.1) by their finite difference approximations, we have

$$\frac{1}{h^2}\left[u_{i-1,j} - 2u_{i,j} + u_{i+1,j}\right] + \frac{1}{h^2}\left[-u_{i,j-1} - 2u_{i,j} + u_{i,j+1}\right] = 0$$

$\Rightarrow$
$$u_{i,j} = \frac{1}{4}\left[u_{i-1,j} + u_{i+1,j} + u_{i,j+1} + u_{i,j-1}\right] \tag{7.5.2}$$

This shows that value of u at any interior mesh point is the average of its values at four neighbouring points to the left, right, above and below. The equation (7.5.2) is known as standard 5-point formula exhibited in figure 7.5.2

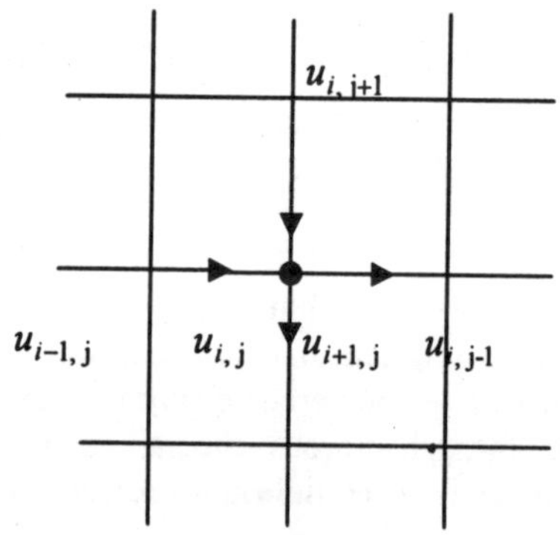

Figure 7.5.2

Sometimes we may use a formula

$$u_{i,j} = \frac{1}{4}\left[u_{i-1,\,j+1} + u_{i+1,\,j-1} + u_{i+1,\,j+1} + u_{i-1,\,j-1}\right] \tag{7.5.3}$$

which is similar to (7.5.2).This shows that the value of $u_{i,j}$ is the average of its values at the four-neighbouring diagonal mesh points and the equation (7.5.3) is called diagonal 5-point formula and is represented in figure 7.5.3.

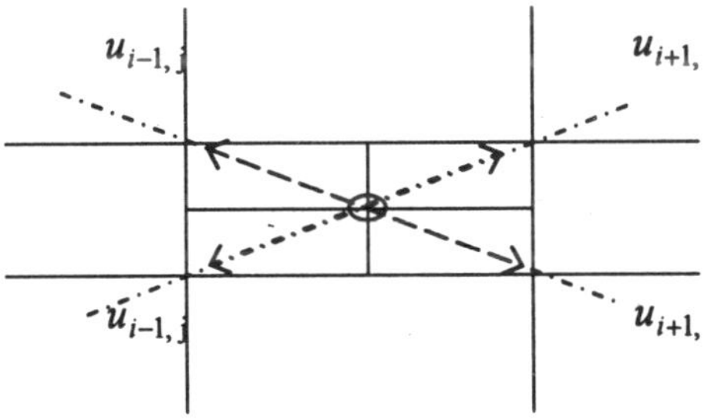

Figure 7.5.3

Although (7.5.3) is less accurate than (7.5.2) but serves as reasonably good approximation for obtaining starting values at the mesh points. We use the diagonal five-point formula (7.5.3) to find the initial values of u at the interior mesh points and compute $u_{3,3}$, $u_{2,4}$, $u_{4,4}$, $u_{4,2}$ *and* $u_{2,2}$ in this order. We have

$$u_{3,3} = \frac{1}{4}\left(b_{1,5} + b_{5,1} + b_{5,5} + b_{1,1}\right)$$

$$u_{2,4} = \frac{1}{4}\left(b_{1,5} + u_{3,3} + b_{3,5} + b_{1,3}\right)$$

$$u_{4,4} = \frac{1}{4}\left(b_{3,5} + b_{5,5} + b_{5,3} + u_{3,3}\right)$$

$$u_{4,2} = \frac{1}{4}\left(u_{3,3} + b_{5,1} + b_{3,1} + b_{5,3}\right)$$

$$u_{2,2} = \frac{1}{4}\left(b_{1,3} + b_{3,1} + u_{3,3} + b_{1,1}\right)$$

The values at the remaining interior points i.e. $u_{2,3}, u_{3,4}, u_{4,3}$ and $u_{3,2}$ are computed by the standard five-point formula (7.5.2). Thus, we obtain

$$u_{2,3} = \frac{1}{4}\left(b_{1,3}, + u_{3,3} + u_{2,4} + u_{2,2}\right)$$

$$u_{3,4} = \frac{1}{4}\left(u_{2,4} + u_{4,4} + b_{3,5} + u_{3,3}\right)$$

$$u_{4,3} = \frac{1}{4}\left(u_{3,3}, + b_{5,3} + u_{4,4} + u_{4,2}\right)$$

$$u_{3,2} = \frac{1}{4}\left(u_{2,2} + u_{4,2} + u_{3,3} + u_{3,1}\right)$$

After determining $u_{i,j}$ once, their accuracy is improved either by using Jacobi's iterative method or using Gauss-Seidal iterative method. The process is repeated till the two consecutive iterations becomes very close i.e. the difference between two consecutive iterations becomes negligibly small in order to achieve the desired level of accuracy. The iterative formula in case of Jacobi method and Gauss-Seidal method are given below.

Jacobi's method

Here we have

$$u_{i,j}^{(n+1)} = \frac{1}{4}\left[u_{i-1,j}^{(n+1)} + u_{i+1,j}^{(n)} + u_{i,j+1}^{(n+1)} + u_{i,j-1}^{(n)}\right] \tag{7.5.4}$$

where $u_{i,j}^{n}$ denotes the *n*th iterative value of $u_{i,j}$. This gives us the improved values of $u_{i,j}$ at the interior mesh points and is called point Jacobi's formula.

Gauss-Seidal method

In this method the iteration formula is

$$u_{i,j}^{n+1} = \frac{1}{4}\left[u_{i-1,j}^{(n+1)} + u_{i+1,j}^{(n)} + u_{i,j+1}^{(n+1)} + u_{i,j-1}^{(n)}\right] \tag{7.5.5}$$

This uses the latest iterative value available and scans the mesh points symmetrically from left to right along successive rows. Gauss-Seidal method is simple and well suited to computer calculations. It being slow, the working is same but lengthy, however it can be shown that the Gauss –Seidal scheme converges twice as fast as Jacobi's method. The accuracy depends on the mesh size i.e. smaller the *h*, better the accuracy. But for too small values of *h*, the round off errors and labour of computations also increase.

Solution of Poisson's equation

Consider the Poisson's equation

$$\frac{\partial^2 u}{\partial x^2} + \frac{\partial^2 u}{\partial y^2} = f(x, y) \tag{7.5.6}$$

the method of solving equation (7.5.6) in similar to that of the La
place equation (7.5.1). Here the standard 5-point formula for (7.5.6) taken the form

$$u_{i-1,j} + u_{i+1,j} + u_{i,j+1} + u_{i,j-1} - 4u_{i,j} = h^2 f(ih, jh) \tag{7.5.7}$$

Using (7.5.6) at each interior mesh point, we arrive at a system of linear equations in the nodal values $u_{i,j}$, which can be solved by Gauss-Seidal method. The error in replacing u_{xx} by the finite difference approximations is of the order $O(h^2)$. Since $k = h$, the error in replacing u_{yy} by the difference approximation is also of the order $O(h^2)$. Hence the error in solving Laplace and Poisson's equations by finite difference method is of the order $O(h^2)$.

Example 7.5.1 Given the values of $u(x, y)$ on the boundary of the square in figure 7.5.4, evaluate the function $u(x, y)$ satisfying the Laplace equation $\nabla^2 u = 0$ at the pivotal points of the figure 7.5.4 below by(a) Jacobi's method (b) Gauss- Seidal method .

Solution: For initial values of u_1, u_2, u_3, u_4 we assume that $u_4 = 0$ then

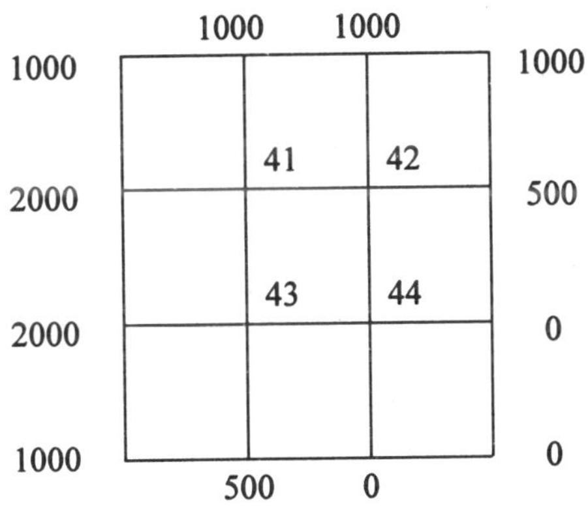

Figure 7.5.4

$$u_1 = \frac{1}{4}(1000 + 0 + 1000 + 2000) = 1000 \text{ (by diagonal formula)}$$

$$u_2 = \frac{1}{4}(1000 + 500 + 1000 + 0) = 625 \text{ (by standard formula)}$$

$$u_3 = \frac{1}{4}(2000 + 0 + 1000 + 500) = 875 \text{ (by standard formula)}$$

$$u_4 = \frac{1}{4}(875 + 0 + 625 + 0) = 375 \text{ (by standard formula)}$$

Jacobi method

The Jacobi's formulae are

$$u_1^{(n+1)} = \frac{1}{4}\left(2000 + u_2^{(n)} + 1000 + u_3^{(n)}\right), \quad u_2^{(n+1)} = \frac{1}{4}\left(u_1^{(n)} + 500 + 1000 + u_4^{(n)}\right)$$

$$u_3^{(n+1)} = \frac{1}{4}\left(2000 + u_4^{(n)} + u_1^{(n)} + 500\right), \quad u_4^{(n+1)} = \frac{1}{4}\left[u_3^{(n)} + 0 + u_2^{(n)} + 0\right].$$

We carry out the successive iteration using above formula as given in the table below.

Iteration (n)	u_1	u_2	u_3	u_4
1	1125	719	969	375
2	1172	750	1000	422
3	1188	774	1024	438
4	1200	782	1032	450
5	1204	788	1038	454
6	1206.5	790	1040	456.5
7	1208	791	1041	458
8	1208	791.5	1041.5	458

Since there is no significant difference between seventh and eight iteration values, therefore we have.

$$u_1 = 1208,\ u_2 = 792, \qquad u_3 = 1042,\ u_4 = 458$$

Gauss-Seidal method

In this case we carry out the successive iterations, using Gauss-Seidal formulae

$$u_1^{(n+1)} = \frac{1}{4}\left[2000 + u_2^{(n)} + 1000 + u_3^{(n)}\right], \quad u_2^{(n+1)} = \frac{1}{4}\left[u_1^{(n+1)} + 500 + 1000 + u_4^{(n+v)}\right]$$

$$u_3^{(n+1)} = \frac{1}{4}\left[2000 + u_4^{(n)} + u_1^{(n+1)} + 500\right], \quad u_4^{(n+1)} = \frac{1}{4}\left[u_3^{(n+1)} + 0 + u_2^{(n+1)} + 0\right]$$

Iteration (n)	u_1	u_2	u_3	u_4
1.	1125	750	1000	438
2.	1188	782	1032	454
3.	1204	789	1040	458
4.	1207	791	1041	458
5.	1208	791.5	1041.5	458.25

Since there is no significant difference between the fourth and fifth iteration values, we have

$$u_1 = 1208,\ u_2 = 792,\ u_3 = 1042,\ u_4 = 458$$

Example 7.5.2 Solve the Poisson equation $\nabla^2 u = -10(x^2 + y^2 + 10)$ over the square with sides $x = 0 = y$, $x = 3 = y$ with $x = 0$ on the boundary and mess length $=1$.

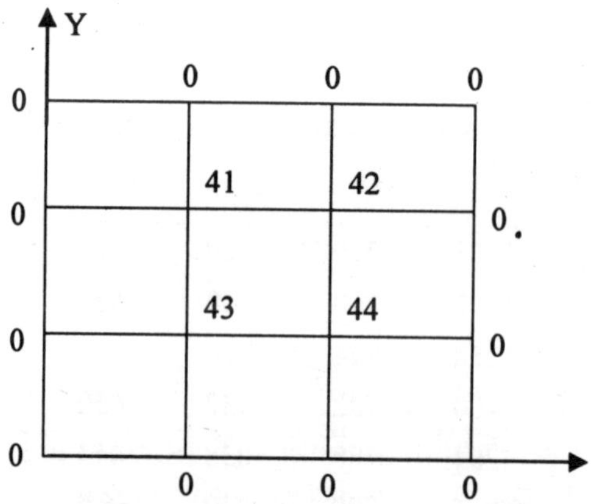

Figure -7.5.5

Solution: Here $h = 1$, therefore the standard 5 point formula for given equation is given by

$$u_{i-1,j} + u_{i+1,j} + u_{i,j+1} + u_{i,j-1} - 4u_{i,j}$$

$$= -10\left(i^2 - j^2 + 10\right) \tag{7.5.8}$$

For $u_1 \left(i = 1, j = 2\right)$ from (7.5.8), we get.

$$0 + u_2 + u_3 - 4u_1 = -10(1 + 4 + 10)$$

$$\Rightarrow \qquad u_1 = \frac{1}{4}(u_2 + u_3 + 150)$$

For $u_2 \left(i = 2, j = 2\right)$, formula (7.5.8) gives us

$$u_2 = \frac{1}{4}(u_1 + u_4 + 180)$$

For $u_3 \left(i = 1, j = 1\right)$, from (7.5.8) we have

$$u_3 = \frac{1}{4}(u_1 + u_4 + 120)$$

For $u_4 \left(i = 2, j = 1\right)$, we get.

$$u_4 = \frac{1}{4}(u_2 + u_3 + 150) = u_1$$

Thus the above equations reduce to

$$u_1 = \frac{1}{4}(u_2 + u_3 + 150), \quad u_2 = \frac{1}{2}(u_1 + 90), \quad u_3 = \frac{1}{2}(u_1 + 60)$$

Now we solve these equations by Gauss-Seidal iteration method and solution is given below:

Iteration (n)	u_1	u_2	u_3
1	37.5	64	49
2	66	78	63
3	73	82	67
4	75	82.5	67.5
5	75	82.5	67.5

Because there is no difference in the values of 4th and 5th iterations, therefore we have

$$u_1 = 75, \quad u_2 = 82.5, \quad u_3 = 67.5, \quad \text{and} \quad u_4 = 75$$

This is the required solution.

7.5.2 Solution of elliptic equations by relaxation method

Consider the Laplace equation

$$\frac{\partial^2 u}{\partial x^2}+\frac{\partial^2 u}{\partial y^2}=0 \tag{7.5.6}$$

We take a square region and divide it into a square net of mesh size *h*. Let value of *u* at *A* be u_0 and its values at four adjacent points be u_1, u_2, u_3 u_4 as shown in figure - 7.5.6

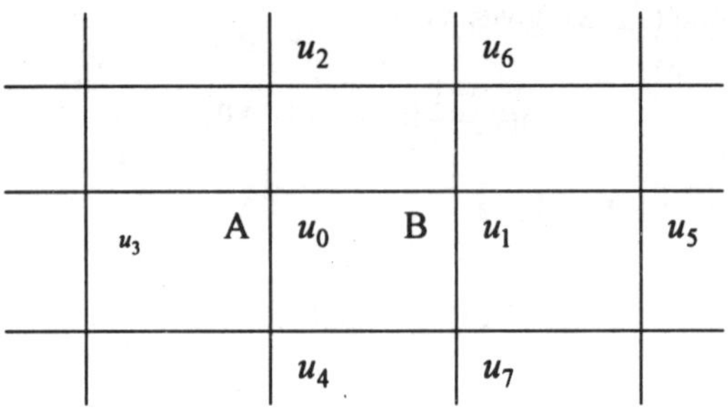

Figure 7.5.6

Then $$\frac{\partial^2 u}{\partial x^2} \simeq \frac{u_1+u_3-2u_0}{h^2} \text{ and } \frac{\partial^2 u}{\partial y^2} \simeq \frac{u_2+u_4-2u_0}{h^2}$$

If (7.5.6) is satisfied at *A*, then we have

$$\frac{u_1+u_3-2u_0}{h^2}+\frac{u_2+u_4-2u_0}{h^2} \simeq 0$$

$$\Rightarrow \quad u_1+u_2+u_3+u_4-4u_0 \simeq 0\ 1$$

Let r_0 be the residual (discrepancy) at the mesh point *A*, then

$$r_0 = u_1+u_2+u_3+u_4-4u_0 \tag{7.5.7}$$

Similarly the residual at the point *B*, is given by

$$r_1 = u_0+u_5+u_6+u_7-4u_1 \tag{7.5.8}$$

and so on. The main aim of the relaxation process is to reduce all the residuals to zero, by making them as small as possible step by step. We, therefore, try to adjust the value of *u* at an internal mush point, so as to make the residual threat zero. When the value of u is changing at a mesh point, the values of the residuals at the neighbouring interior points will also be changed. If u_0 is given an increment 1, then (i) (7.5.7) shows that r_0 is changed by -4 and (ii) (7.5.8) shows that r_1 is changed by 1.The relaxation pattern is shown in figure 7.5.7. In general equation (7.5.5) of Gauss –Seidal method can be written as

$$u_{i,j}^{(n+1)} = u_{i,j}^{(n)} + \frac{1}{4}\left[u_{i-1,j}^{(n+1)} + u_{i+1,j}^{(n)} + u_{i,j-1}^{(n)} + u_{i,j+1}^{(n)} - 4u_{i,j}^{(n)}\right] = u_{i,j}^{(n)} + \frac{1}{4}R_{i,j}$$

which shows that $\frac{1}{4}R_{i,j}$ is the change in the value of $u_{i,j}$ for one Gauss –Seidal iteration. In successive over relaxation method, large changes than this is given to $u_{i,j}^{(n)}$ and the iteration formula is written as

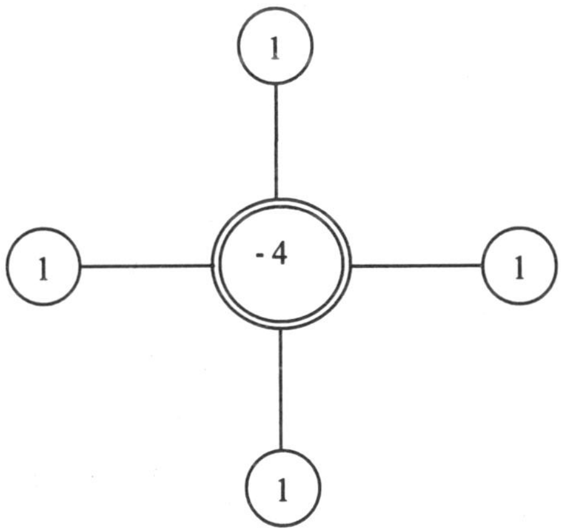

Figure-7.5.7

$$u_{i,j}^{(n+1)} = u_{i,j}^{(n)} + \frac{1}{4}wR_{i,j} = \frac{1}{4}w\left[u_{i-1,j}^{(n+1)} + u_{i+1,j}^{(n)} + u_{i,j-1}^{(n+1)} + u_{i,j+1}^{(n)}\right] + (1-w)u_{i,j}^{(n)} \tag{7.5.9}$$

The rate of convergence of (7.5.9) depends on the choice of w, which is called the accelerating factor and lies between 1 and 2. It was shown by *B.A.* Carrie that $w = 1.875$ the rate of convergence of (7.5.9) is twice as fast as that when w=1 and for $w = 19$ the rate of convergence is 40 times greater than that for $w = 1$. However in general, it is difficult to estimate the best value of w.

Algorithm 7.5.1: To solve an equation by relaxation method

1. Write down by trial, the initial values of u at the interior mesh points by diagonal - averaging or cross averaging.

2. Calculate the residuals at each of their points by (7.5.7) above. The application of this formula at a point near the boundary, chopped off one or more end points, since there is no residuals at boundary.

3. Write the residuals at a mesh point on the right of this point and the value of u on its left.

4. Obtain the solution by reducing the residuals to zero one by one, by giving suitable increment to u and using figure 7.5.7. At each step, we reduce numerically largest residual to zero and record the increment of u in the left and the modified residual on the right.

5. When a round of relaxation is completed, the value of u and its increments added at each point. Using these values, calculate all the residuals afresh. If some of the recalculated residuals are large, we liquidate these again.

6. Stop the relaxation process, when the current values of the residuals are quite small. The current value of u at each of the nodes gives us the solution.

This method combines simplicity with the speed of convergence but has its only drawback that it is unsuitable for computer calculations.

Example *7.5.3* Solve by relaxation method, the Laplace equation

$$\frac{\partial^2 u}{\partial x^2}+\frac{\partial^2 u}{\partial y^2}=0$$

inside the square bounded by $x=0, x=4, y=0, y=4,$ given that $u=x^2y^2$ on the boundary.

Solution: We take $h=1$ and find u on the boundary from $u=x^2y^2$. The initial values of u at the nine mesh-points are estimated to be 24, 56, 104, 16, 32, 56, 8, 16, and 24, as shown on the left of each point in figure 7.5.8 below.

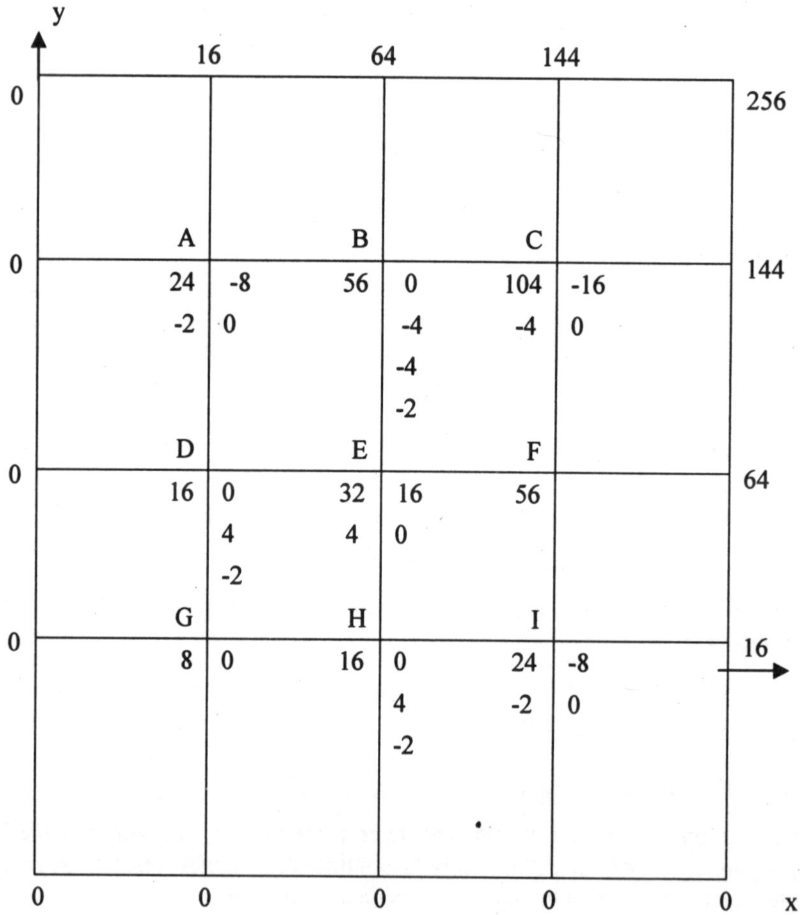

Figure 7.5.8

Therefore, residual at A i.e $r_A=0+56+16+16-4\times 24=-8$

Similarly $r_B=0, r_C=-16$ $, r_D=0,\ r_E=16\ ,\ r_H=0,\ r_I=-8\ , r_F=0,\ r_G=0,$

i The numerically largest residual is $r_E = 16$.To liquidate it, we increase *u* by 4 so that the residual becomes zero and the residual at neighbouring nodes get increased by 4.

ii The next, numerically largest-residual is $r_C = -16$. To reduce it to zero, we increase *u* by – 4 so that the residual at the adjacent nodes are increased by –4.

iii Again, the numerically largest residual is $r_A = -8$. To reduce it to zero, we increase *u* by -2 so that the residuals at the adjacent nodes are increased by –2

iv Finally, the largest residual is $r_I = -8$ and to liquidate it we increase *u* by –2 so that the residuals at the adjacent points are increased by –2.

v The numerically largest current residual being 2, we stop the relaxation process. Thus the final values of *u* are

$$u_A = 22, \quad u_B = 56, \quad u_C = 100$$

$$u_D = 16, \quad u_E = 36, \quad u_F = 56$$

$$u_G = 8, \quad u_H = 16, \quad u_I = 22$$

which is the required solution.

Example 7.5.4 Solve the equation $\frac{\partial^2 u}{\partial x^2} + \frac{\partial^2 u}{\partial y^2} = 0$ in the domain of figure 7.5.9, below by (a) Jacobi's method, (b) Gauss-Seidal method and (c) S.O.R. Method.

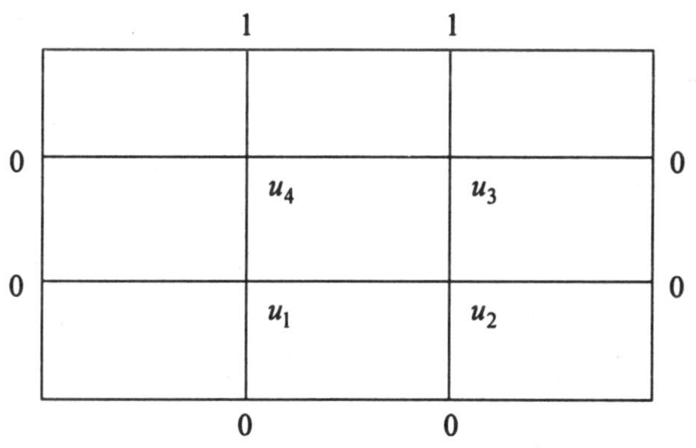

Figure 7.5.9

Solution: Jacobi's method

To start Jacobi's process, we obtain the approximate values of u_1, u_2, u_3 and u_4 as

$$u_1^{(1)} = \frac{1}{4}(0+0+0+1) = 0.25, \quad u_2^{(1)} = \frac{1}{4}(0+0+0+1) = 0.25$$

$$u_3^{(1)} = \frac{1}{4}(1+1+0+0) = 0.5, \quad u_4^{(1)} = \frac{1}{4}(1+1+0+0) = 0.5$$

The iteration formula has been continued by using the formula. (7.5.4) and the results are givn below.

Iteration (n)	u_1	u_2	u_3	u_4
1	0.1875	0.1875	0.4375	0.4375
2	0.15625	0.15625	0.40625	0.40625
3	0.14062	0.14062	0.39062	0.39062
4	0.13281	0.13281	0.38281	0.38281
5	0.12891	0.12891	0.37891	0.37891
6	0.12695	0.12695	0.37695	0.37695
7	0.12598	0.12598	0.37598	0.37598

Gauss–Seidal method

Using the iteration formula (7.5.5) the five successive iterations are given below.

Iteration (n)	u_1	u_2	u_3	u_4
1	0.25	0.3125	0.5625	0.46875
2	0.21875	0.17187	0.42187	0.39844
3	0.14844	0.13672	0.38672	0.38086
4	0.13086	0.12793	0.37793	0.37646
5	0.12646	0.12573	0.37573	0.37537

Relaxation (SOR) method

With $w = 1.1$, three successive iterated values obtained by using the formula (7.5.9) are given below.

Iteration (n)	u_1	u_2	u_3	u_4
1	0.275	0.35062	0.35062	0.35062
2	0.16534	0.10683	0.38183	0.37432
3	0.11785	0.12181	0.37216	0.37341

The ADI method

This is an efficient method for the numerical solution of elliptic partial differential equations and was proposed by Peaceman and Rachford. We demonstrate its applicability with reference to the Laplace equation two in dimensions. We consider Laplace's equation in two dimensions

$$\frac{\partial^2 u}{\partial x^2} + \frac{\partial^2 u}{\partial y^2} = 0 \tag{7.5.10}$$

and the standard five-point formula

$$u_{i-1,j} + u_{i+1,j} + u_{i,j-1} + u_{i,j+1} - 4u_{i,j} = 0 \tag{7.5.11}$$

The use of formula (7.5.11) involves the solution of a system of algebraic equations, whose coefficients matrix, for $n = 6$, is of the form.

$$A = \begin{bmatrix} -4 & 1 & 0 & 1 & 0 & 0 \\ 1 & -4 & 1 & 0 & 1 & 0 \\ 0 & 1 & -4 & 0 & 0 & 1 \\ 1 & 0 & 0 & -4 & 1 & 0 \\ 0 & 1 & 0 & 1 & -4 & 1 \\ 0 & 0 & 1 & 0 & 1 & -4 \end{bmatrix} \tag{7.5.12}$$

The general form of such a system is given by

$$B = \begin{bmatrix} T & 1 & & & & \\ & & & 0 & & \\ 1 & T & 1 & & & \\ \cdot & \cdot & \cdot & \cdot & \cdot & \cdot \\ & & 1 & T & & 1 \\ & 0 & & 1 & & T \end{bmatrix} \tag{7.5.13}$$

where T is a tridiagonal matrix of the form

$$T = \begin{bmatrix} -4 & 1 & & & & \\ 1 & -4 & 1 & O & & \\ \cdot & \cdot & \cdot & \cdot & \cdot & \cdot \\ & O & & 1 & -4 & 1 \\ & & & & 1 & -4 \end{bmatrix} \tag{7.5.14}$$

System A is called a block tridiagonal system and such systems can be solved by elimination or Gauss–Seidal methods. Peaceman and Rachford showed that it is possible to obtain directly tridiagonal systems in numerical solution of Laplace equation, which are much easier to solve than back tridiagonal systems. Their method of procedure, called the alternating direction implicit–method (**ADI** method) is described below. We rearrange formula (7.5.11) in either of two the ways:

$$u_{i-1,j} - 4u_{i,j} + u_{i+1,j} = -u_{i,j-1} - u_{i,j+1} \tag{7.5.15}$$

and

$$u_{i,j-1,} - 4u_{i,j} + u_{i,j+1} = -u_{i-1,j} - u_{i+1,j} \tag{7.5.16}$$

The ADI is an iteration method and formula (7.5.15) and (7.5.16) are used as iteration formulae. Therefore we have

$$u_{i-1,j}^{(r+1)} - 4u_{i,j}^{(r+1)} + u_{i+1,j}^{(r+1)} = -u_{i,j-1}^{(r)} - u_{i,j+1}^{(r)} \tag{7.5.17}$$

and

$$u_{i,j-1}^{(r+2)} - 4u_{i,j}^{(r+2)} + u_{i,j+1}^{(r+2)} = -u_{i-1,j}^{(r+1)} - u_{i+1,j}^{(r+1)} \tag{7.5.18}$$

The formulae (7.5.17) and (7.5.18) are used to compute function values at all interval mesh points along rows and columns respectively. For $j = 1,2,3,\ldots(n-1)$, equation (7.5.17) yields a tridiagonal system of equations and equation (7.5.17) so does for $i = 1,2,3,\ldots,(n-1)$. In ADI method, formulae (7.5.17) and (7.5.18) are used alternatively. For example, for the first row $j = 1$, formula (7.5.17) gives

$$u_{i-1,1}^{(r+1)} - 4u_{i,1}^{(r+1)} + u_{i+1,1}^{(r+1)} = -u_{i,0}^{(r)} - u_{i,2}^{(r)},\ i = 1,2,3,\ldots(n-1) \qquad (7.5.19)$$

Equations (7.5.19) along with boundary conditions represent a tridiagonal system of equations and are easily solved for $u_{i,1}^{(r+1)}$. We next put j=2 and obtain $u_{i,2}^{(r+1)}$ on the second row. The process is repeated for all the rows viz up to $j = n-1$. Next, we alternate the direction i.e. we use formula (7.5.18) to compute $u_{i,j}^{(r+2)}$. It is easy to see that at every stage we are encountered to solve a tridiagonal system of equations.

Example: 7.5.5 Solve Laplace equation $\frac{\partial^2 u}{\partial x^2} + \frac{\partial^2 u}{\partial y^2}$ =0 in the domain of figure 7.5.10

Solution: We apply formulae (7.5.17) and (7.5.18) after relabelling the interval mesh points as in figure 7.5.10.

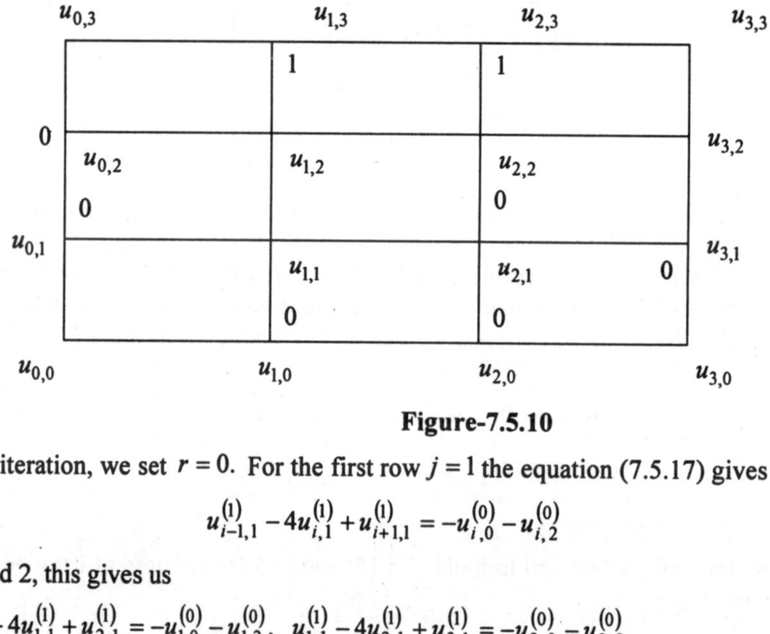

Figure-7.5.10

To start the iteration, we set $r = 0$. For the first row $j = 1$ the equation (7.5.17) gives us

$$u_{i-1,1}^{(1)} - 4u_{i,1}^{(1)} + u_{i+1,1}^{(1)} = -u_{i,0}^{(0)} - u_{i,2}^{(0)}$$

with $i = 1$ and 2, this gives us

$$u_{0,1}^{(1)} - 4u_{1,1}^{(1)} + u_{2,1}^{(1)} = -u_{1,0}^{(0)} - u_{1,2}^{(0)},\ u_{1,1}^{(1)} - 4u_{2,1}^{(1)} + u_{3,1}^{(1)} = -u_{2,0}^{(0)} - u_{2,2}^{(0)}$$

Using boundary conditions and assuming that $u_{1,2}^{(0)} = 1$ and $u_{1,2}^{(0)} = 1$, the above equations yield

$$u_{1,1}^{(0)} = u_{2,1}^{(0)} = \frac{1}{3} = 0.3333.$$

Again for $j = 2$, equation (7.5.17) leads to

$$u_{i-1,2}^{(1)} 4u_{i,2}^{(1)} + u_{(i+1,2)}^{(1)} = -u_{i,1}^{(0)} - u_{i,3}^{(0)}$$

with $i=1$ and 2,this equation gives us

$$u_{0,2}^{(1)} - 4u_{1,2}^{(1)} + u_{2,2}^{(1)} = -u_{1,1}^{(0)} - u_{1,3}^{(0)}$$

and

$$u_{1,2}^{(1)} - 4u_{2,2}^{(1)} + u_{3,2}^{(1)} = -u_{2,1}^{(0)} - u_{2,3}^{(0)}$$

Solving theses equation along with boundary conditions, we get

$$u_{1,2}^{(1)} = u_{2,2}^{(1)} = \frac{1}{3} = 0.3333.$$

Now, we alternate with direction and compute the function values on the columns, starting with the first one. For this, we use formula (7.5.18) with $r=0$. Setting $i=1$, formula (7.5.18) becomes

$$u_{1,j-1}^{(2)} - 4u_{1,j}^{(2)} + u_{1,j+1}^{(2)} = -u_{0,j}^{(1)} - u_{2,j}^{(1)}$$

Putting $j=1$ and 2 in the above equation and using the boundary values to solve the resulting equations, we get

$$u_{1,1}^{(2)} = \frac{8}{45} = 0.1778 \text{ and } u_{1,2}^{(2)} = \frac{17}{45} = 0.3778$$

To compute the value on the second column, we set $i=2$ in (7.5.18) to get

$$u_{2,j-1}^{(2)} - 4u_{2,j}^{(2)} + u_{2,j+1}^{(2)} = -u_{i,j}^{(1)} - u_{3,j}^{(1)}$$

Putting $j=1$ and 2 in the above equation and solving the resulting equations by using boundary equations, we get

$$u_{2,1}^{(2)} = 0.1778, \quad u_{2,2}^{(2)} = 0.3778$$

The iterations are continued to improve the function values obtained first on the rows, then on the columns, and so on.

7.6 Parabolic equations

7.6.1 Solution of one dimensional heat equation

Consider the one dimension heat equation

$$\frac{\partial u}{\partial t} = c^2 \frac{\partial^2 u}{\partial x^2} \tag{7.6.1}$$

where $c^2 = K/\rho C$ is the diffusivity of the substance $(cm^2/\sec)$. We can solve equation (7.6.1) by the following three methods.

(1) Schmidt method

(2) Crank- Nicholson method

(3) Iterative methods

Schmidt method

We consider a rectangular mesh in the $x-t$ plane with spacing h along x direction and k along time, 't' direction. Denoting a mesh point $(x, t) = (ih, jk)$ as simply i, j, we have

$$\frac{\partial u}{\partial t} = \frac{u_{i,j+1} - u_{i,j}}{k}, \quad \frac{\partial^2 u}{\partial x^2} = \frac{u_{i-1,j} - 2u_{i,j} + u_{i+1,j}}{h^2}$$

Using these in (7.6.1), we obtain

$$u_{i,j+1} - u_{i,j} = \frac{kc^2}{h^2}\left[u_{i-1,j} - 2u_{i,j} + u_{i+1,j}\right]$$

or $$u_{i,j+1} = ru_{i-1,j} + (1-2r)u_{i,j} + ru_{i+1,j} \quad (7.6.2)$$

where $r = kc^2/h^2$ is the mesh ratio parameter. The relation (7.6.2) is a relation between the function values at the two time levels $j+1$ and j and hence is called 2-level formula. This formula enables us to determine the value of u at the $(i, j+1)^{th}$ mesh point in is terms of the known function values at the points x_{i-1}, x_i and x_{i+1} at the instant t_j. The schematic form of (7.6.2) is shown in figure-7.6.1

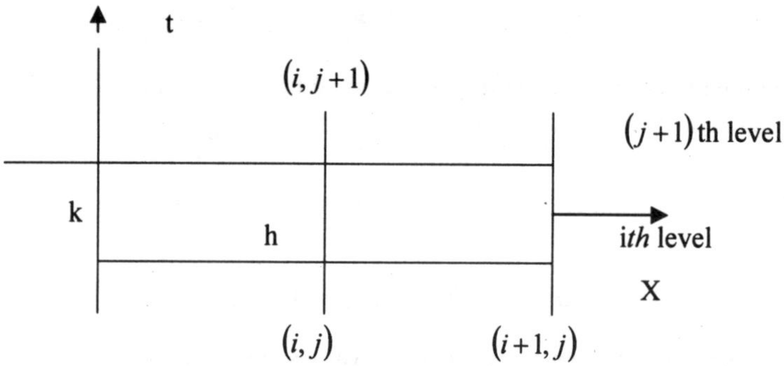

Figure-7.6.1

Hence (7.6.2.) is called the Schmidt explicit formula which is valid only for $0 < r \le \frac{1}{2}$. In particular when $r = \frac{1}{2}$, (7.6.2) reduces to $u_{i,j+1} = \frac{1}{2}(u_{i-1,j} + u_{i+1,j})$ (7.6.3)

which shows that the value of u at x_i at time t_{j+1} is the mean of the u-values at x_{i-1} and x_{i+1} time t_j. This relation (7.6.3), known as Bender-Schmidt recurrence relation, gives the values of u at the internal points with the help of boundary conditions.

Crank -Nicholson method

Crank and Nicholson proposed a method in 1947 according to which $\frac{\partial^2 u}{\partial x^2}$ is replaced by the average of the finite–difference approximations on the j_{th} and $(j+1)$ the rows. Thus

$$\frac{\partial^2 u}{\partial x^2} = \frac{1}{2}\left[\frac{u_{i-1,j} - 2u_{i,j} + u_{i+1,j}}{h^2} + \frac{u_{i-1,j+1} - 2u_{i,j+1} + u_{i+1,j+1}}{h^2}\right]$$

and hence equation (7.6.1) is replaced by

$$\frac{1}{k}\left[u_{i,j+1} - u_{i,j}\right] = \frac{c^2}{2h^2}\left[u_{i-1,j} - 2u_{i,j} + u_{i+1,j} + u_{i-1,j+1} - 2u_{i,j+1} + u_{i+1,j+1}\right]$$

which gives on re-terms the arranging

$$-ru_{i-1,j+1} + (2+2r)u_{i,j+1} - ru_{i+1,j+1} = ru_{i-1,j} + (2-2r)u_{i,j} + ru_{i+1,j} \tag{7.6.4}$$

where $r = kc^2/h^2$. On the left hand side of (7.6.4) we have three unknowns and on the right hand side all the three quantities are known. Equation (7.6.4), which is an implicit scheme, is called Crank–Nicholson formula and it is convergent for all finite values of r. If there are N internal mesh points on each row, then formula (7.6.4) gives *N* simultaneous equations for the N unknowns in terms of the given boundary values. Similarly, the internal mesh points on all rows can be calculated. The computational model of this method is given below in figure 7.6.2.

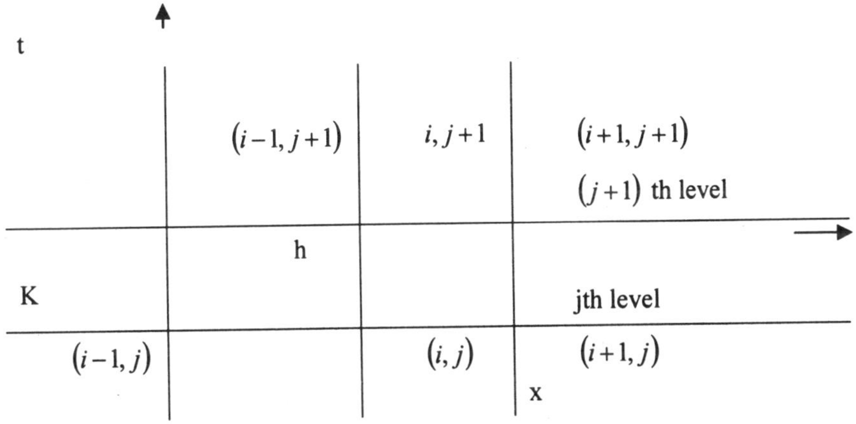

Figure-7.6.2

Iterative methods

The iterative methods discussed in section 7.5 can also be applied to solve the finite-difference equations obtained in the preceding section. In the Crank – Nicholson method the partial differential equation (7.6.1) is replaced by the finite difference equation

$$(1+r)u_{i,j+1} = u_{i,j} + \frac{1}{2}r\left(u_{i-1,j+1} + u_{i+1,j} + u_{i+1,j+1} + u_{i-1,j} - 2u_{i,j}\right)$$

where $r = kc^2/h^2$ In this equation, the unknowns are $u_{i,j+1}$, $u_{i-1,j+1}$ *and* $u_{i+1,j+1}$ and all others are known since they were already computed at the j_{th} step. Hence, dropping that j's and setting.

$$c_i = u_{i,j} + \frac{1}{2} r\left(u_{i-1,j} - 2u_{i,j} + u_{i+1,j}\right) \tag{7.6.5}$$

the above equation can be written as

$$u_i = \frac{r}{2(1+r)}\left[u_{i-1} + u_{i+1}\right] + \frac{c_i}{1+r} \tag{7.6.6}$$

From equation (7.6.6), we obtain the iteration formula

$$u_i^{(n+1)} = \frac{r}{2(1+r)}\left[u_{i-1}^{(n)} + u_{i+1}^{(n)}\right] + \frac{c_i}{1+r} \tag{7.6.7}$$

which expresses the $(n+1)^{th}$ iteration in terms of *n*th iteration and is known as *Jacobi's iteration* formula. It can be seen form equation (7.6.7) that at the line of computing $u_i^{(n+1)}$, the latest value of u_{i-1}, namely $u_{i-1}^{(n+1)}$, is already available. Hence, the convergence of Jacobi's iteration can be improved by replacing $u_{i-1,}^{(n+1)}$ in (7.6.7) by its latest values available, namely by $u_{i-1,}^{(n+1)}$. Accordingly, we obtain the formula.

$$u_i^{(n+1)} = \frac{r}{2(1+r)}\left[u_{i-1}^{(n+1)} + u_{i+1}^{(n)}\right] + \frac{c_i}{1+r} \tag{7.6.8}$$

which is called the Gauss–Seidal iteration formula. It can be shown that the scheme (7.6.8) converges for all finite values of *r* and that it converges twice as fast as Jacobi's scheme. This method is valid for all finite values of *r*. The equation (7.6.8) can be rewritten as

$$u_i^{(n+1)} = u_i^{(n)} + \left[\frac{r}{2(1+r)}\left\{u_{i-1}^{n+1} + u_{i+1}^{(n)}\right\} + \frac{c_i}{(1+r)} - u_i^{(n)}\right]$$

from which it is clear that expression with in the square brackets is the difference between the *n*th and $(n+1)$ iterations. If we take the difference to be *w* times this expression, then

$$u_i^{(n+1)} = u_i^{(n)} + w\left[\frac{r}{2(1+r)}\left\{u_{i-1}^{(n+1)} + u_{i+1}^{(n)}\right\} + \frac{c_i}{(1+r)} - u_i^{(n)}\right] \tag{7.6.9}$$

This is called the successive over–relaxation method or SOR method. Here w is called the relaxation factor and it generally, lies between 1 and 2.

DuFort and Frankel method

If we replace the derivatives in (7.6.1) by the central-difference approximations,

$$\frac{\partial u}{\partial t} = \left(u_{i,j+1} - u_{i,j-1}\right)/2k$$

and

$$\frac{\partial^2 u}{\partial x^2} = \left(u_{i-1,j} - 2u_{i,j} + u_{i+1,j}\right)/h^2$$

we obtain

$$u_{i,j+1} - u_{i,j-1} = \frac{2kc^2}{h^2}\left[u_{i-1,j} - 2u_{i,j} + u_{i+1,j}\right]$$

$$\Rightarrow \qquad u_{i,\,j+1} = u_{i,\,j-1} + 2r\left[u_{i-1,\,j} - 2u_{i,\,j} + u_{i+i,\,j}\right] \qquad (7.6.10)$$

where $r = \dfrac{kc^2}{h^2}$ This formula (7.6.10) is called the Richardson scheme and is a 3-level method. If we replace $u_{i,j}$ by the mean of the values $u_{i,j-1}$ and $u_{i,\,j+1}$ i.e. $u_{i,\,j} = \dfrac{1}{2}\left(u_{i,\,j-1} + u_{i,\,j+1}\right)$ in (7.6.10), we get

$$u_{i,\,j+1} = u_{i,\,j-1} + 2r\left[u_{i-1,\,j} - \left(u_{i,\,j-1} + u_{i,\,j+1}\right) + u_{i+1,\,j}\right]$$

which on simplification becomes

$$u_{i,\,j+1} = \frac{1-2r}{1+2r}u_{i,j-1} + \frac{2r}{2+2r}\left\{u_{i-1,\,j} + u_{i+1,\,j}\right\} \qquad (7.6.11)$$

This difference scheme is called DuFort–Frankel method that is a 3-level explicit method and its computational model is given in figure-7.6.3.

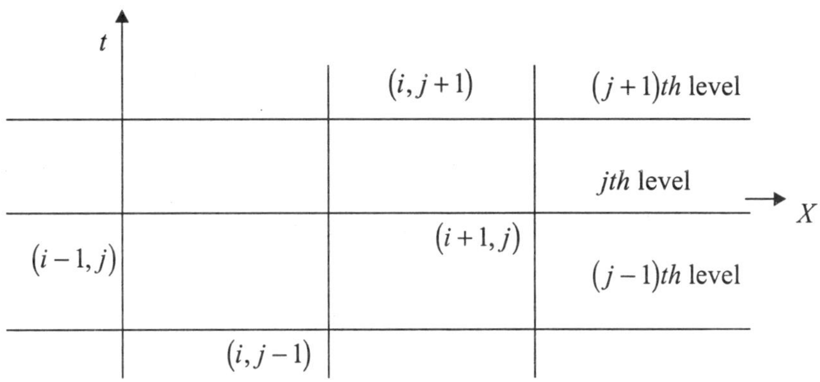

Figure-7.6.3

Example 7.6.1 Solve the partial differential Equation

$$\frac{\partial u}{\partial t} = \frac{\partial^2 u}{\partial x^2}$$

subject to the conditions $u(x,0) = \sin \pi x, 0 \le x \le 1;\ u(0,t) = u(1,t) = 0$, using (a) Schmidt method, (b)Crank–Nicholson method, (c) DuFort – Frankel method. Carry out computations for the two levels, taking $h = \dfrac{1}{3}, k = \dfrac{1}{36}$

Solution: Here $c^2 = 1,\ h = \dfrac{1}{3},\ k = \dfrac{1}{36}$ so that $r = \dfrac{kc^2}{h^2} = \dfrac{1}{4}$. Also

$$u_{1,0} = \sin\frac{\pi}{3} = \frac{\sqrt{3}}{2},\quad u_{2,0} = \frac{2\pi}{3} = \sin\frac{\sqrt{3}}{2}$$

and all boundary values are zero as shown in figure 7.6.4 below

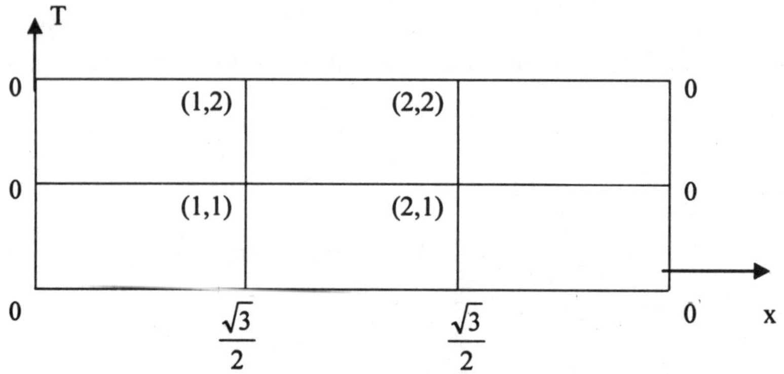

Figure-7.6.4

(a) Schmidt's method

The Schmidt's formula (7.6.2) in this case becomes

$$u_{i,j+1} = \frac{1}{4}\left(u_{i-1,j} + 2u_{i,j} + u_{i+1,j}\right)$$

For $i = 1,2,\ j = 0$:

$$u_{1,1} = \frac{1}{4}\left[0 + 2\frac{\sqrt{3}}{2} + \frac{\sqrt{3}}{2}\right] = 0.65$$

$$u_{2,1} = \frac{1}{4}\left[\frac{\sqrt{3}}{2} + 2\frac{\sqrt{3}}{2} + 0\right] = 0.65$$

For $i = 1,2,\ j = 1$:

$$u_{1,2} = \frac{1}{4}\left[0 + 2(0.65) + 0.65\right] = 0.49$$

$$u_{2,2} = \frac{1}{4}\left[0.65 + 2(0.65) + 0\right] = 0.49$$

(b) Crank–Nicholson method

The Crank-Nicholson formula (7.6.4) is becomes

$$-\frac{1}{4}u_{i-1,j+1} + \frac{5}{2}u_{i,j+1} - \frac{1}{4}u_{i+1,j+1} = \frac{1}{4}u_{i-1,j} + \frac{3}{2}u_{i,j} + \frac{1}{4}u_{i+1,j}$$

For $i = 1,2;\ j = 0$: $10u_{1,1} - u_{2,1} = 7\frac{\sqrt{3}}{2}$ and $-u_{1,1} + -10u_{2,1} = 7\frac{\sqrt{3}}{2}$

Solving these equations, we get

$$u_{1,1} = u_{2,1} = 0.67$$

For $i = 1,2,\ j = 1$: we have

$$10u_{1,2} - u_{2,2} = 4.69$$

$$-u_{1,2} + 10u_{2,2} = 4.69$$

Solving these equations, we obtain

$$u_{1,2} = u_{2,2} = 0.52$$

(c) DuFort-Frankel method

The Du Fort-Frankel formula (7.6.11) here becomes

$$u_{i,j+1} = \frac{1}{3}\left(u_{i,j-1} + u_{i-1,j} + u_{i+1,j}\right)$$

To start the computations we require $u_{1,1}$ and $u_{2,1}$. So we may take $u_{1,1} = u_{2,1} = 0.65$ form the Schmidt method. For $i = 1,2, j = 1$ $j = 1$, we get

$$u_{1,2} = \frac{1}{3}\left(\frac{\sqrt{3}}{2} + 0 + 0.65\right) = 0.5, \qquad u_{2,2} = \frac{1}{3}\left(\frac{\sqrt{3}}{2} + 0.65 + 0\right) = 0.5$$

Hence the required solution of the question

Example 7.6.2: Solve the partial differential equation $\frac{\partial u}{\partial t} = \frac{\partial^2 u}{\partial x^2}$ subject to the initial condition $u = \sin \pi x$, at $t = 0$ for $0 \le x \le 1$ and $u = 0$ at $x = 0$ and $x = 1$ for $t > 0$ by Gauss-Seidal method.

Solution: We choose h= 0.2 and $k = 0.02$ so that $r = \frac{k}{h^2} = \frac{1}{2}$. The formula (7.6.8.) therefore becomes

$$u_i^{(n+1)} = \frac{1}{6}\left[u_{i-1}^{(n+1)} + u_{i+1}^{(n)}\right] + \frac{2}{3}c_i \qquad (7.6.12)$$

Let the values of u at the interior mesh points on the row corresponding to $t = 0.02$ be u_1, u_2, u_3, u_4 as shown in figure 7.6.5.

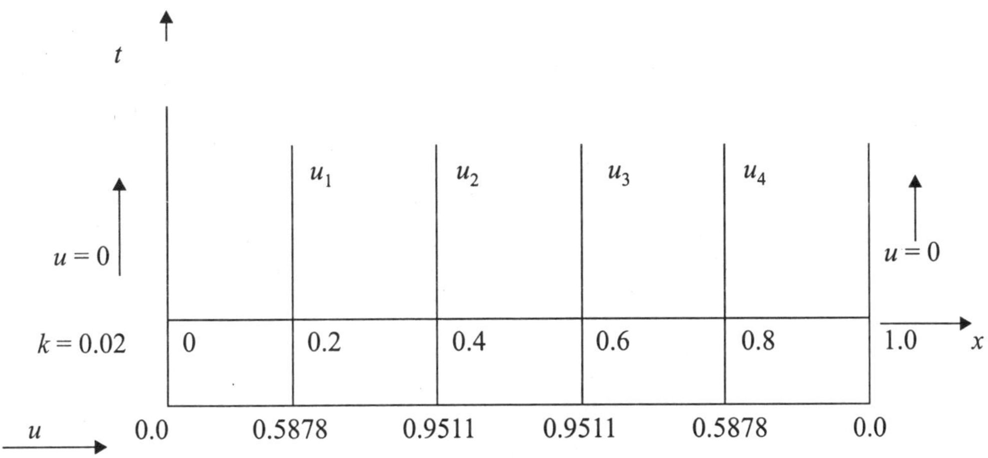

Figure-7.6.5

Using the formula (7.6.12), we obtain

$$u_1^{(n+1)} = \frac{1}{6} u_2^{(n)} + 0.3544 \qquad \text{(i)}$$

$$u_2^{(n+1)} = \frac{1}{6}\left[u_1^{(n+1)} + u_3^{(n)}\right] + 0.5736 \quad \text{(ii)}$$

$$u_3^{(n+1)} = \frac{1}{6}\left[u_2^{(n+1)} + u_4^{(n)}\right] + 0.5736 \quad \text{(iii)}$$

$$u_4^{(n+1)} = \frac{1}{6}\ u_3^{(n+1)} + 0.3544 \qquad \text{(iv)}$$

Equations (*i*) , (*ii*), (*iii*), and (*iv*) can now be used to find better approximations for $u_1, u_2, u_3,$ and u_4 respectively. The table below gives us the successive iterations of $u_1, u_2, u_3,$ and u_4 corresponding to t =0.02

X	0.0	0.2	0.4	0.6	0.8	1.0
$u(x)$	0.0	0.5878	0.9511	0.9511	0.5878	0.0
$n=0$	0.0	0.5878	0.9511	0.9511	0.5878	0.0
$n=1$	0.0	0.5129	0.8176	0.8078	0.4890	0.0
$n=2$	0.0	0.4907	0.7900	0.7868	0.4855	0.0
$n=3$	0.0	0.4861	0.7858	0.7855	0.4853	0.0
$n=4$	0.0	0.4854	0.7854	0.7854	0.4853	0.0
$n=5$	0.0	0.4853	0.7854	0.7854	0.4853	0.0

The symmetry of the solution about x= 0.5 is quite clear from the above table. The analytic solution of the problem is given by $u = e^{-\pi^2 t} \sin \pi x$ and the exact values of u for x=0.2 and x=0.4 are respectively 0.4825 and 0.7807. The percentage error in both the solutions is approximately 0.6% and this error can be reduced by taking a finer mesh.

7.7 Hyperbolic equations

The wave equation

$$\frac{\partial^2 u}{\partial t^2} = c^2 \frac{\partial^2 u}{\partial x^2} \qquad (7.7.1)$$

is the simplest example of hyperbolic partial differential equations. Its solution is the displacement function $u(x,t)$ defined for values of x from 0 to l and for t from 0 to ∞ , satisfying the initial and boundary conditions. Such equations arise from convective type of problems in vibrations, wave mechanics, gas dynamics, elasticity, electromagnetic and seismology.

We consider the boundary value problem defined by

$$\frac{\partial^2 u}{\partial t^2} = c^2 \frac{\partial^2 u}{\partial x^2}, \quad 0 \angle\ x\ \angle 1 \qquad (7.7.2)$$

subject to the conditions

$$\left.\begin{aligned} u(x,0) &= f(x) \\ \frac{\partial u}{\partial t}(x,0) &= g(x) \end{aligned}\right\}, \quad 0 \le x \le 1 \tag{7.7.3}$$

$$\left.\begin{aligned} u(0,t) &= \phi(t) \\ u(1,t) &= \phi(t) \end{aligned}\right\} \quad 0 \le t \le T \tag{7.7.4}$$

which models the transverse vibrations of a stretched string. As previously, we use the difference approximations for the derivatives

$$\frac{\partial^2 u}{\partial x^2} = \frac{1}{h^2}\left(u_{i-1,j} - 2u_{i,j} + u_{i+1,j}\right) + O\left(h^2\right) \tag{7.7.5}$$

$$\frac{\partial^2 u}{\partial t^2} = \frac{1}{k^2}\left(u_{i,j-1} - 2u_{i,j} + u_{i,j+1}\right) + O\left(k^2\right) \tag{7.7.6}$$

where $x = ih,\ \ i = 0,1,2,...$ and $t = jk,\ \ j = 0,1,2,3...$.

The derivative $\frac{\partial u}{\partial t}$ is approximated as

$$\frac{\partial u}{\partial t} = \frac{u_{i,j+1} - u_{i,j-1}}{2k} + O\left(k^2\right) \tag{7.7.7}$$

Using (7.7.3) and (7.7.4) in (7.7.2), we obtain

$$\frac{1}{k^2}\left(u_{i,j-1} - 2u_{i,j} + u_{i,j+1}\right) = \frac{c^2}{h^2}\left(u_{i-1,j} - 2u_{i,j} + u_{i+1,j}\right)$$

$$\Rightarrow u_{i,j+1} = -u_{i,j-1} + \alpha^2\left(u_{i-1,j} + u_{i+1,j}\right) + 2(1-\alpha^2)u_{i,j}$$

where $\alpha = \frac{ck}{h}$. The formula (7.7.8) shows that the function values at the jth and $(j-1)^{th}$ levels are required in order to determine those at the $(j+1)^{th}$ level. Such difference schemes are called three level difference schemes compared to the two level schemes derived in the parabolic equations case. By expanding the terms in (7.7.8) as Taylor's series and simplifying, it can be shown that the truncation error in (7.7.8) is $O\left(k^2 + h^2\right)$ and the formula (7.7.8) holds good if $\alpha < 1$, which is the condition for stability. There also exist implicit finite difference schemes for the equation (7.7.2). Two such schemes are

$$\frac{u_{i,j+1} - 2u_{i,j} + u_{i,j-1}}{k^2} = \frac{c^2}{2h^2}\left[u_{i+1,j+1} - 2u_{i,j+1} + u_{i-1,j+1}\right] + \left[u_{i+1,j-1} - 2u_{i,j-1} + u_{i-1,j-1}\right] \tag{7.7.9}$$

and

$$\frac{u_{i,j+1} - 2u_{i,j} + u_{i,j-1}}{k^2} = \frac{1}{4h^2}\left[\left(u_{i+1,j+1} - 2u_{i,j+1} + u_{i-1,j+1}\right)\right.$$

$$\left. + 2\left(u_{i+1,j} - 2u_{i,j} + u_{i-1,j}\right) + \left(u_{i,j-1} - 2u_{i,j-1} + u_{i-1,j-1}\right)\right] \tag{7.7.10}$$

The formulae (7.7.9) and (7.7.10) hold good for all values of $\frac{k}{h}$. The use of formula (7.7.8) is demonstrated in the following example.

Example 7.7.1 Solve the equation $\frac{\partial^2 u}{\partial t^2} = 16\frac{\partial^2 u}{\partial x^2}$, taking on $\Delta x = 1$ up to $t = 1.25$ and subject to the boundary and initial conditions $u(0, t) = u(5, t) = \frac{\partial u}{\partial t}(x, 0) = 0$ and $u(x, 0) = x^2(5 - x)$.

Solution: Here $c^2 = 16$, therefore the difference equation for the given equation is

$$u_{i,j+1} = 2\left(1-16\alpha^2\right)u_{i,j} + 16\alpha^2\left(u_{i-1,j} + u_{i+1,j}\right) - u_{i,j-1} \qquad (7.7.11)$$

where $\alpha \frac{k}{h}$. Taking h=1 and choosing k so that the coefficient of $u_{i,j}$ vanishes, we have

$$16\alpha^2 = 1 \Rightarrow \frac{k^2}{h^2} = \frac{1}{16} \Rightarrow k = \frac{h}{4} = \frac{1}{4}$$

Therefore (7.7.11) reduces to

$$u_{i,j+1} = u_{i-1,j} + u_{i+1,j} - u_{i,j-1} \qquad (7.7.12)$$

which gives us a convergent solution as $\frac{k}{h} < 1$.

Now since $u(0,t) = u(s,t) = 0$

$\therefore \quad u_{0,j} = 0$ and $u_{5,j} = 0$, for all j

Also $u(x,0) = x^2(5-x)$ implies that

$$u_i,0 = i^2(5-i) = 4,12,18,16 \text{ for } i=1,1,2,3,4 \text{ at } t=0$$

finally since $\frac{\partial u}{\partial E}(x,0) = 0$, therefore we have $\frac{u_{i,j} + 1 - u_{i,j}}{k} = 0$ for $j = 0$ *i.e.* $u_{i,0} = u_{i,0}$ which means the entries of the second sow are same those of the first row putting $j = 1$, in (7.7.12)we get $u_{i,2} = u_{i-1,1} + u_{i+1,1} - u_{i,0}$ Now taking $i = 1,2,3,4$ successively, we obtain

$u_{1,2} = 0 + 12 - 4 = 8, u_{2,2} = 4 = 18 - 12 = 10, u_{3,2} = 12 + 16 - 18 = 10, u_{4,2} = 18 + 0 - 16 = 2$.

Putting $j = 2,3,4$ successively in (7.7.12) the entries of the remaining forth, fifth and sixth row and obtained the value of the $u_{i,j}$ are given in the following table

i / j	0	1	2	3	4	5
0	0	4	12	18	16	0
1	0	4	12	18	16	0
2	0	8	10	10	2	0
3	0	6	6	-6	-6	0
4	0	-2	-10	-10	-8	0
5	0	-16	-18	-12	-4	0

Exercises

1. Solve the Laplace equation $u_{xx} + u_{yy} = 0$ for the square mesh with boundary value as shown below:

(i)

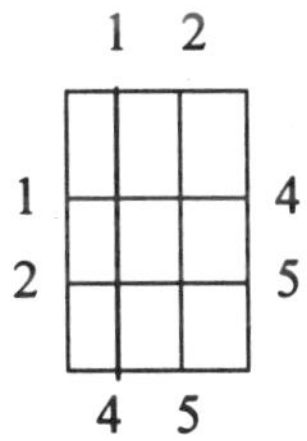

(ii)

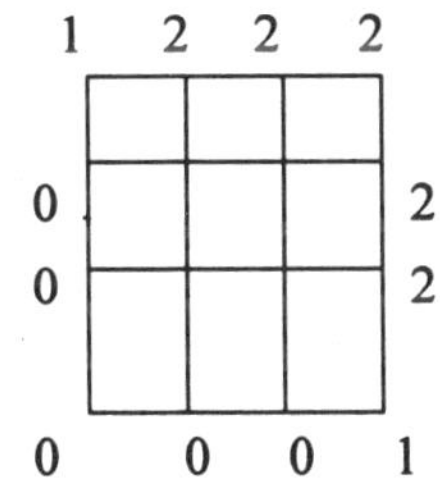

2. Use central difference approximation to solve $\nabla^2 u = 0$ at the nodal points of the square given bellow by using the boundary values indicated.

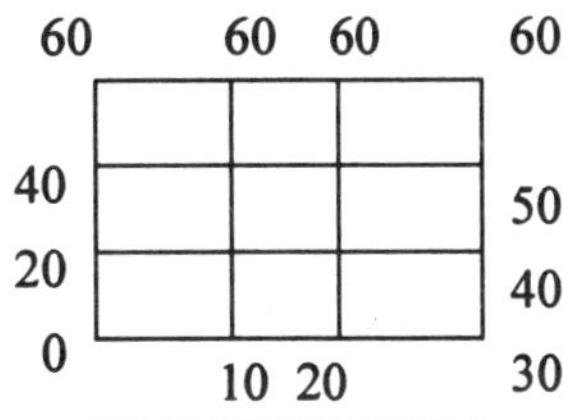

3. Solve $u_{xx} + u_{yy} = 0$ in the domain given below by (i) Jacobi's method (ii) Gauss-Seidal method

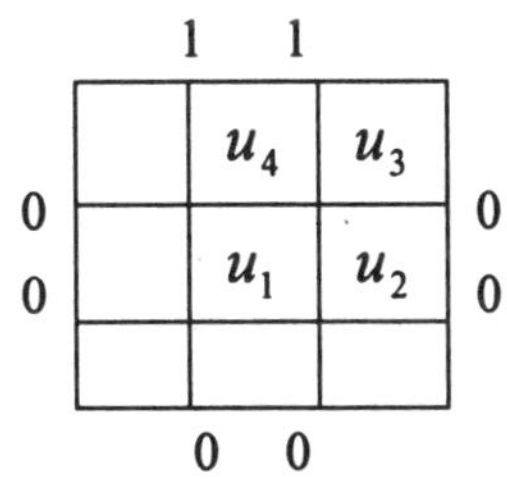

4. Solve the Laplace equation $\nabla^2 u = 0$ in the domain

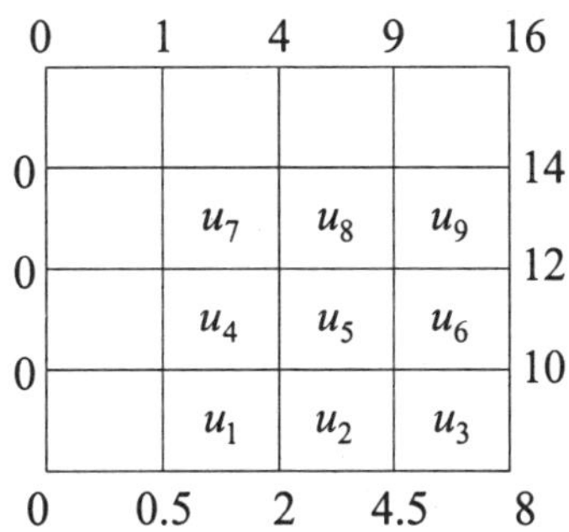

5. Solve the Poisson's equation $\nabla^2 u = 8x^2 y^2$ for the square mesh given below with $u(x, y) = 0$ on the boundary and mesh length=1

y ↑

$$\begin{array}{ccc} u_1 & u_2 & u_1 \\ u_2 & u_3 & u_2 \\ u_1 & u_2 & u_1 \end{array}$$

x →

6. Find the solution of the parabolic equation $u_{xx} = 2u_t$ given that $u(0,t) = u(4,t) = 0$ and $u(x,0) = x(4-x)$, taking $h = 1$. the value up to $t = 5$

7. Solve the equation $\frac{\partial^2 u}{\partial x^2} = \frac{\partial u}{\partial t}$ with $u(0,t) = 0$, $u(x,0) = x(1-x)$ and $u(1,t) = 0$ Assume $h = 0.1$ Tabulate u for $t = k$, $2k$ and $3k$ choosing an appropriate value of k.

8. Solve the boundary value problem $u_t = u_{xx}$ under the conditions $u(0,t) = u(1,t) = 0$ and $u(x,0) = \sin \pi x$, $0 \le x \le 1$ by Gauss-Seidel method. Take $h = 0.2$ and $\alpha = \frac{1}{2}$.

9. Solve the 2D heat equation $\frac{\partial u}{\partial t} = \frac{\partial^2 u}{\partial x^2} + \frac{\partial^2 u}{\partial y^2}$ satisfying conditions: $u(x, y, 0) = \sin \pi x \sin \pi y$, $0 \le x,\ y \le 1$ and boundary conditions: $u = 0$ at $x = 0$ and $x = 1$ for $t > 0$. Obtain the solution up to two time levels with $h = \frac{1}{3}$ and $\alpha = \frac{1}{8}$

10. Solve $u_{tt} = u_{xx}$ given that $u(0,t) = u(1,t) = 0$, $u(x,0) = \frac{1}{2}x(1-x)$ and $u(x,0) = 0$ taking $h = k = 0.1$ for $0 \le t \le 0.4$. Compare your solution with the exact solution at $x = 0.5$ and $t = 0.3$

11. The transverse displacement u of a point at a distance x from one end and at any time t of a vibrating string satisfies the equation $u_{tt} = 4u_{xx}$ with $u = 0$ at $x = 0, t > 0$, $u = 0$ at $x = 4, t > 0$ and the initial conditions $u = x(4-x)$ and $\frac{\partial u}{\partial t} = 0$ at $t = 0$, $0 \le x \le 4$. Solve this equation numerically for $0 \le t \le 2$, taking $h = 1$ and $k = 0.5$.

12. The function x satisfying the equation $\frac{\partial^2 u}{\partial t^2} = \frac{\partial^2 u}{\partial x^2}$ and the conditions : $u(x,0) = \frac{1}{8}\sin \pi x$, $u_t(x,0) = 0$, for $0 \le x \le 1$, $u(0,t) = x(1,t) = 0$ for $t \ge 0$. Use the explicit scheme to calculate u for $x = 0(0.1)1$ and $t = 0(0.1)0.5$.

13. Solve $u_{tt} = u_{xx}, 0 < x < 1$ subject to the conditions $u(0,t) = 0, u(1,t) = 0, \frac{\partial u}{\partial t}(x,0) = 0$ and $u(x,0) = \sin \pi x$. compare your result with the exact solution given by $u(x,t) = \sin \pi x \cos \pi t$.

14. Solve $u_{tt} = u_{xx}$ with boundary condition $u(0,t) = 0, u(1,t) =, \frac{\partial u}{\partial t}(x,0) = 0$ and $u(x,0) = x - x^2$. Take $h = 0.25$ and $k = 0.2$

15. Solve the heat conduction equation subject to the boundary conditions $u(0,t) = u(1,t) = 0$ and $u(x,0) = x - x^2$. Take $h = \frac{1}{4}$ and $k = 0.025$.

16. Solve the Laplace equation $\nabla^2 u = 0$ at all points within the square mesh given below by ADI method

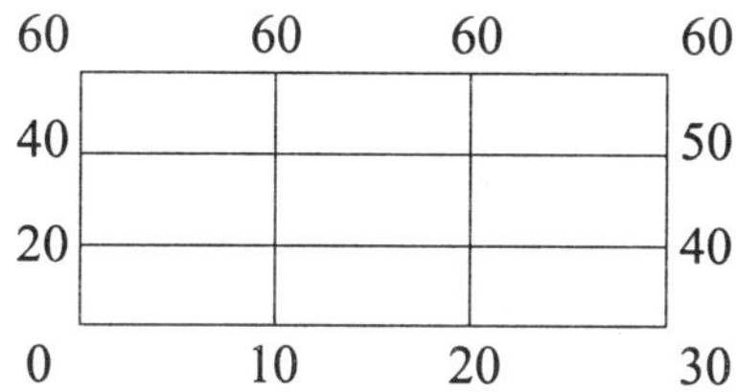

17. Solve the partial differential equation $u_{xx} + u_{yy} = -10(x^2 + y^2 + 10)$ over a square region with sides $x = 0, y = 0, x = 3, y = 3$ taking the boundary condition $u = 0$ on all the boundary sides and mesh size $\Delta x = \Delta y = 1$.

CHAPTER 8

INTEGRAL EQUATIONS

8.1 Introduction

An equation involving unknown function under the integral sign is known as an integral equation. Integral equation may involve any number of variables. If the unknown function appears linearly i.e. in first degree in the integral equation, the equation is called linear integral equation. The most general form of linear integral equation is

$$h(s)g(s) = f(s)+\lambda\int_a^b k(s,t)g(t)\,dt \tag{8.1.1}$$

where $h(s), f(s), \mathrm{k(s,t)}$ are given functions and $g(s)$ is the unknown function. The limits of integration *a*, *b* may be finite or infinite, constants or variables.

8.2 Classification of integral equations

8.2.1 Fredholm integral equations

If the limits a and b are constants, the linear integral equation is called Fredholm integral equation and (8.1.1) can be classified as follows:

(i) Fredholm Integral equation of first kind

In this case $h(s)=0$ and the integral equation (8.1.1) has the form

$$f(s)+\lambda\int_a^b k(s,t)g(t)dt = 0 \tag{8.2.1}$$

(ii) Inhomogeneous Fredholm Integral equation of Second kind

In this case $h(s)=1$ and $f(s)\neq 0$, for all $s\in[a,b]$ and from (8.1.1), we have

$$g(s)=f(s)+\lambda\int_a^b k(s,t)g(t)dt \tag{8.2.2}$$

(iii) Homogeneous Fredholm integral equation of second kind

In this case $h(s)=1, \mathrm{f(s)}=0,$ for all $s\in[a,b]$ and from equation (8.1.1) we have

$$g(s)=\lambda\int_a^b k(s,t)g(t)dt \tag{8.2.3}$$

8.2.2 Voltera integral equations

The integral equation (8.1.1) is said to be Voltera integral equation if the upper limit of the integral (i.e. b) is a function of s. In this case the integral equation (8.1.1) is classified as follows:

(a) Voltera Integral equation of first kind

In this case $h(s) = 0$ and the equation (8.1.1.) takes the form

$$f(s) + \lambda \int_a^b k(s,t)\ g(t)\ dt\ = 0 \tag{8.2.4}$$

(b) In homogenous Voltera equation of second kind

In this case $h(s) = 1, f(s) \neq 0,$ for $s \in [a,b]$ and the equation (8.1.1) becomes

$$g(s) = f(s) + \lambda \int_a^b k(s,t) g(t) dt \tag{8.2.5}$$

(c) Homogenous Voltera equation of second kind

In this case $h(s) = 1, f(s) = 0,$ for all $s \in [a,b]$ and equation (8.1.1) reduces to

$$f(s) + \lambda \int_a^b k(s,t) g(t) dt = 0 \tag{8.2.6}$$

The parameter λ in equation (8.2.1) to (8.1.6) may be real or complex and is independent of the variables and t. The function $k(s,t)$ occurring in this equation is called kernel of the integral equations.

Definition 8.2.1 The kernel $k(s,t)$ is said to be symmetric of $k(s,t) = k(t,s)$ and the corresponding integral equation is called symmetric integral equation.

Definition 8.2.2 The kernel $k(s,t)$ is said to be degenerate (or separable) kernel, if it can be written as

$$k(s,t) = \sum_{i=1}^{n} a_i(s) b_i(t) \tag{8.2.7}$$

Definition 8.2.3 An integral equation is said to be singular if (i) either one or both of the limits a and b become infinite or (ii) the kernel $k(s,t)$ becomes infinite at certain points of the interval $[a,b]$ i.e. for s and $t \in [a,b]$.

Theorem 8.2.1: The Non-homogenous Fredholm integral equation of second kind in case of separable kernel can be reduced to a system of linear algebraic equations. Hence, such an equation may have a unique solution or an infinite number of solutions.

Proof: Consider the Fredholm integral equation of second kind

$$g(s) = f(s) + \lambda \int_a^b k(s,t)\ g(t)\ dt \tag{8.2.8}$$

where $s, t \in [a,b]$ and a, b are constants. Let for any finite n, the kernel $k(s,t)$ be

$$k(s,t)=\sum_{i=1}^{n} a_i(s)b_i(t) \tag{8.2.9}$$

Upon using (8.2.9) in (8.2.8), we get

$$g(s)=f(s)+\lambda\int_a^b\sum_{i=1}^{n} a_i(s)b_i(t)g(t)dt$$

$$=f(s)+\lambda\sum_{i=1}^{n}[a_i(s)]\int_a^b b_i(t)g(t)dt]$$

$$=f(s)+\lambda\sum_{i=1}^{n} c_i a_i(s) \tag{8.2.10}$$

where $c_i = \int_a^b b_i(t)g(t)dt$, is an unknown constant. Multiplying $(8.2.10)$ by $b_i(s)$ and integrating with respect to s from $s=a$ to $s=b$, we get

$$\int_a^b g(s)b_1(s)ds$$

$$=\int_a^b b_i(s)f(s)ds+\lambda\sum_{j-1}^{n} c_j \int_a^b a_j(s)b_i(s)ds.$$

This implies that

$$c_i = f_i + \lambda\sum_{d=1}^{n} a_{ij}c_j \tag{8.2.11}$$

where $f_i = \int_a^b b_i(s)f(s)ds$, and $a_{ij} = \int_a^b b_i(s)a_j(s)ds$ are unknown constants.

The equation $(8.2.11)$ can be rewritten as

$$c_i - \sum_{j=1}^{n} a_{ij}c_j = f_i \tag{8.2.12}$$

This is a system of *n*-linear algebraic equations in n- unknown constants $c_1, c_2, c_3, \ldots c_n$. Solving equations $(8.2.12)$ for $c_1, c_2, c_3, \ldots c_n$ and substituting in $(8.2.10)$, we get the required solution. The system $(8.2.12)$ can be written as

$$\begin{aligned}
(1-\lambda a_{11})c_1 - \lambda a_{12}c_2 - \lambda a_{13}c_3 - \cdots - \lambda a_{1n}c_n &= f_1 \\
-\lambda a_{21}c_1 + (1-\lambda a_{22})c_2 - \lambda a_{23}c_3 - \cdots - \lambda a_{2n}c_n &= f_2 \\
\vdots \qquad \vdots \qquad \vdots \qquad \vdots \quad &\vdots \\
-\lambda a_{n1}c_1 - \lambda a_{n2}c_2 - \lambda a_{n3}c_3 - \cdots + (1-\lambda a_{nn})c_n &= f_n
\end{aligned}$$

or

$$AC = F \tag{8.2.13}$$

where

$$A_{nxn} = (\delta_{ij} - \lambda a_{ij}),\ C = \begin{bmatrix} c_1 \\ c_2 \\ c_3 \\ \vdots \\ c_n \end{bmatrix},\ F = \begin{bmatrix} f_1 \\ f_2 \\ f_3 \\ \vdots \\ f_n \end{bmatrix} \text{ and } \delta_{ij} = \begin{cases} 1, i = j \\ 0, i \neq j \end{cases} \tag{8.2.14}$$

Equation (8.2.8) will have a unique solution if the equation (8.2.12) or (8.2.13) has a unique solution. Therefore, the condition for unique solution is that $|A| \neq 0$.

$$D(\lambda) = |A| = \det(\delta_{ij} - \lambda a_{ij}) \tag{8.2.15}$$

The values of λ for which $D(\lambda) = 0$, the integral equation (8.2.8) either does not possess a solution or has infinitely many solutions depending upon the rank of matrix A. These values of λ are called eigenvalues of the integral equation (8.2.8). This completes the proof of the theorem.

Theorem 8.2.2 : An homogeneous Fredholm integral equation of second kind with a separable kernel can be reduced to a system of linear homogeneous set of algebraic equations.

Proof: Consider a homogeneous Fredholm integral equation of second kind

$$g(s) = \lambda \int_a^b k(s,t) g(t) dt \tag{8.2.16}$$

Since the Kernel is separable, so

$$k(s,t) = \sum_{i=1}^{n} a_i(s) b_1(t),\quad n \in N$$

Therefore,

$$g(s) = \lambda \int_a^b \sum_{i=1}^{n} a_i(s) b_i(t) g(t)\ \mathrm{dt} = \lambda \sum_{i=1}^{n} a_i(s) c_i \tag{8.2.17}$$

where $c_i = \int_a^b b_i(t) g(t) dt$ is a known constant. Multiplying (8.2.17) with $b_i(s)$ and integrating w.r.t, s from a to b we get

$$c_i = \int_a^b b_i(s)g(s)ds = \lambda\sum_{j=1}^{n} c_j \int_a^b a_j(s)b_i(s)ds$$

$$= \lambda\sum_{j=1}^{n} c_j a_{ij},$$

where $= \int_a^b b_i(s)a_i(s)ds$. Therefore,

$$c_i - \lambda\sum_{j=1}^{n} a_j c_j = 0, i = 1,2,3,\ldots\ldots.n \tag{8.2.18}$$

is the required system of n homogenous linear algebraic equations. In the matrix form (8.2.18) can be written as

$$AC = 0 \tag{8.2.19}$$

where

$$A = \begin{bmatrix} 1-\lambda a_{11} - \lambda a_{12} - \cdots - \lambda a_{1n} \\ \lambda a_{11} - \lambda a_{12} - \cdots \quad\quad - \lambda a_{1n} \\ \vdots \\ -\lambda a_{n1} \quad -\lambda a_{n2} - \cdots \quad -\lambda a_{nn} \end{bmatrix}, C = \begin{bmatrix} c_1 \\ c_2 \\ \vdots \\ c_n \end{bmatrix} \tag{8.2.20}$$

The system of equations (8.2.19) will have a non-trivial solution only if

$$|\ A\ | = 0$$

i. e. determinant of $D(\lambda) = 0$

It rank of $A = r \le n$, then (8.2.19) has $n - r$ linearly imported solutions corresponding to each set of solutions of (8.2.19) we get a solution of integral equation (8.2.16) by putting the values of

$$c_i, \text{i} = 1,2,3,\ldots,\text{n}, \text{in } \text{g}(\text{s}) = \lambda\sum_{i=1}^{n} a_i(s)c_i.$$

This completes the proof of the theorem. Thus we have seen that for non-homogeneous equations λ is a numerical parameter where as for homogeneous equations it is an eigenvalue parameter because in such a case the integral equation presents an eigenvalue problem in which the objective is to determine those values of λ, called the eigenvalues for which the integral equation possesses non-trivial solution called eigen functions. To solve an integral equation of any type is to find the unknown function satisfying the equation. In most cases, however, the solution of an integral equation by analytic techniques is not possible, and a straight forward numerical approach is to replace the integral equation by a set of linear algebraic equations and solve them by any of the methods described in chapter 3.

Fredholm integral equations, especially of the second kind, occur quite frequently in practice and hence to a brief treatment of these equations we discuss some of the recent methods for their numerical solutions in the following sections.

8.3 Picard's method of successive approximations

8.3.1 Fredholm integral equations

Consider the Fredholm integral of equation second kind

$$g(x) = f(x) + \lambda \int_a^b k(x,t) g(t) dt \tag{8.3.1}$$

Define a sequence of continuous functions $\{g_m(x)\}$, $m = 0,1,2,.........$ and $x \in [a,b]$ such that

$$g_0(x) = f(x) \text{ and } \quad g_m(x) = f(x) + \lambda \int_a^b k(x,t) g_{m-1}(t) dt$$

Therefore, $g_1(x) = f(x) + \lambda \int_a^b k(x,t) g_0(t) dt = g_0 + \lambda \int_a^b k(x,t) f(t) dt$

$$g_2(x) = g_0 + \lambda \int_a^b k(x,t) g_1(t) dt = g_0 + \lambda \int_a^b k(x,t) \left\{ g_0 + \lambda \int_a^b k(t,s) f(s) ds \right\} dt$$

$$= g_0 + \lambda \int_a^b k(x,t) g_0(t) dt + \lambda^2 \int_a^b \int_a^b k(x,t)(k(t,s) f(s) ds\, dt = (1 + \lambda k + \lambda^2 k^2) f$$

since $k f = \int_a^b k(s,t) f(t) dt$ and $g_0(t) = f(t)$

Similarly, $g_3(x) = (1 + \lambda\kappa + \lambda^2\kappa^2 + \lambda^3\kappa^3) f$, and so on. In general, we have.

$$g_n(x) = (1 + \lambda\kappa + \lambda^2 k^2 + \cdots + \lambda^n \kappa^n) f \tag{8.3.2}$$

This function (8.3.2) is called the nth approximation to the solution of the integral equation (8.3.1). If the $\lim_{n \to \infty} g_n(x)$ exists and is equal to $g(x)$, then $g(x)$ is the exact solution of integral equation (8.3.1). In that case, $g(x) = \sum_{n=0}^{\infty} \lambda^n k^n f$. If we denote the kernel of the operator k^n by $k_n(x,s)$ then $k^n f = \int_a^b k_n(x,s) f(s) ds$, for n = 1, 2, 3, …, such that $k_1 = k$, $k_2 = \int_a^b k(x,t) k(t,s) dt$ and so on. In general, we have

$$k_n(x,s) = \int_a^b k(x,t) k_{n-1}(t,s) dt, \tag{8.3.3}$$

Here $k_n(x,s)$ is called an nth iterated kernel and thus Equation (8.3.2) become

$$g_n(x) = f(x) + \sum_{m=1}^{n} \lambda^m \int_a^b k_m(x,t) f(t) dt. \tag{8.3.4}$$

8.3.2 Voltera integral equations

Consider the integral equation

$$g(x) = f(x) + \lambda \int_a^x k(x,t) g(t) dt \tag{8.3.5}$$

Zeroth order approximation is given by $g_0(x) = f(x)$. First order approximation is given by

$$g_1(x) = f(x) + \lambda \int_a^x k(x,t) g_0(t) dt = f(x) + \lambda \int_a^x k(x,t) f(t) dt$$

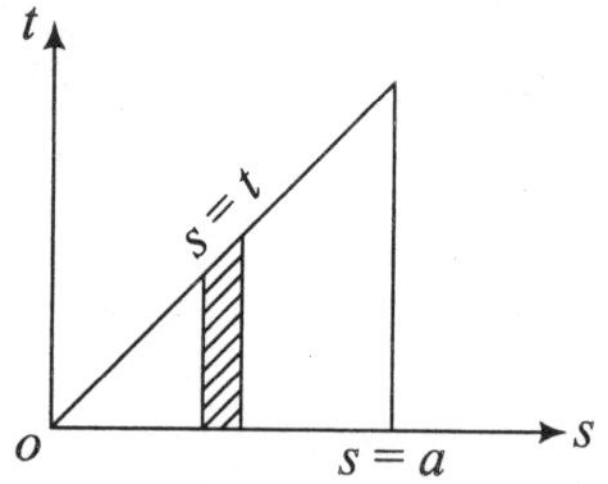

Fig. 8.3.1

Second order approximation is given by

$$g_2(x) = f(x) + \lambda \int_a^x k(x,t) g_1(t) dt$$

$$= f(x) + \lambda \int_a^x k(x,t) f(t) f(x) dt + \lambda^2 \int_a^x k(x,t) \int_a^t k(t,s) f(s) ds dt$$

Now the region of integration in the double integral is the st-plane indicated by the triangle ABC as shown in figure (8.3.1). Changing the order of integration, we have.

$$g_2(x) = f(x) + \lambda \int_a^x k(x,t) f(t) dt + \lambda^2 \int_a^x f(s) ds \int_a^t k(x,t) k(t,s) dt$$

$$= f(x) + \lambda \int_a^x k(x,t) f(t) dt + \lambda^2 \int_a^t k_2(x,s) f(s) ds$$

$$= f(x) + \lambda \int_a^x k(x,t) f(t) dt + \lambda^2 \int_a^x k_2(x,t) f(t) dt$$

where $$k_2(x,t) = \int_t^x k(x,s) k(s,t) ds.$$

Continuing in this way, we find that

$$g_n(x) = f(x) + \sum_{m=1}^{n} \lambda^m \int_a^b k_m(x,t) f(t) dt,$$

where $$k_m(x,t) = \int_a^x k_{m-1}(x,s) k(s,t) ds$$

Thus
$$g_n(x)=f(x)+\lambda\int_a^x\sum_{m=1}^{n}\lambda^{m-1}k_m(x,t)f(t)dt \tag{8.3.6}$$

Therefore the Voltera integral equation will have a solution if the sum of the series

$$\sum_{m=1}^{\infty}\lambda^{m-1}k_m(x,t)$$

exists and if it is denoted by $\Gamma(x,t,\lambda)$ then the solution is given by

$$g(x)=f(x)+\lambda\int_a^x\Gamma(x,t,\lambda)f(t)dt \tag{8.3.7}$$

where
$$\Gamma(x,t,\lambda)=\sum_{m=0}^{\infty}\lambda^{m-1}k_m(x,t), \tag{8.3.8}$$

is called Neumann series. and

$$k_m(x,t)=\int_t^x k_{m-1}(x,s)k(s,t)ds \tag{8.3.9}$$

Example 8.3.1 Apply the method of successive approximations to solve the integral equation

$$g(x)=e^x-x-\int_0^1 x\left(e^{xt}-1\right)g(t)dt \tag{8.3.10}$$

Solution: Here

$$k(x,t)=x\left(e^{xt}-1\right)$$

$$=x\left(\left(1+xt+\frac{x^2t^2}{1.2}+\frac{x^3t^3}{1.2.3}+\cdots+\frac{x^nt^n}{1.2.3.\cdots n}+\cdots\right)-1\right)$$

$$=x^2t+\frac{x^3t^2}{1.2}+\frac{x^4t^3}{1.2.3}+\cdots+\frac{x^{n+1}t^n}{1.2.3\cdots n}+\cdots$$

We take $k(x,t)\cong x^2t+\dfrac{x^3t^2}{1.2}+\dfrac{x^4t^3}{1.2.3}$ (up to first three terms) and then

$$g(x)=e^x-x-\int_0^1(x^2t+\frac{x^3t^2}{2!}+\frac{x^4t^3}{3!})g(t)dt$$

$$=e^x-x-c_1x^2-c_2\frac{x^3}{2}-c_3\frac{x^4}{6}$$

where

$$c_1 = \int_0^1 g(t)dt = \int_0^1 te^x dt - \int_0^1 t^2 dt - c_1 \int_0^1 t^3 dt - \frac{c_2}{2}\int_0^1 t^4 dt - \frac{c_3}{6}\int_0^1 t^5 dt$$

$$= 1 - \frac{1}{3} - \frac{c_1}{4} - \frac{c_2}{10} - \frac{c_3}{36}$$

This implies that

$$\frac{5}{4}c_1 + \frac{c_2}{10} + \frac{c_3}{36} = \frac{2}{3} \tag{8.3.11}$$

$$c_2 = \int_0^1 t^2 g(t)dt = \int_0^1 t^2 e^t dt - \frac{1}{4} - \frac{c_1}{5} - \frac{c_2}{12} - \frac{c_3}{42}$$

or

$$\frac{c_1}{5} + \frac{13}{12}c_2 + \frac{c_3}{42} = e - \frac{9}{4} \tag{8.3.12}$$

Also

$$c_3 = \int_0^1 t^3 g(t)dt = \int_0^1 t^3 e^t dt - \frac{1}{5} - \frac{c_1}{6} - \frac{c_2}{14} - \frac{c_3}{48}$$

$$\frac{c_1}{6} + \frac{c_2}{14} + \frac{49}{48}c_3 = -\frac{29}{5} - 2e \tag{8.3.13}$$

Solving equation (8.3.11) to (8.3.13) for c_1, c_2 and c_3 we get the required solution.

Example 8.3.2 Solve the integral equation

$$g(x) = 1 + \lambda\int_0^1 (1 - 3xt)g(t)dt.$$

Use Picard's method to find first three approximations to the solution of this equation.

Solution: The given equation is

$$g(x) = 1 + \lambda\int_0^1 (1 - 3xt)g(t)dt$$

$$= 1 + \lambda\int_0^1 g(t)\,dt - 3\lambda x\int_0^1 t\,g(t)\,dt$$

$$= 1 + \lambda c_1 - 3\lambda\, x\, c_2 \tag{8.3.14}$$

where

$c_1 = \int_0^1 g(t)\,dt$ and $c_2 = \int_0^1 tg(t)dt$. Therefore,

$g(x) = 1 + \lambda c_1 - 3\lambda\, x\, c_2$

Therefore,

$$c_1 = \int_0^1 g(t)\,dt = \int_0^1 \left(1+\lambda c_1 + 3t\lambda c_2\right)dt = 1+\lambda c_1 + \frac{3\lambda c_2}{2}$$

$$c_2 = \int_0^1 tg(t)dt = \int_0^1 (1+\lambda c_1)tdt - 3\lambda c_2 \int_0^1 t^2 dt$$

$$= \frac{1+\lambda c_1}{2} - \frac{3\lambda c_2}{3} = \frac{1}{2} + \frac{\lambda}{2}c_1 - \lambda c_2$$

=> $$2(1-\lambda)c_1 + 3\lambda c_2 - 2 = 0 \tag{8.3.15}$$

and $$\lambda c_1 - 2(1+\lambda)c_2 + 1 = 0 \tag{8.3.16}$$

Solving equation (8.3.15) and (8.3.16), we get

$$c_1 = \frac{1}{2-\lambda} \text{ and } c_2 = \frac{2}{4-\lambda^2}$$

Using these of c_1 and c_2 in equation (8.3.14) we get

$$g(x) = 1 + \frac{\lambda}{2-\lambda} - \frac{6\lambda x}{4-\lambda^2} \; \frac{2[(\lambda(1+3x))-2]}{\lambda^2-4}, \; \lambda \neq 2$$

which is the required solution of the given integral equation.

Picard's method

Since $g_{0(x)} = 1$, therefore we have

$$g_1(x) = 1 + \lambda \int_0^1 (1-3xt)dt = 1 + \lambda - \frac{3\lambda x}{2}$$

$$g_2(x) = 1 + \lambda \int_0^1 (1-3xt)\left\{1+\lambda - \frac{3\lambda x}{2}\right\}dt = 1+\lambda+\frac{\lambda^2}{4} - \frac{3\lambda x}{2}(1-\frac{\lambda}{2})$$

$$g_3(x) = 1 + \lambda \int_0^1 (1-3xt)\left\{1+\lambda\left(1+\frac{\lambda}{4}\right) - \frac{3}{2}\lambda(1-\frac{\lambda}{2})t\right\}dt$$

$$= 1+\lambda+\frac{\lambda^2}{4}+\frac{5}{8}\lambda^3 - \frac{3\lambda}{2}\left(1+\frac{3}{4}\lambda^2\right)x,$$

This is the required solution of the given integral equation.

8.4 Finite difference method

We consider the integral equation

$$g(s) = f(s) + \int_a^b k(s,t)g(t)dt \tag{8.4.1}$$

We know that a definite integral can be closely approximated by a quadrate formula of the form

$$\int_a^b F(x)dx = \sum_{m-1}^{n} A_m F(x_m) \tag{8.4.2}$$

where A_m and x_m are the weights and abscissa, respectively. Consequently (8.4.1) can be written as

$$g(s) - \sum_{m=1}^{n} A_m\, k(s,t_n)\, g(t_m) = f(s) \tag{8.4.3}$$

where $t_1, t_2, t_3, \ldots t_n$ are points in which the interval (a, b) is subdivided. Further, equation (8.4.3.) must hold for values of s in the interval (a, b) and hence it must hold for $s = t_1, s = t_2, \ldots. s = t_n$. Thus we option.

$$g(t_i) - \sum_{m=1}^{n} A_m\, k(t_i,t_m) g(t_m) = f(t_i),\ i = 1,2,3,\ldots,n \tag{8.4.4}$$

which is a system of n-linear equations in $g(t_i), g(t_2), \cdots, g(t_n)$. When the unknowns $g(t_i)$ are determined, equation (8.4.3.) gives an approximation for $g(s)$. Clearly, different types of quadrature formulae can be employed for the approximation of the integral in (8.4.1).

Example 8.4.1 Consider the integral equation.

$$f(s) - \int_0^1 (s+t) f(t)dt = \frac{3}{2}s - \frac{5}{6} \tag{8.4.5}$$

It can be easily seen that the analytic solution of (8.4.5) is given by $f(s) = s-1$. For numerical solution, we divide the range $[0,1]$ into two equal subintervals so that $h = \frac{1}{2}$. Applying trapezoidal rule to approximate the integral in (8.4.5), we obtain

$$f(s) - \frac{1}{4}\left[s f_o + 2\left(s + \frac{1}{2}\right) f_1 + (s+1) f_2\right] = \frac{3}{2}s - \frac{5}{6} \tag{8.4.6}$$

where $f_i = f(t_i)$. Setting $s = t_i$, where $t_o = 0, t_1 = \frac{1}{2}$ and $t_2 = 1$ we obtain the system of equations

$$12 f_o - 3 f_1 - 3 f_2 = -10$$

$$-3 f_o + 12 f_1 - 9 f_2 = -2$$

$$-3 f_o - 9 f_1 + 6 f_2 = 8$$

The solution of this system of equation is given by

$$f_0 = -\frac{1}{2}, f_1 = -\frac{5}{6}, f_2 = \frac{1}{2}.$$

Upon using there values in (8.4.6) we get $f(s)=\frac{4}{3}s-=\frac{11}{12}$ as the solution.

It we use Simpson's rule to approximate the integral in (8.4.5), we get

$$f(s)-\frac{1}{6}\left[s\ f_o+4\left(s+\frac{1}{2}\right)f_1+(s+1)\ f_2\right]=\frac{3}{2}\ s-\frac{5}{6} \qquad (8.4.7)$$

Setting s= t_i, we get

$$\begin{aligned} 6f_o-2f_1-f_2&=-5\\ -f_o+4+1-3t_2&=-1\\ -f_o-6f_1+4f_2&=4 \end{aligned}$$

Which implies that $f_o=-1,\ f_1=\frac{-1}{2}$ and $f_2=0$. Using these values in equation (8.3.6), we get

$$f\ (s)=\frac{1}{6}\left[-s+4\left(s+\frac{1}{2}\right)\left(-\frac{1}{2}\right)\right]+\frac{3}{2}\ s-\frac{5}{6}=\mathrm{s}-1$$

which is the exact solution . Thus the Simpson's rule gives exact result in this case because the integrand is a second degree polynomial in t.

Example 8.4.2 Consider the Love's integral equation

$$y(x)+\int_{-1}^{1}k(x,s)y(s)ds=1 \qquad (8.4.8)$$

where

$$k(x,s)=\frac{1}{\pi}\left[\frac{1}{1+(x-s)^2}\right] \qquad (8.4.9)$$

This equation occurs in an electrostatic problem considered by Love in 1949. The analytic method suggested by Love is laborious and hence various numerical methods were proposed. The simplest is to approximate the integral term in (8.4.8.) by trapezoidal rule. We divide the interval (-1, 1) into n smaller intervals of width h such that $s_i=-1+ih,\ i=0,1,2,\ldots\mathrm{n}$ and $nh=2$. The equation (8.4.8) gives us

$$y_i=\sum_{j=0}^{n-1}\int_{s_i}^{s_{i+1}}k(x_i,s)y(s)ds=1$$

Upon using the trapezoidal rule, the above equation becomes

$$y_i+\sum_{j=0}^{n-1}\frac{h}{2}\left[k(x_i,s_j)y\left(s_j\right)+k\left(x_i,s_{j+1}\right)y\left(s_j+1\right)\right]=1$$

which can be rewritten as

$$y_1+\frac{h}{2}k(x_i,s_o)y_o+\frac{h}{2}k(x_i,s_n)y_n+h\sum_{j+1}^{n-1}k(x_i,s_j)y_j=1,\quad i=0,1,2,\cdots,n \qquad (8.4.10)$$

Equation (8.4.10) represents a system of $n+1$ linear equations in n+1unknowns viz. $y_o, y_1, y_2 \cdots y_n$, and can be solved on a computer. The solution is symmetric and the computed values of $y(x)$ at $x = 0$ and $x = 1$ are given below:

x	Exact $y(x)$	n	Computed $y(x)$	error
0.0	0.65741	4	0.66026	0.00285
		8	0.65812	0.00071
		16	0.00759	0.00018
		32	0.65746	0.00005

x	Exact $y(x)$	n	Computed $y(x)$	error
1.0.	0.75572	4	0.75452	0.00120
		8	0.75542	0.00030
		16	0.75564	0.00008
		32	0.75570	0.00002

8.5 Chebyshev polynomials

The Chebyshev polynomial of degree n over the interval [-1, 1] is defined by the relation

$$T_n(x) = \cos(n\cos^{-1} x) \tag{8.5.1}$$

This implies that

$$T_n(x) = T_{-n}(x) \tag{8.5.2}$$

Let $\cos^{-1} x = \theta$ so that $x = \cos\theta$ and equation (8.5.1) gives us

$$T_n(x) = \cos n\theta$$

Therefore, $T_0(x) = 1$ and $T_1(x) = x$.

Using the trigonometric identity

$$\cos(n-1)\theta + \cos(n+1)\theta = 2\cos n\theta \cos\theta$$

we obtain $T_{n-1}(x) + T_{n+1}(x) = 2xT_x(x)$

$$\Rightarrow T_{n+1}(x) = 2xT_x(x) - T_{n-1}(x) \tag{8.5.3}$$

This is the recurrence relation which can be used to compute successively all $T_n(x)$ since we know $T_0(x)$ and $T_1(x)$. The first six Chebyshev polynomials are

$$T_0(x)=1,\ T_1(x)=x,\ T_2(x)=2x^2-1,$$

$$T_3(x)=4x^3-3x, \qquad T_4(x)=4x^4-8x^2+1,$$

$$T_5(x)=16x^5-20x^3+5x,\ T_6(x)=32x^6-48x^4+18x^2-1 \tag{8.5.4}$$

It can be easily seen that the coefficient of x^n in $T_n(x)$ is always 2^{n-1}. Again, if we set $y=T_n(x)=\cos n\theta$, we get

$$\frac{dy}{dx}=\frac{n\sin\theta}{\sin\theta} \text{ and } \frac{d^2y}{d^2x}=\frac{-n^2\cos n\theta+n\sin\theta.\cot\theta}{\sin^2\theta}=\frac{-n^2y+x\dfrac{dy}{dx}}{1-x^2}$$

Thus $T_n(x)$ satisfies the differential equation

$$(1-x^2)\frac{d^2y}{d^2x}-x\frac{dy}{dx}+n^2y=0. \tag{8.5.5}$$

The equations (8.5.5) provide us

$$1=T_0(x),\quad x=T_1(x),\quad x^2=\frac{T_0(x)+T_2(x)}{2}, x^3=\frac{3T_1(x)+T_3(x)}{8},$$

$$x^4=\frac{3T_0(x)+4T_2(x)+T_4(x)}{8}, x^5=\frac{10T_1(x)+5T_3(x)+T_5(x)}{16},$$

$$x^6=\frac{10T_0(x)+15T_2(x)+6T_4(x)+T_6(x)}{32}, \tag{8.5.6}$$

and so on. An important property of $T_n(x)$ is given by

$$\int_{-1}^{1}\frac{T_m(x)T_n(x)}{\sqrt{1-x^2}}dx=\begin{cases}0, & m\neq n\\ \dfrac{\pi}{2}, & m=n\neq 0\\ \pi, & m=n=0\end{cases} \tag{8.5.7}$$

This means that the polynomials $T_n(x)$ are orthogonal with the function $\dfrac{1}{\sqrt{1-x^2}}$. In order to prove (8.5.7) we put $x=\cos\theta$. We have,

$$\int_0^{\pi}T_m(\cos\theta)T_n(\cos\theta)d\theta=\int_0^{\pi}\cos m\theta\cos n\theta\,d\theta$$

$$= \left[\frac{\sin(m+n)\theta}{2(m+n)} + \frac{\sin(m-n)\theta}{2(m-n)} \right]_0^\pi = \begin{cases} 0, & m \neq n \\ \frac{\pi}{2}, & m = n \neq 0 \\ \pi, & m = n = 0 \end{cases}$$

In approximation theory, we use monic polynomials i.e., Chebyshev polynomials in which coefficient of x^n is unity. If $p_n(x)$ is a monic polynomial, we can write

$$p_n(x) = 2^{1-n} T_n(x), \quad n \geq 1. \tag{8.5.8}$$

Chebyshev polynomials have the property that of all monic polynomials $p_n(x)$ of degree n with leading coefficient unity, the polynomial $2^{1-n} T_n(x)$ has the least upper bound for its absolute value in the range (-1, 1). Because $|T_n(x)| \leq 1$, the upper bound referred above is 2^{1-n}. Thus, in Chebyshev approximation, the maximum error is kept down to a minimum and is often referred to as minimax principle and the polynomial (8.5.8) is called the minimax polynomial. This process leads to the best lower-order approximation, called the minimax approximation, to a given polynomial. This is illustrated with the help of following example.

Example 8.5.1: Find the best lower-order approximation to the cubic polynomial $2x^3 + 3x^2$.

Solution: Using the relation (8.5.6), we have

$$2x^3 + 3x^2 = \frac{2}{4}[T_3(x) + 3T_1(x)] + 3x^2,$$

$$= 3x^2 + \frac{3}{2}T_1(x) + \frac{1}{2}T_3(x),$$

$$= 3x^2 + \frac{3}{2}x + \frac{1}{2}T_3(x), \qquad (\because T_1(x) = x)$$

The polynomial $3x^2 + \frac{3}{2}x$ is the required lower-order approximation to the given cubic with a maximum error $\pm\frac{1}{2}$ in the range (-1, 1).

8.5.1 Lanczos process of economization of power series

We consider the power series expansion of f(x) in the form

$$f(x) = A_0 + A_1 x + A_2 x^2 + \cdots + A_n x^n, \qquad -1 \leq x \leq 1 \tag{8.5.9}$$

Using the relations given in (8.5.6), we convert (8.5.1) into an expansion in Chebyshev polynomials. We obtain

$$f(x) = B_0 + B_1 T_1(x) + B_2 T_2(x) + \cdots + B_n T_n(x), \tag{8.5.10}$$

For large number of functions, the expansion (8.5.10) converges more rapidly than the power series given by (8.5.9). This is known as economization of the power series.

Example 8.5.2: Economize the power series

$$\sin x = x - \frac{x^3}{6} + \frac{x^5}{120} - \frac{x^7}{5040} + \cdots$$

Solution: Since $\frac{1}{5040} = 0.000198\cdots$, the truncated series

$$\sin x = x - \frac{x^3}{6} + \frac{x^5}{120},$$

will produce a change in the fourth decimal place only. Writing in terms of Chebyshev polynomials, we get

$$\sin x \cong T_1(x) - \frac{1}{24}[3T_1(x) + T_3(x)] + \frac{1}{120 \times 16}[10\,T_1(x) + 5T_3(x) + T_5(x)]$$

$$= \frac{169}{192}T_1(x) - \frac{5}{128}T_3\,(x) + \frac{1}{1920}T_5(x) \approx \frac{169}{192}T_1(x) - \frac{5}{128}T_3(x)$$

Since $\frac{1}{1920} = 0.00052$ so truncated series will produce a change in the fourth decimal place only. Thus the economized series is

$$\sin x = \frac{169}{192}x - \frac{5}{128}(4x^3 - 3x) \qquad = \frac{383}{384}x - \frac{5}{32}x^3$$

8.6 Chebyshev series method

Consider the Fredholm integral equation in the form

$$y(x) + \int_{-1}^{1} k(x,s)y(s)ds = f(x) \tag{8.6.1}$$

where $-1 \le x\,;\, s \le 1,$ and we write

$$y(x) = \sum_{r=0}^{n} a_r T_r(x) \tag{8.6.2}$$

$$f(x) = \sum_{r=0}^{n} f_r T_r(x) \tag{8.6.3}$$

where $T_r(x)$ is the rth Chebyshev polynomial. Substituting (8.6.2) and (8.6.3) in (8.6.1), and interchanging the order of integration and summation in the integral term we obtain

$$\sum_{r=0}^{N} a_r\, T_r(x) + \sum_{j=0}^{N} a_j \int_{-1}^{1} k(x,s)\, T_j(s) ds = \sum_{r=0}^{N} f_r\, T_r(x) \quad (8.6.4)$$

Now, we can find the expansion

$$\int_{-1}^{1} k(x,s)\, T_j\,(s)\, ds = \sum_{r=0}^{n} b_{jr}\, T_r\,(x) \quad (8.6.5)$$

and use it in (8.6.4). Then we can equate corresponding coefficients of each $T_r(x)$ and both sides to obtain a set of $(N+1)$ equations in $(N+1)$ unknowns as given below.

$$a_r + \sum_{j=0}^{n} a_j b_{ir} = f_r \,, r = 0,1,2, \ldots.N \quad (8.6.6)$$

The method provides the solution as a Chebyshev expanses, is due to Elliott [1963]. The variation of this method based on Clenshaw and Curtis [1960] quadrature scheme which gives better accuracy and due to EI – Gendi [1969] is discussed below.

$$\int_{-1}^{1} f(x) dx \quad (8.6.7)$$

Consider the numerical quadrature of the definite integral where $f(x)$ is defined and well behaved in the interval $-1 \le x \le 1$. The Clenshaw and Curtis quadrature is based on the approximation

$$f(x) = \sum_{r=0}^{N} a_r\, T_r(x) \quad (8.6.8)$$

where $$\mathrm{a_r} = \frac{2}{N} \sum \sum_{J=0}^{N} f + \left(\cos\frac{j\pi}{N}\right) \cos\frac{rj\pi}{N} \quad (8.6.9)$$

and double primer denotes that the first and last terms have to be halved. The equations (8.5.7) to on simplification lead to

$$\int_{-1}^{1} T_{2j}(x) dx = \frac{2}{1-4j^2} \quad (8.6.10)$$

We can write $$\int_{-1}^{1} f(x) dx = \sum_{s=0}^{N} b_{NS}\, f_s \quad (8.6.11)$$

where for even N, $$b_{NS} = \frac{4}{N} \sum_{j=0}^{N/2} \frac{1}{1-4j^2} \cos\frac{2j\pi s}{N} \,, s = 1,2,3,\ldots,N-1 \quad (8.6.12)$$

and $$b_{N0} = b_{NN} = \frac{1}{N^2-1} \quad (8.6.13)$$

Using the procedure to approximate (8.6.1), we get

$$[I + A]\,[y] = [f] \tag{8.6.14}$$

where the elements of A are defined as

$$a_{ij} = b_{Nj} k_{ij}, \tag{8.6.15}$$

$$k_{ij} = k\left(-\cos\frac{i\pi}{N},\ -\cos\frac{j\pi}{N}\right), i, j = 0,1,2,\ldots, N \tag{8.6.16}$$

and

$$y_i = y\left(-\cos\frac{i\pi}{N}\right) \tag{8.6.17}$$

The system (8.6.14) can new be solved to obtain directly the values of y.

Example 8.6.1 Solve the Love's integral equation $y(x) + \int_{-1}^{1} k(x,s)y(s)ds = 1$, where $k(x,s) = \frac{1}{\pi}\left[\frac{d}{d^2 + (x-s)^2}\right]$, for d = 1.0 by using Chebyshev method.

Solution: The analytical solution of the given equation in given by

$$y(x) = 1 + \sum_{n=1}^{\infty} (-1)^n \int_{-1}^{1} k_n(x,s)ds,$$

where the iterated kernels $k_n(x,s)$ are given by

$$k_1(x,s) = \frac{1}{\pi[1 + (x-s)^2]} \text{ and } k_n(x,s) = \int_{-1}^{1} k_{n-1}(x,t)k(t,s)ds$$

The analytical method of solution is some what laborious and numerical solutions are preferred. The following table gives the results obtained by the Chebyshev series method with N = 8 and d =1.0. Chebyshev series solution of Love's equation with d = 1.0

$x_j = -\cos\frac{i\pi}{N}$	$y(x_j)$
0.0	0.65740981
0.38268	0.67248912
0.70711	0.70866017
0.92388	0.74265684
1.0	0.75571801

It is clear that this method gives better accuracy than the trapezoidal method, but it too gives inaccurate results for small values of d.

8.7 Method of cubic spline

We know that in the interval $x_{j-1} \le x \le x_j$, S(x) is given by

$$S(x) = M_{j-1}\frac{(x_j - x)^3}{6h} + M_j\frac{(x - x_{j-1})^3}{6h} + \left(y_{j-1} - \frac{h^2}{6}M_{j-1}\right)\left(\frac{x_j - x}{h}\right) + \left(y_j - \frac{h^2}{6}m_j\right)\frac{(x - x_{j-1})}{h} \quad (8.7.1)$$

where $M_j = S''(x_j)$, $y_j = y(x_j)$ and $x_j = x_0 + jh, j = 0,1,2,3,...N$.

Using (8.7.1) in (8.6.1) we obtain

$$y(x_i) + \sum_{j=1}^{N} k(x_i, s)\left[M_{j-1}\frac{(s_j - s)^3}{6h} + M_j\frac{(s - s_{j=1})^3}{6h}\right] + \left[\left(y_{j-1} - \frac{h^2}{6}M_{j-1}\right)\frac{(s_j - s)}{h} + \left(y_j - \frac{h^2}{6}M_j\right)\frac{(s - s_{j-1})}{h}\right]ds$$

$$= f(x_i), i = 0,1,2,3,..., N \quad (8.7.2)$$

Putting $s = s_{j-1} + ph$, the equation (8.7.2) simplifies to

$$y(x_i) + h\sum_{j=1}^{N}\int_0^1 k\left(x_i, -s_{j-1} + ph\right)\left[M_{j-1}\frac{(1-p)^3 h^2}{6} + M_j\frac{p^3 h^2}{6}\right]$$

$$+\left(y_{j-1} - \frac{h^2}{6}M_j - 1\right)\left[(1-p) + \left(y_j - \frac{h^2}{6}Mj\right)p\right]dp = f(x_i), \quad (8.7.3)$$

$$i = 0,1,2,3,..........N.$$

The integrals $\int_0^1 k\left(x_i, s_{j-1} + ph\right)p^m\,dp,\ m = 0,1,2,\ and\ 3$ in equation (8.7.3) have to be evaluated either analytically or by numerical techniques. Then the equation (8.7.3) together with the relations

$$\frac{h}{6}M_{j-1} + \frac{2h}{3}M_j + \frac{h}{6}M_{j+1} = \frac{y_{j-1} - 2y_j + y_{j-1}}{h},\ j = 1,2,3...N-1, M_o = M_N = 0 \quad (8.7.4)$$

will form a set of $(2N+2)$ linear algebraic equations in $(2N+2)$ unknowns viz., $y_0, y_1, y_2, ..., y_n, m_0, m_1 m_2, ..., m_n$.

Example 8.7.1 Solve the Love's integral equation

$$y(x)+\int_{-1}^{1} k(x\ ,s\ )y(s)ds = 1, \quad \text{where} \quad k(x,s)=\frac{d}{\pi[d^2+(x-s)^2]}, \quad \text{using cubic spline method.}$$

Solution: The method of cubic spline can be applied to solve Love's equation even the values of d are small. For $m = 0$, the integral $\int_0^1 k(x_i, s_{j-1}+ph)dp$ in equation (8.7.3.) provides us

$$X_o = \int_0^1 k(x_i, s_{j-1}+ph)dp$$

$$= \frac{1}{\pi}\int_0^1 \frac{d}{d^2+(x_i - s_{j-1} - ph)^2}\ dp$$

Putting $x_i = -1+ih$ and $s_{j-1} = -1+(j-1)h,$ we obtain

$$X_0 = \frac{1}{\pi h}\tan^{-1}[\frac{h/d}{1+\frac{h^2}{d^2}(i-j)(i-j+1)}]$$

Similarly, we get

$$X_1 = \int_o^1 k(x_i, s_{j-1}+ph)p\,dp = \frac{d}{2\pi h^2}\left[\log\ \frac{d^2+h^2(i-j)^2}{d^2+h^2(i-j+1)^2}\right]+(i-j+1)X_0$$

$$X_2 = \int_0^1 k(x_i, s_{j-1}+ph)p^2 dp = \frac{d}{\pi h^2} - \left[\frac{d^2}{h^2}+(i-j+1)^2\right]X_o + 2x_1(i-j+1)$$

and $$X_3 = \int_0^1 k(x_i, s_{j-1}+ph)p^3 dp = \frac{d}{2\pi h^2}[5+4(i-j)] + \left[3(i-j+1)^2 - \frac{d^2}{h^2}\right]X_1$$

$$-2(i-j+1)\left[\frac{d^2}{h^2} + (i-j+1)^2\right]X_o$$

This system of equations was solved by the Gauss- Seidal iteration method and a subroutine was used for this. It was found that method is unsuitable for finding the solution for larger values of d as the convergence is rather slow. The cubic spline solutions of Love's equation are given below.

x	$y(x)$		
d=0.1	d=0.01	d=0.001	
0.0	0.51261	0.50146	0.50015
0.2	0.51470	0.50158	0.50016
0.4	0.51858	0.50187	0.50019
0.6	0.52876	0.50261	0.50026
1.8	0.60688	0.51713	0.50271
1.0	0.78627	0.69641	0.67179

For $d = 1.0$ the value obtained with 500 interiors for $n - 1.0$ is o.80692 as compared to the true value 0.75572. Thus the spline method for the numerical solution of Fredholm integral equations is potentially useful. The solution obtained by the spline method can be improved upon by regard it as the initial iterates in an iterative method of higher order of convergence.

8.8 Method of generalized quadrature

Consider the integral equation

$$f(x)+\int_a^b k(x,t)f(t)dt=\phi(x),\quad a\le x\le b \qquad (8.8.1)$$

Let $b-a=nh$ and $t_j=a+jh, j=0,1,2,...n$ so that $t_o=a$ and $t_n=b$. The (8.8.1) becomes

$$f(x)+\sum_{j=o}^{n-1}\int_{t_j}^{t_{j+1}} k(x,t)f(t)dt=\phi(x) \qquad (8.8.2)$$

We replace $f(t)$ in the integral by the linear interpolating function $f_n(t)$ given by

$$f_n(t)=\frac{1}{h}\left[(t_{j+1}-t)f(t_j)+(t-t_j)f(t_j+1)\right] \qquad (8.8.3)$$

and we obtain

$$f(x)+\frac{1}{h}\sum_{j=0}^{n-1}\int_{t_j}^{t_{j+1}} k(x,t)\left[(t_{j+1}-t)f(t_j)+(t-t_j)f(t_{j+1})\right]dt=\phi(x)$$

We set to $t=t_j+ph$ get

$$f(x)+h\sum_{j=o}^{n-1}\int_0^1\left[(1-p)f(t_j)+p\,f(t_{j+1})\right]k(x,t_{j+ph})dp=\phi(x)$$

If we put $x=x_i, i=0,1,2,3,...,n$, we get

$$f_i+\sum_{j=0}^{n-1}\left[\alpha_{ij}f_j+\beta_{ij}f_{j+1}\right]=\phi_i,\qquad i=0,1,2,3,...,n \qquad (8.8.4)$$

Where

$$\alpha_{ij}=h\int_0^1(1-p)k(x_i,t_{j+ph})dp,$$

$$\beta_{ij}=h\int_0^1 p\,k(x_i,t_j+ph)dp \text{ and } f_i=f(x_i)=f(a+ih)dp.$$

Example 8.8.1: Solve Love's integral equation

$$y(x)+\int_0^1 k(x,s)y(s)ds=1,\quad k(x,s)=\frac{1}{\pi[1+(x-s)^2]}$$

Solution: The integrals α_{ij} and β_{ij} in (8.8.5) are here given by

$$\alpha_{ij} = h\left[X_0(i,j) - X_1(i,j)\right]$$

and $$\beta_{ij} = h.X_1(i,j),$$

where

$$X_0(i,j) = \int_0^1 k\left(x_i, t_{j+ph}\right)dp$$

$$= \frac{1}{\pi}\int_0^1 \frac{dp}{1+h^2(i-j-p)^2}$$

$$= \frac{1}{\pi h}\tan^{-1}\left(\frac{h}{1+h^2(i-j)(i-j-1)}\right),$$

and

$$X_1(i,j) = \int_0^1 pk\left(x_i, t_j + ph\right)dp = \frac{1}{\pi}\int_0^1 \frac{pdp}{1+h^2(i-j-p)^2}$$

$$= \frac{1}{2\pi h^2}\log\left[\frac{1+h^2(i-j-1)^2}{1+h^2(1-j)^2}\right] + (i-j)X_1(i,j)$$

With $n = 4$, we obtain the equations

$$1.076 f_0 + 0.126 f_1 + 0.081 f_2 + 0.050 f_3 + 0.018 f_4 = 1.0$$

$$0.071 f_0 + 1.153 f_1 + 0.126 f_2 + 0.081 f_3 + 0.029 f_4 = 1.0$$

$$0.047 f_0 + 0.126 f_1 + 1.153 f_2 + 0.126 f_3 + 0.047 f_4 = 1.0$$

$$0.029 f_0 + 0.081 f_1 + 0.126 f_2 + 0.153 f_3 + 0.071 f_4 = 1.0$$

$$0.018 f_0 + 0.050 f_1 + 0.081 f_2 + 0.126 f_3 + 1.076 f_4 = 1.0$$

The solution of this Centro –symmetric system was obtained on a digital computer and the results obtained as given in the table below.

x	$y_e(x)$	n	$y_c(x)$	Error
0.0	0.65741	4	0.65609	0.00132
		8	0.65708	0.00033
		16	0.65733	0.00008
		32	0.65739	0.00002
1.0	0.75572	4	0.75484	0.00088

8	0.75550	0.00022
16	0.75566	0.00006
32	0.75570	0.00002

Comparison with the results obtained by trapezoidal rule shows that the rule gives better accuracy and rate of convergence is h^2.

Exercises

1. Solve the integral equations

(i) $g(x) = e^x - x - \int_0^1 x\left(e^{xt} - 1\right)g(t)dt,$ (ii) $g(x) = 1 + \lambda\int_0^1 (1 - 3xt)g(t)dt$

(iii) $\phi(x) = \sin x - \frac{x}{4} + \frac{1}{4}\int_0^{\frac{\pi}{2}} x\xi\phi(\xi)d\xi,$ (iv) $\phi(x) = 1 + \lambda\int_0^{\pi}\sin(x + t)\phi(t)dt$

(iv) $f(x) = \frac{3}{2}e^x - \frac{1}{2}xe^x - \frac{1}{2} + \frac{1}{2}\int_0^1 t\, f(t)dt$

2. Solve the Voltera integral equations

$$g(x) = 1 + x + \lambda\int_0^x (x - t)g(t)dt \quad , g(x) = f(x) + \lambda\int_0^x e^{x-t}g(t)dt,$$

and find the resolvant kernel. $g(x) = \sinh(x) + \int_0^x e^{t-x}g(t)dt,$

$$g(x) = 29 + 6x + \int_0^x (6x - 6t + 5)g(t)dt \quad f(x) = 1 + \int_0^x xtf(t)dt.$$

3. Solve the integral equation with degenerate kernel

(i) $f(x) - \lambda\int_{\frac{-\pi}{4}}^{\frac{\pi}{4}} \tan s\ f(s)ds = \cos x,$ (ii) $f(x) - \lambda\int_0^{\frac{\pi}{2}} \sin x\cos t\, f(t)dt = \sin x$

(iii) $f(x) = \cos x + \lambda\int_0^x \sin(x - u)f(u)du$

4. Solve the integral equations

(i) $\phi(x) = \dfrac{15x-2}{18} + \dfrac{1}{3}\int_0^1 (x+t)\phi(t)dt$, (ii) $f(x) = \dfrac{3}{2}x - \dfrac{5}{6} + \int_0^1 (x+t)f(t)dt$

(iii) $f(x) = \dfrac{3}{2}e^x - \dfrac{1}{2}e^x - \dfrac{1}{2} + \dfrac{1}{2}\int_0^1 t\, f(t)dt$,

by using (i) trapezoidal and (ii) cubic spline method.

5. Prove that

   (i) $T_n(x)$ is a polynomial in x of degree x, (ii) The coefficient of x^n in $T_n(x)$ is 2^{n-1}

   (iii) $x^2 = \dfrac{1}{2}[T_0(x) + T_2(x)]$

6. Express in x:

   (i) $T_0(x) + 2T_1(x) + T_2(x)$, (ii) $2T_0(x) - \dfrac{1}{4}T_2(x) + \dfrac{1}{8}T_4(x)$

7. Express the following polynomials in terms of Chebyshev polynomials

   (i) $1 + x - x^2 + x^3$

   (ii) $1 - x^2 + 2x^4$

8. Obtain the least lower-degree approximation to the cubic $x^3 + 2x^2$

9. Economize the series

   (i) $\sinh x = x + \dfrac{x^3}{6} + \dfrac{x^5}{120} + \dfrac{x^7}{5040} + \cdots$, on [-1, 1] allowing a tolerance of 0.0005

   (ii) $f(x) = 1 - \dfrac{x}{2} - \dfrac{x^2}{8} - \dfrac{x^7}{16}$

CHAPTER 9

C LANGUAGE AND PROGRAMS

9.1 Introduction

Computers understand only binary codes consisting of numbers 0 and 1 which is also known as machine language. Because the machine language is difficult to understand by the users, computer professionals have devised symbolic languages like FORTRAN, COBOL, Pascal, C, C^{++} etc., which are known as *high level* languages. There are special programs known as compilers and interpreters which translate the programs written in high level languages to machine (*low level*) language. The language C is a general purpose programming language, originally designed by Dennis Ritchie in 1972 at Bell laboratories and standardized by the American National Standards Institute (ANSI) in 1988 and named as ANSI C. A powerful language C is used for many purposes like writing operating systems, business and scientific applications and even the C compiler itself. It does not dependent on a particular hardware and the programs written in C are portable across any system. The programs written in C are efficient, fast and user friendly. This is the reason that C is one of the most popular computer languages today. In this chapter we give an overview of the features of C language followed by programs of some standard numerical methods in C.

9.2 Features of C language

In this section we list out all the basic features of C language in order to understand it.

9.2.1 Constants and variables

Constants

The C constants are numbers which do not change during execution of a program. These may be of three types namely, *integer*, *floating point (real)* and *string* such as 1897, -1589, 1.728, -0.2317, and "Enter the data" etc., respectively. It may be noted that the string constants are enclosed in double quotes ('').

Variables

The variables can contain different C constants during the execution of the program. These are declared in C programs by first specifying the type *int* for an integer and *float* for the floating point and then the variable names separated by commas. The general format is

type *list of variables.*

For example,
```
int     a, b, c;
float   a, b, c;
float   a=3.4;   etc.
```

Rules for naming C variables

(*i*) A variable name may contain only alphabets, digits and underscore (–).

(*ii*) It must begin with a alphabet or an underscore (–).

(*iii*) It can be as long as you wish, but on some C systems only first 31characters are considered.

(*iv*) Lower and upper case alphabets are treated as different in C. For example, Num and num are two entirely different variable names. As a matter of convention, lower case alphabets are used.

9.2.2 Operators, expressions and statements

Arithmetic operators

The arithmetic operators used in C for different operations are as under:

Symbol	Operator
+	Addition
-	Subtraction
*	Multiplication
/	Division

While using these operators, following order of precedence is adopted

(*i*) *, / (*ii*) +, −

If the order of operators is to be different, then circular brackets are used. There is no exponentiation operator in C, but various library functions are available for the same. For example to calculate the square root *sqrt* function is used.

Mathematical expressions

The mathematical expressions in C consist of a sequence of arithmetic operators and variable names. For example

$a + b,\ \ a/b + c,\ \ a/(b + c),\ \ b * b - 4 * a * c,$ (beta +gamma) and exp$(b * \ln(a))$

are the arithmetic expressions written for

$$a + b, \frac{a}{b} + c, \frac{a}{b + c}, \sqrt{b^2 - 4ac},\ \ \alpha(\beta + \gamma) \text{ and } a^b$$

in the C language. The presence of multiplication operator [*] is never assumed in C, but it has to be written explicitly. Here *exp* and *ln* are library functions

Arithmetic statements

The arithmetic statements in C are of the form *var = exp;* where *var* is an integer or a floating point variable and *exp* is a mathematical expression written in C format. The = sign tells C to calculate the value of *exp* and assign it to *var.* For example $n = i * i;$ calculates the value of $i * i$ and assigns the result to the variable $n.$ If $i = 5,$ then n gets the value 25. Following points are kept in mind while writing arithmetic statement in C

(i) A statement in C always terminates by a semi colon (;) and

(ii) C also permits statements of the type $k = n = i * i;$

This is equivalent to the statements $n = i * i; k = n;$

(iii) To test the equality of two expressions in C, it uses '=='

Shorthand assignment operators

In addition to the assignment operator =, C also supports certain shorthand assignment operators such as (++, −−, + =, − =, * =, / =). The use of shorthand assignment operators results in concise and more efficient programs. Their use is illustrated below.

Statement with assignment operator	Statement with shorthand assignment operator
$a = a + 1$	$a{+}{+}$ or $+{+}a$
$a = a - 1$	$a{-}{-}$ or $-{-}a$
$a = a + 4$	$a+ = 4$
$a = a - 4$	$a- = 4$
$a = a * 4$	$a* = 4$
$a = a / 4$	$a / 4$

Relational and logical operators

The relational operators available in C are:

Mathematical operator	C operator
$>$	>
$\geq$	>=
$<$	<
$\leq$	<=
$=$	= =
$\neq$	!=

Logical expressions

The Logical expressions are mathematical expressions connected by relational operators. Their value is either true or false. For example

Assuming $i = 2, j = 3$

$i < j$ is true

$i == j$ is false

$(i * j) > (i + j)$ is true.

The result of a logical expression in C is an integer. Here 0 is taken as false, and non zero integer is taken as true.

Logical operators

The logical operators are used to test more than one condition *i.e.*, to combine more than one logical expression. Following are the logical operators available in C:

Logical operator	C symbol
AND	&&
OR	II
NOT	!

The following tables illustrate their use.

Logical expression 1	Logical expression 2	**AND**	**OR**
True	True	True	True
True	False	False	True
False	True	False	True
False	False	False	False

Logical expression	NOT
True	False
False	True

9.2.3 Arrays , user defined types and directives

Arrays

An aggregate of variables of the same type is called an array and variables are called elements of the array. The Rules for naming of arrays are same as those for variable names. For example int $b[10]$; creates a one-dimensional array $b[0]$, b[1],...[9] and float $c[2][2]$; creates two-dimensional array $C[0][0]$, $C[0][1], C[1][0], C[1][1]$ in C. The subscripts always start from zero in C.

Initialization of arrays at the time of declaration

At the time of declaration the arrays can also be initialized. The syntax is

$$type\ array - name[size] = \{list\ of\ values\}$$

For example, int $a[2][2] = \{\{3,\ 1\},\ \{1,\ 2\}\}$; initializes

$a[0][0]$ to 3, $a[0][1]$ to 1, $a[1][0]$ to 2 and $a[1][1]$ to 1.

User defined types

In addition to the built in types *int* and *float,* C allows users to define an identifier that can represent an existing data type. The syntax is

$$typedef \quad type \qquad identifier$$

For example the statement *typedef int number;* defines number to mean the same as *int* and *typedef float matrix [2] [2];* and defines matrix to mean same as 2×2 array of float. These two statements enable declarations of the form *number* $a,\ b,\ c;$ which declares three integers a, b, c and *matrix x;* declares a two dimensional array x having 4 elements $x[0][0]$, $x[0][1]$, $x[1][0]$ and $x[1][1]$.

Preprocessor directives

The lines in a C program that begin with a hash (#) sign are called preprocessor directives. The two most commonly used such directives are *# define* and *# include*. There is no semi colon (;) after the directive.

(i) **# define:** The form of its syntax is **# define** <name replacement >

It instructs the computer to replace all occurrences of name with the replacement even before the program is processed *i.e.*, checked for syntax. For example in the statements

define N 2 and int $a[N]$;

Before the program is processed by the compiler, the second statement *i.e.*, int $a[N]$; is changed to int $a[2]$; and the first one is removed. Therefore, the resulting processed statement is int $a[2]$.

(ii) **# include:** It has syntax of the form: **# include** < header-file-name>

This instructs the computer to insert the contents of the mentioned header file at the place where the directive appeared. The header file contains declarations of various functions and many preprocessor directives.

9.2.4 Input and output statements

The input statement is *scanf* and it has syntax of the form

scanf ('control string", & variable 1, & variable 2, ……);

The *control string* contains the format of the data being input by the user and it contains

$\% d$, for an integer and $\% f$, for a floating point type.

The ampersand (&) symbol is necessary before the integer or floating point type variable. Its significance is discussed in the following section under *functions*. For example we assume the declarations int c; and float $a[3][3]$;

The statement scanf ("$\%d$ $\%f$", $\&c$, $\&$ $a[1][2]$); takes input from the user and stores it in the corresponding variables, $\%d$ corresponds to c, and $\%f$ corresponds to $a[1][2]$.

Output statement

The output statement is *printf* and it has syntax as under:

printf ("control string", argument 1, argument 2, ……) ;

The arguments can be C constants or variables and the control string values can consist of

(i) The characters that will be printed as such

(ii) Format specifications for variables

(iii) Escape sequences.

Format specifications

The various format specifications in C are $\% w d$, for integers where w specifies the minimum width ($i.e.$, number of digits) for output and % w pf, for floating point where p is number of digits to be displayed after the decimal point, after rounding off, if necessary, and w also includes the decimal point. If width of the number is less than the specified width, the number is right justified in that width. However, if width of the number is greater than the specified width, it will be printed in full.

Escape sequences

These are sequences of two characters meant for performing special tasks. The first character is always a back slash (\). The most commonly used escape sequence is $\backslash\ n$ and it causes the output to start from next line. An example of the *printf* statement is: The declarations

Int $i = 1$; and float $a = 32.57$; with *printf* ("$\%3d \backslash n$ $\%8.2f \backslash n$", i, a);

will give output as $\not{b}\not{b}1$ and $\not{b}\not{b}\not{b}$ 32.57, where $\not{b}$ is a blank space.

9.2.5 Decision making statements and loops

Following are the decision making statements and loops available in C language.

If statement

The syntax of if statement is

if (Lexp)

{T statements}
else
{F statements}

Here *Lexp* is a logical expression, T statements are C statements executed when value of *Lexp* is True, and F statements are C statements executed when Lexp has value False. The else part is optional.

Loops

The various types of loops used in C language are given below.

(*i*) **While Loop:** The syntax of *while loop* is

(a) **while** (Lexp)
{statements}

(b) **do**
{statements}
while (Lexp)

Both these forms of the while loop cause execution of statements while value of Lexp is true. The difference between the two forms is that in the latter, the statements are executed at least once irrespective of the value of Lexp.

(ii) **For loop:** The syntax of *for loop* is

for (initialization statement ; Lexp; increment statement)
{statements}

This loop is best explained by a flow chart below:

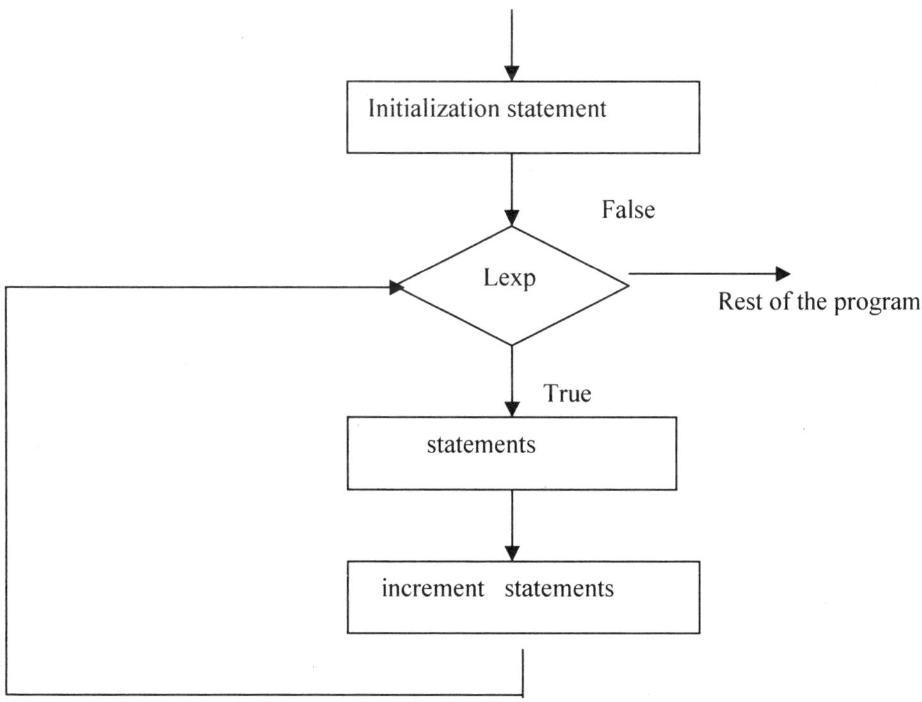

9.2.6 Comments and Break statements

The comments in C start with "/*" and end with "*/". For example,

/*Picard's Method */.

Break statement

If a break statement is encountered inside a loop, then loop is exited, irrespective of the value of *Lexp*, and the program continues with the statement immediately following the loop.

9.2.7 Functions

These are the basic block of a C program. *Functions* contain *Statements* that specify what is to be done. Every program must contain a function named *main.* The program begins execution at the first statement of *main.* Apart from *main* the C functions are classified into

- Library functions
- User defined functions

These functions are called from main to perform various tasks.

(i) *Library functions* such as printf, scanf, sqrt, cos, sin, fabs (used to get the absolute value of a floating point variable) etc., are already available and we just have to use them

(ii) User defined functions have to be written by the user in the program and have syntax of the form:

```
return-type function-name (Argument-list)
{
statements
}
```

9.2.8 Sample program

Here Line numbers are being added for reference purpose and are not part of the program.

```
1.  / * Sample program to illustrate the terms and concepts * /
2.  # include < stdio.h>
3.  float add (float a, float x );
4.  void half (float * x).
5.  {
6.  * x/ = 2;
7.  return
8.  }
9.  main ( )
10. {
11. float a = 2, b = 2,c;
12. c = add (a, b);
13. print ("%f %f %f \n", a, b, c);
14. half (&a); half (&b);
15. printf ("&f &f \n", a, b, c);
```

```
16. }
17. float add (float a, float x)
18. {
19. float sum ;
20. sum = a + x ;
21. a = 20, x = 20; /* changing the formal arguments*/
22. return sum ;
23. }
```

Discussion of sample program

- **Declaration and definition**

(i) Line 3 is the declaration of the function named add. It indicates that there is a function *add* which takes two arguments a and b both of type *float* and returns a *float* $i.e.$, the argument list is *float* a, float b and return-type is float. This can be *void* if function doesn't return anything as can be seen from line 4. It also indicates that the function is defined later in the program.

(ii) Lines 17-23 are the definitions of the function add and they define how the function will make use of the arguments it received and return the required sum.

(iii) Lines 4-8 constitute both the declaration and definition of the function half.

- **Calling a function**

The Line 12 calls the function *add* with a and b as arguments and stores the value returned by it into the variable c.

- **Actual and Formal arguments**

(i) The variables a and x in the declaration of function add in line 3 are called the formal arguments.
(ii) The variables a and b in the call to the function (Line 12) are the actual arguments.

- **Call by value/Call by reference**

The values of actual arguments in C language are *always copied* to the formal arguments when a function is called. This way of passing arguments is called *call* by value. In this any change made to the formal arguments in the function does not affect the value of actual arguments, in contrast to other languages especially, FRTRAN.

- **Simulating a call by reference in C**

The thumb rule that can be followed to make the formal argument refer to the actual argument and not just receive a copy of it is as under:
"Precede the actual argument with an ampersand & (Line 14) and precede the formal argument with an asterisk *(Line 4)".
The actual working of this involves the concepts of pointers, which are another data type in C.

- **Return statement**

The return statement in a function passes the control back to the calling function along with the calculated value (Line 22). Its syntax is *return expression;* . The expression can be omitted; in this case the return statement causes the function to just terminate then and there and pass control back to the calling function (Line 7). In case no return statement is present in a function, an implicit return takes place on encountering the right curly braces }. For example, in the function *main*, the control passes back to the caller, the operating system in this case after line 16.

9.3 C TO FORTRAN CONVERSION

The C programs in the text may be translated almost line by line using FORTRAN statements equivalently as given below.

C statements	FORTRAN equivalents
int	INTEGER
float, double	REAL
# define N 50	PARAMETERS (N = 50)
# include < Math. h>	INTRISIC
scanf ( ) statement	READ (*, *) statement
printf ( ) statement	WRITE (*, *) statement
Preprocessor macro definition	Function sub programs
Label name	Statement number
go to < label >	GOTO < number >
exit ( ) function	GOTO inside a program
if (---) ---;	STOP inside a program
if (---) --- else	IF (---) THEN ---
for (---) {---}	DO --- CONTINUE
}	END

9.4 PROGRAMS OF SOME NUMERICAL METHODS

Based on the algorithm and flow chart, we write the instructions in a code which the computer can understand. A set of such instructions is called a *program*. If there are errors in the program these will be pointed out by the computer during compilation process. Such errors are known as *syntax* errors. After correcting the compilation (syntax) errors, the program is executed with the input data to check for logical errors which may be due to misinterpretation of the algorithm or due to incorrect usage of computer language. These errors are fatal and may stop the working of the computer sometimes or lead to incorrect results and conclusions. Such execution time errors are also known as *symntax* errors. The process of detecting and correcting errors is called *debugging* of the program. While writing a program, our aim should be that the same program is able to run on any computer conveniently with the minimum number of modifications. In this section we give C programs of some commonly used numerical methods.

- **Bisection method**

```
/* Bisection Method * /
 # include < stdio. h >
# include <math. h>
float f(float x)
{
return    f(x);                    (f(x) to be defined here by the user)
 }
void bisect (float *x, float a, float b, int * itr)
{
*x = (a+ b) / 2;
 + +(*itr) ;
 printf (" Iteration no. %3d  X= %7.5f \ n",*itr, *x) ;
}
 main ( )
{
int   itr =  0, maxitr ;
 float x, a, b, aerr, x1;
 printf (" Enter the values of a, b,"   "allowed error, maximum iterations\ n  ");
scanf ("%f %f %f %d", & a,  & b, & aerr, &maxitr) ;
 bisect (&x, a, b,  & itr) ;
  do
 {
if   (f (a) *f (x) < 0)
b = x;
 else
 a = x ;
bisect  (&x1, a, b, & itr);
 if (fabs (x1 – x) < aerr)
 {
 printf (" After % d iterations, root "  "= %6.4f \ n", itr,  x1);
 return  0;
 }
  x = x1;
 } while (itr < maxitr);
 printf ( "solution does not converge,"   " iterations not sufficient");
return 1;
}
```

- **Regula-Falsi Method**

```
/* Regula Falsi Method */
 # include < stdio. h >
 # include <math. h>
float f(float x)
```

```
{
return f(x);                (f(x) to be defined here by the user)
}
void regula (float *x, float x0, float x1,  float fx0, float fx1, int  *itr)
{
*x =x0-((x1 – x0) / (fx1-fx0)) *fx0 ;
++(*itr) ;
printf (“iteration no.   %3d  X= %7.5f  \ n”, *itr, *x);
}
main ( )
{
int   itr =0, maxitr;
float x0,  x1,  x2,  x3, aerr;
printf (" Enter the values for x0, x1,” allowed error,  maximum iterations \n");
scanf ("%f %f %f %d",    & x0,    &x1,     &aerr,    &maxitr);
regula (  &x2,  x0,  x1,   f(x0),  f(x1 ),   &itr);
do
{
if (f(x0) *f(x2) < 0)
x1 = x2 ;
else
x0 = x2 ;
regula  (&x3 , x0, x1, f(x0), f(x1), &itr) ;
if (fabs (x3 –x2 ) < aerr)
{
printf ("After %d iterations,"   "root  = % 6.4f\ n",  itr,   x3 ) ;
return 0;
}
x2 =x3 ;
} while (itr < maxitr);
printf ("Solution does not converge,"   "iterations not sufficient \ n") ;
return 1;
}
```

- **Newton - Raphson method**

```
/ * Newton Raphson Method */
# include < stdio. h >
# include <math. h>
float f(float x)
{
return f(x);                (f(x) to be defined here by the user)
}
float df( float x)
{
return df(x);               Here  df(x)= derivative of  f(x)
```

```
}
 main ( )
{
int  itr, maxitr ;
float h, x0, x1,  aerr ;
 printf ("Enter x0, allowed error,"  "maximum iterations\n");
scanf ("%f %f %d",& x0, &aerr, &maxitr) ;
for (itr =1; itr <maxitr ; itr++)
{
 h = f(x0) / df (x0) ;
x1 =x0 – h;
 printf ("Iteration no.%3d,"  "x = %9.6f\ n", itr, x1);
 if (fabs (h) < aerr)
  {
  printf (" After  % 3d iterations ,"  "root = %8.6f \ n", itr, x1) ;
return 0 ;
}
 x0  = x1;
 }
printf ("Iterations not sufficient ,"  "solution does not converge \ n");
return 1;
}
```

- **Muller's Method**

```
/* Muller's Method */
#include <stdio. h>
#include <math. h>
#define I 2
float y(float x)
{
 return f(x);                     (f(x) to be defined here by the user)
}
main ( )
{
 int i, itr, maxitr;
float x[4], hi,  di,  mu ,  s,  h,  aerr;
printf "Enter the initial"  "approximatons \ n");
 for (i =I-2;  i <3;  i++)
scanf ("%f",  &  x[i ];
printf (" Enter allowed error ,"  "maximum iterations \ n");
scanf ("%f %d" ,&aerr, &maxitr);
 for (itr = 1;  it r <= maxitr;   itr++)
{
hi = (x[ I ]-x[ I-1])/(x[I-1]-x[I-2]) ;
di = (x[ I ]-x[ I-2])/(x[I-1]-x[I-2]) ;
```

```
mu = y (x[I –2]) * hi*hi - y(x [I-1])*di*di +   y(x[ I ]*(di+hi) ;
s  = sqrt ( ( mu*mu –4*y (x[ I ]) *di *hi *(y(x[I-2] ) *hi-y(x[I-1]) *di +y(x[ I ]))));
if (mu < 0)
h = (2*y(x[ I ])* di)/(-mu+ s) ;
else
h = (2*y(x[ I ])* di)/(-mu- s) ;
x[I+1] =x [ I ]+h*(x[ I] – x[I-1]) ;
printf (" Iteration no. % 3d, "  "x = % 7.5f \ n", itr, x [I+1] ;
if (fabs (x[ I+1] –x [ I ]) < aerr)
{
printf ("After % 3d iterations," "the solution is % 6.4f\ n", itr, x[I+1]);
return 0 ;
}
for (i=I-2; i <3 ; i++)
x[ i ] =x[i+1];
}
printf ("Iterations not sufficient,"  "solution does not converge \ n") ;
return 1 ;
```

- **Gauss elimination method**

```
/* Gauss elimination method */
# include < stdio. h>
#define N 4
main ( )
{
float a[N] [N+1], x[N] ,  t, s;
int i,  j,  k ;
printf (" Enter the elements of the"  "augmented matrix rowwise\ n");
for (i=0;  i<N;  i ++)
for (j=0 ;  j<N+1;  j++)
scanf ("%f",  &a [i ][ j ]);
for (j = 0;  j <N-1;   j++)
for (i=j+1;   i<N;   i++)
{
t =  a[ i ] [j ] /a[ j ] [ j ] ;
for  (k = 0;   k<N+1; k++)
a[i] [k] - = a [ j ][k] *t ;
}
/* now printing the upper triangular matrix */
printf (" The upper triangular matrix"  "is : - \ n");
for (i = 0 ; i<N; i++)
{
for ( j =0; j < N+1; j++
printf ("%8.4f",a[ i ][ j ]);
printf ("\ n");
```

```
}
/* now performing back substitution */
for ( i = N-1; i > = 0 ; i- -)
{
 s = 0 ;
 for ( j = i +1; j < N; j++)
  s + =a [ i ] [ j ] *x [ j ];
 x[ i ]= (a [ i ][ N ] – s) / a [ i ] [ i ] ;
}
/ * now printing the results */
printf("The solution is : - \ n");
for (i =0 ;  i<N; i++)
printf (" x[%3d] = %7.4f \ n",i+1, x[ i ]);
}
```

- **Gauss – Jordan method**

```
/* Gauss - Jordan method */
#include <stdio.h>
#define N 3
main ( )
{
 float a[N] [N+1], t;
 int  i, j, k;
 printf (" Enter the elements of  the"  "augmented matrix rowwise \ n");
for (i=0; i<N+1; i++)
 for ( j =0 ; j<N+1; j++)
  scanf ("%f", & a[i][j];
/* now calculating the values of x1,  x2 ,…., xN */
for ( j=0;  j <N;  j++)
for (i=0;  i<N;   i++)
 if (i !=j)
 {
 t = a[ i ] [ j ]/ a[ j ] [ j ];
 for (k =0;   k<N+1;   k++)
 a[ i ] [ k ] - = a[ j ] [ k ] *t;
  }
/* now printing the diagonal matrix */
printf (“The diagonal matrix is :- \ n”) ;
for (i=0; i<N; i++)
{
 for ( j =0; j<N+1; j++)
  printf (“%9.4f ” , a[ i ] [ j ]);
printf (“\n”);
}
/* now printing the results */
printf (“The solution is :- \n”);
for (i = 0 ; i <N ; i++)
```

```
printf ("x[%3d] = % 7.4f \n", i+, a[ i ] [N] /a [ i ] [ i ]) ;
}
```

- **Crout's method**

```
/* Crout method */
#include <stdio. h>
# define N 4
typedef float matrix [N] [N];
matrix l, u, a;
float b [N], x [N], v[N] ;
void urow (int i)
{
float s;
int j, k;
for (j = i; j <N ; j++)
{
s = 0;
for (k = 0 ; k <N-1; k++)
s + = u[k] [ j] *1[ i] [ k ] ;
u [ i] [ j ] = a [ i ] [ j ] –s ;
}
}
void 1col (int j)
{
float s ;
int i, k;
for (i =j+1; i < N : i ++)
{
s = 0 ;
for (k = 0 ; k < = j – 1, k++)
s + = u[ k ] [ j ] * 1 [ i ] [ k ] ;
1[ i] [ j ] = ( a [ i ] [ j ] –s) /u [ j ] [ j ] ;
}
}
void printmat (matrix x)
{
int i, j ;
for (i =0 ; i<N; i++)
{
for (j = 0; j < N; j++)
{
for (j = 0 ; j <N; j++)
printf (" % 8.4 f", x [ i ] [ j ]) ;
printf ("\n");
}
}
main ( )
```

```
{
 int i, j, m;
 float s ;
printf (" Enter the elements of augmented "  " matrix rowwise \ n");
for (i=0 ; i < N ; i++)
{
for (j = 0;  j < N;  j + +)
scanf ("%f",  & a [i] [ j ] ) ;
scanf "%f",  & b [i] ) ;
 }
  /* now calculating the elements of 1 and u */
 for (i =0;  i<N;  i++)
{
 urow(m)
if (m < N – 1) lcol (m) ;
l[i] [i] = 1.0;
 for (m = 0; m < N; m + +)
}
/ * now printing the elements of 1 and u */
 printf  ("\t \ tL \ n");  printmat  (1) ;
printf  ("\t \ tU \ n"); printmat  (u) ;
 /* now solving LV = B by forward substitution */
  for (i = 0 ; i< N; i++)
 {
 s = 0 ;
for ( j = 0 ; j < =i –1 ; j ++)
 s += 1 [ i ] [ j ] * v [ j ] ;
 v[ i ] = b [i ] –s;
}
/ * now solving UX = V by backward substitution */
 for (i = N  – 1; i >=  0; i  )
{
s = 0;
 for (j = i + 1; j < N ; j + + )
s+= u [ i ] [ j ] * x [ j ] ;
x [ i] = (v[ i ] – s ) / u [i ] [ i ] ;
}
/  *  printing the results */
 printf (" The solution is  :- \ n");
for (i = 0 ; i< N; i++)
 printf (" x [ % 3d] = %6.4f \n", i+1, x [ i ] );
}
```

- **Gauss -Seidal method**

```
/* Gauss -Seidal method */
#include <stdio. h>
#include <math. h>
main ( )
{
 float a[ N ] [N+1], x[N] ,aerr, maxerr, t, s, err ;
int i, j, itr, maxitr;

/* first initializing the array x */
for (i=0; i<N; i++) x[ i ] = 0;
printf(" Enter the elements of the" ) "augmented matrix row wise\ n");
for ( i =0 ; i<N; i++)
 for ( j =0; j<N+1 ; j++)
  scanf ("%f", &a [ i ] [ j ]) ;
printf ("Enter the allowed error," "maximum iterations \n");
scanf ("%f %d , & aerr, & maxitr);
printf ("Iteration x[ 1] x[ 2 ]" " x[ 3 ] \n");
for (itr =1; itr<maxitr; itr++)
{
 maxerr = 0 ;
for (i =0 ; i < N ; i++)
 {
s = 0;
 for ( j = 0 ; j <N; j++)
  if ( j ! = i) s += a[ i] [ j ] *x [ j ] ;
t = (a [ i ][N] –s) / a [ i ] [ i ];
err = fabs (x [ i ] – t) ;
if (err maxerr) maxerr = err;
x [ i ] = t;
}
printf (" %5d ", itr) ;
for ( i = 0 ; i <N; i++)
printf ("%9.4f " ,x[ i ]) ;
printf ( "\n") ;
if (maxerr<aerr)
{
  printf ("converges in %3d" "iterations \n" ,itr) ;
 for (i = 0; i <N; i++)
  printf ("x[%3d ] = %7.4f \n", i+1, x[ i] );
  return 0;
  }
 }
printf ("Solution does not converge," "iterations not sufficient \n") ;
return 1;
}
```

- **Power method**

```
/* Power method for finding largest eigen value */
#include <stdio.h>
#include <math.h>
#define N 3
typedef float array [ N ];
void findmax (float *max, array x)
{
 int i;
*max = fabs (x[ 0 ] ) ;
for ( i=1;  i<N;  i++ )  if  ( fails ( x[i] > * max  )
* max= fabs (x [i] );
}
main ( )
{
  float a[N] [N] ,x[N] , r[N] ,maxe, err, errv, aerr, e,s,t;
int i, j, k, itr, maxitr;
printf (“Enter the matrix rowwise\n”) ;
for (i =0;  i<N;  i++)
for ( j =0;  j<N;  j++)
scanf  (“%f”, &a[ i ] [ j ] ) ;
printf (“Enter the initial approximation”  “to the eigen vector\n”) ;
for (i = 0;  i<N;  i++)
scanf (“ %f ” , &x[ i ]) ;
printf (“Enter the allowed error,”  “maximum iterations \n”) ;
scanf (“ %f  %d, &aerr, &maxitr) ;
printf (“Itr no. Eigenvalue”  “ EigenVector\n”);
/* now finding the largest eigenvalue in the initial approx. to eigen vector */
findmax ( & e, x) ;
/* now starting the iterations */
for (itr = 1;  itr<=maxitr; itr++)
{
 /*loop to multiply the matrices a and x */
 for ( i = 0;  i <N;  i ++)
{
 s = 0 ;
for (k = 0 ; k<N; k++)
s + = a [ i ] [ k ]*x[ k ] ;
r [ i ] = s ;
}
 findmax (&t, r) ;
for (i =0 ; i <N; i++)  r[i ] /= t;
  maxe = 0 ;
  for (i =0;i<N;i++)
 {
  err = fabs (x[ i ] - r[ i ] ) ;
 if (err > maxe) maxe  = err;
  x[ i ] = r[ i ];
```

```
}
 errv = fabs (t-e) ;
  e = t;
printf ("% 4d % 12 .4 f " , itr,e);
for (i = 0; i < N ; i + +)
 printf  ("% 9.3 f ", x(i));
printf ("\n")  ;
 if ( (errv <= aerr) && (maxe <= aerr))
{
 printf ("Converges in %d"  "iterations \n" , itr) ;
 printf ("Largest eigen value"    "= %6.2f\n",e) ;
printf ("Eigen vector:-\n") ;
for (i= 0 ; i <N; i++)
printf ("x[ %3d] = %6.2f \n", i+1,x[ i ]) ;
printf ("\n") ; return 0;
}
}
printf ("Solution does not converge",  "iterations not sufficient \n");
return 1;
}
```

- **Least squares method**

```
/* Parabolic fit by least squares */
#include <stdio.h>
main ( )
{
 float augm [ 3 ] [4 ] ={ {0, 0, 0, 0}, (0, 0, 0, 0}, {0, 0, 0, 0}};
 float t, a, b, c, x, y, xsq;
Int i, j, k, n;
 puts ("Enter the no. of pairs of"  "observed values:");
scanf (" %d", &n );
 augm [ 0 ] [ 0 ] = n;
  for (i =0; i<n;i ++)
  {
 printf ("pair no. %d\n",  i+1);
 scanf ("%f %f", &x, &y);
xsq = x*x;
augm [ 0 ] [ 1] += x;
augm [ 0 ] [ 2 ] += xsq;
augm [ 1 ] [ 2 ] += x*xsq;
augm [ 2 ] [ 2 ] += xsq*xsq;
augm [ 0 ] [ 3 ] += y ;
augm [ 1 ] [ 3 ] += x*y;
augm [ 2]  [ 3 ] +=xsq*y;
}
augm [ 1 ] [ 1 ] = augm [ 0 ] [ 2 ];
augm [ 2 ] [ 1 ] = augm [ 1 ] [ 2 ];
augm [ 1 ] [ 0 ] = augm [ 0 ] [ 1 ];
```

```
augm [ 2 ] [ 0 ] = augm [ 1 ] [ 1 ] ;
puts ("The augmented matrix is  : -") ;
for ( i = 0 ; i<3; i++)
{
  for ( j=0; j<4; j++)
printf (" %9.4f " ,augm [ i ] [ j ]) ;
 printf ("\n") ;
}
/* Now solving for a, b, c by Gauss Jordan Method */
for ( j = 0 ; j<3 ; j++)
   for (i = 0 ; i<3; i++)
 if ( i !=j )
 {
   t = augm [ i ] [ j ] / augm [ j ] [ j ];
  for (k =0 ; k<4;k++)
   augm [ i ] [ k ] -= augm [ j ] [ k ] * t;
 }
a = augm [ 0 ] [ 3 ] /augm [ 0 ] [ 0 ];
b = augm [ 1 ] [ 3 ] /augm [ 1 ] [ 1 ] ;
c = augm [ 2 ] [ 3 ] /augm [ 2 ] [ 2 ] ;
printf ( "a = %8.4f  b = %8.4f"   " c = % 8.4f\ n", a, b, c) ;
}
```

- **Newton forward difference formula**

```
/* Newton's forward interpolation */
# include <stdio.h>
# define MAXN 100
#define ORDER 4
main ( )
{
 float as [MAXN +1], ay [MAXN+1],  diff [MAXN +1 ] [ ORDER +1] ,
   nr = 1. 0, dr = 1. 0, x, p, h, yp;
int n, i, j, k;
printf ("Enter the value of n\n");
scanf ("%d", &n) ;
printf ("Enter the values in form x, y\n");
for (i =0 ; i < =n; i++)
  scanf ( " %f  %f  ",&ax [ i ], & ay [ i ]) ;
printf ("Enter the values of x"   "for which value of y is wanted \n");
scanf ( " %f  ", &x) ;
h = ax [1] –ax [ 0 ] ;
/* now making the diff. table */
/* calculating the 1st  order differences */
for ( i =0 ; i<=n- 1; i++)
  diff [ i ] [1] =ay [i+1]-ay [ i ];
/* calculating the second &  higher order differences .*/
for ( j = 2 ; j< =ORDER ; j++)
for (i = 0; i<=n-j; i ++ )
diff [ i ] [ j ]= diff [i +1][ j – 1] - diff [ i ] [ j –1];
```

```
/* now finding x0 */
i = 0;
while (!(ax[ i ]> x)) i++;
/* now ax [i ] is x0 & ay [ i ] is y0 */ i - ;
p = (x-ax [ i ] /h; yp = ay [ i ];
/* Now carrying out interpolation */
for (k=1 ; k<=ORDER ; k++)
{
nr * = p-k+1;  dr *=k;
 yp + = (nr/dr)*diff [ i ] [ k ] ;
}
printf (“when x =%6.1f, y = %6.2 f \ n”,  x,  yp);
}
```

- **Lagrange’s interpolation method**

```
/*Lagrange’s Interpolation */
#include <stdio.h>
#define MAX 100
main ( )
{
  float ax  [MAX+1], ay[MAX+1], nr, dr, x,y = 0;
 int i, j, n;
printf (“Enter the value of n\n”) ;
scanf (“ %d ”, &n) ;
printf (“Enter the set of values\n”);
for  (i = 0 ; i <=n;  i++)
scanf ( “ %f , %f , & ax [i], say [i] ;
puts ( “Enter the value of x for which”   “value of  y is wanted” );
scanf (“%d”, & x) ;
for  (i = 0 ; i <=n; i++)
{
  nr = dr = 1;
 for ( j = 0 ; j<n ; j++)
 if ( j ! = i )
 {
  nr  * = x - ax[ j ];
 dr * =ax [ i ] – ax [ j ] ;
 }
   y + = (nr/dr)*ay[ j ] ;
}
printf ( “ When x=%4. If   y=%7. if \n” ,x, y );
}
```

- **Trapezoidal rule**

```
/* Trapezoidal rule. */
#include <stdio.h>
float y(float x)
{
  return  f(x);                    (f(x) to be defined here by the user)
}
main ( )
{
 float x0, xn, h, s;
int i, n;
puts ("Enter x0, xn, no. of subintervals") ;
scanf ("%f  %f  %d", &x0, &xn, &n );
h = (xn-x0)/n;

s = y(x0)+y(xn) ;
for (i=1;i<=n-1;i++)
s += 2*y(x0+i*h) ;
printf ("Value of integral is % 6.4f\n", (h/2)*s) ;
}
```

- **Simpson's rule**

```
/* Simpson's rule */
#include <stdio.h>
float y(float x)
{
return f(x);                       (f(x) to be defined here by the user)
}
main ( )
{
  float x0, xn, h, s;
  int i, n;
  puts ("Enter x0, xn.  no.of subintervals");
 scanf ("%f  %f  %d", &x0, &xn, &n) ;
  h = (xn-x0) /n;
  s = y(x0)+y(xn)+4*y(x0+h);
  for (i=3;i<=n-1;i+=2)
    s += 4*y(x0+i*h)+2*y(x0+(i-1)*h);
printf("Value of integral is %6.4f\n", (h/3) *s);
}
```

- **Euler method**

```
/*Euler's Method*/
#include <stdio.h>
float df(float x, float y)
```

```
{
 return (f(x);    (f(x) to be defined here by the user)

}
main ( )
{
  float x0, y0, h, x, x1, y1;
 puts ("Enter the values of x0, y0, h, x");
 scanf ( "%f  %f  %f ",  &x0, &y0,  &h,  &x );
x1 =x0;  y1= y0;
while(1)
{
   if (x1>x) return;
   y1 += h*df(x1, y1);
  x1 += h;
printf ( " When x = %3.1f  "   "y = %4.2f\n", x1, y1);
 }
}
```

- **Runge-Kutta method**

```
/* Runge -Kutta Method */
#include <stdio.h>
float f(float x, float y)
{
 return  f (x,y) ;  (f(x,y) to be defined here by the user)
}
main ( )
 {
 float x0, y0, h, xn, x, y, k1, k2, k3, k4, k ;
 printf ( " Enter the values of x0, y0,"  "h, xn\n " );
scanf  ( " %f  %f   %f   %f", &x0, &y0, &h, &xn );
x =x0;  y = y0;
while (1)
{
if ( x == xn ) break;
k1 = h*f(x,y);
k2 = h*f( x+h/2, y+k1/2 );
k3 = h*f ( x+h/2, y+k2/2 ) ;
k4 = h*f (x+h, y+k3 ) ;
k   = (k1+( k2+k3 )*2+k4 )/6;
x + = h; y + = k;
printf ( " When x = %8.4f"  "y = %8.4f\n", x, y ) ;
 }
}
```

- **Miline method**

```
/*Miline predictor corrector*/
# include <stdio.h >
#include <math.h >
float x[5], y[ 5 ], h;
float f(int i)
{
 return x[ i ] –y[ i ] *y[ i ];
}
void corect ( )
{
 y [ 4 ] = y[ 2 ] +(h/3) * (f(2)+4*f(3)+f( 4));
printf (“%23s %8.4f %8.4f \n”, “ ”,y[ 4 ], f( 4 )) ;
}
main ( )
{
  float xr, aerr, yc;
 int i;
  puts (“Enter the values of x0,xr,h”, “allowed error”);
 scanf (“ %f %f %f %f ”, &x[ 0], &xr, &h, &aerr);
puts (“Enter the value of y[ i ], i= 0,3”);
for (i=0; i,<=3; i++) scanf ( “ %f ”, &y[ i ] );
for (i =1; i<=3; i++) x[ i ] = x[ 0 ] +i*h;
puts ( “ x   Predicated ”  “Corrected ” );
puts( “      y            f, ”       y           f” );
while ( 1)
{
if (x[3 ]= xr) return;
x [4 ] =x[3]+h;
y[ 4 ] = y [ 0 ]+ (4*h/3)*(2*(f(1)+f(3))-f(2) ) ;
  printf ( “ %6.2f  %8.4f  %8.4f\n ”, x[4 ], y [4 ], f ( 4 ) ) ;
  corect ( );
  while ( 1 )
  {
yc =y[4 ];
corect ( ) ;
 if (fabs (yc-y[4 ] ) <=aerr) break;
  }
for (i =0; i<=3; i++)
{
  x [ i ] = x[ i+1];
 y[ i ] = y [ i+1];

  }
  }
}
```

- **Elliptic partial differential equations**

```
/* Solution of Laplace's Equation */
#include <stdio.h>
#include <math. h>
#define SQR 4
typedef float array [SQR+1] [SQR+1];
void getrow (int i, array u)
{
  int j;
printf ( " Enter the values of u[%d, j ]," "j=1,%d\n", I, SQR ) ;
 for ( j =1; j<=SQR; j++)
 scanf ( " %f ", &u[ i ] [ j ] );
 }
 void getcol ( int j, array u)
{
 int i;
 printf ( " Enter the values of u[i,%d ],"  " i=2, %d\n", j, SQR-1 ) ;
for (i=2; i<=SQR-1; i++)
scanf ( " %f ", &u[ i ] [ j ] );
}
void printarr (array u,int width, int precision)
{
int i, j;
 for ( i =1; i<=SQR; i++)
 {
for ( j=1; j <=SQR; j++)
printf ( " %*.*f", width, precision, u[ i ] [ j ] );
printf( " \n" );
}
}
main ( )
{
array u;
float maxerr, aerr, err, t;
int i,j,itr,maxitr;
for (i=1; i<=SQR; i++)
 for ( j=1; j<=SQR; j++) u[ i ] [ j ]=0 ;
puts ( " Enter the boundary conditions " );
getrow ( 1,u ); getrow (SQR, u);
getcol (1,u); getcol (SQR, u );
puts ( " Enter allowed error,"  "maximum iterations" );
scanf ( " %f  %f ", &aerr, & maxitr);
for ( itr=1; itr<=maxitr; itr++)
{
 maxerr = 0;
 for (i=2; i<=SQR-1; i++)
  for ( j=2; j<=SQR-1; j++)
  {
 t=(u[ i-1] [ j ] +u[ i+1][ j ]+ u[ i ] [ j+1]+u[ i ] [ j-1])/4;
```

```
err =fabs (u[ i ][ j ] –t);
if (err > maxerr)
maxerr =err;
 u [ i ] [ j ]=t;
}
printf( " Iteration no. %d \n ", itr );
printarr (u, 9, 2);
if (maxerr <= aerr)
{
 printf  ( "After %d iterations \n"    " The solution :-\n ",  itr  );
printarr (u, 8, 1);
 return 0;
  }
}
puts ( " Iterations not sufficient ." );
return 1;
}
```

- **Parabolic partial differential equations**

```
/*Solution of diffusion equations by Bendre Schmidt method*/
#include  <stdio.h>
#define  XEND 8
#define TEND  5
 float  f (int x)
{
  return f(x);                        (f(x) to be defined here by the user)
 }
main ( )
{
  float u[XEND+1][TEND+1], h=1.0,k=0.125,  csqr, alpha, ust, uet;
int i, j;
puts ( "Enter the square of 'c'  " );
scanf (" %f ",  &csqr);
alpha =(csqr*k)/ (h*h);
puts ("Enter the value of u[0,t]");
scanf ( " %f , &ust ) ;
printf ("Enter the value of u[%d, t]\n",  XEND) ; scanf ( " %f  ", &uet );
for ( j=0 ;  j<=TEND;  j++)
 u [ 0 ] [ j ] =u [XEND ] [ j ]=ust;
for ( i =1;  i<=XEND-1;  i++)
u[ i ] [ j +1]=  alpha *u[i-1] [ j ]  + (1-2*alpha)*u[ i ] [ j ] + alpha*u[ i+1] [ j ];
printf ( " The value of alpha is %4.2f\n ",    alpha);
puts ( " The values of u[ i, j ] are:-" ); for ( j =0;  j<TEND;  j++)
{
for (i =0 ;  i<XEND;  i++)
 printf ( " %7.4f ", u[ i ] [ j ] ) ;
   printf ( "\ n" );
  }
}
```

- **Hyperbolic partial differential equations**

```
/* Solution of wave equation */
#include <stdio.h>
#define XEND 5
#define TEND 5
float f(int x)
{
 return f(x);                    (f(x) to be defined here by the user)
}
main ( )
{
 float u[XEND +1] [TEND+1], csqr,ust,uet;
int i, j;
  puts ( "Enter the square of 'c' " );
  scanf ( " %d ",  &csqr);
 printf ( "Enter the value of u[ 0 ] [t ] \n" );
  scanf ( " %f  ",  &ust );
  printf ( "Enter the value of u[ %d ] [ t ]\n", XEND ); scanf ( " %f ", &uet);
for ( j =0 ;  j<=TEND;  j++)
  {
  u[ 0 ] [ j ] = ust;  u[XEND ] [ j ] =uet;
  }
  for (i =1;  i<=XEND –1;  i++)
 u[ i ] [1 ]= u[ i ] [ 0 ] =f( i );
for  ( j =1;  j<=TEND-1;  j++)
 for ( i=1;  i<=XEND-1;  i++)
u[ i ] [ j+1] = u[i-1] [j ] +u [i+1] [ j ] -u[ i ] [ j-1] ;
puts ( " The values of u[ i ] [ j ]are :-" );
for ( j =0;  j <=TEND;  j++)
{
for (i=0 ;  i<=XEND;  i++)
 printf (  " %6. If " , u[ i ] [ j ]);
   printf ( "\n" );
 }
}
```

****

ANSWERS TO KEY PROBLEMS

CHAPTER 1

1. 38.46, 3.26, 4986000, 0.0003222, 19.24 **2.** 0.077

3. 0.04963 **4.** 600.05, 1000.05, 5.766 x 10^2

5. 1.149 x 10^{-4}, 4.836 x 10^{-4} **6.** 0.0025 **10.** 140.4, 13%

12. $\sin t + \cos t = 1 + t - \dfrac{t^2}{2} - \dfrac{t^3}{6} + \dfrac{t^4}{24} + \dfrac{t5}{125} + o\left(t^6\right)$, $\sin t \cos t = t - \dfrac{2}{3}t^3 + \dfrac{2}{15}t^{15} + o\left(t^6\right)$

CHAPTER 2

1. (i) -2.105 (ii) 2.943 (iii) 2.706 (iv) 2.375 (v) 2.621
(vi) 0.657 (vii) 2.279 (viii) 1.8125 (ix) 2.7406 (x) 0.682

2. (i) 2.7065 (ii) 1.4036 (iii) 2.621 (iv) 1.466 (v) 1.328
(vi) 0.853 (vii) 0.6071 (viii) 2.7406 (ix) 1.8438

3. (i) -2.1051 (ii) 2.621 (iii) 1.4036 (iv) 0.567 (v) 0.607 (vi) 2.6875

4. (i) 0.7548 (ii) 0.6071 (iii) 1.516 (iv) 1.088 (v) 0.0913 (vi) 1.068
(vii) 1.466 (viii) 0.5314 (ix) 0.1514 (x) 4.4346 (xi) 4.3311 (xii) -0.70346

5. (i) 2.375 (ii) 0.5110 (iii) 1.834 (iv) 1.532 (v) 1.967 (vi) 2.198
(vii) -2.533 (viii) 2.908 (ix) 1.662 (x) 2.798 (xi) 0.739 (xii) 1.896
(xiii) 1.8438 (xiv) 1.756 (xv) 0.5177 (xvi) 0.5886

6. (i) 1.4142 (ii) 6.4031 (iii) 2.1547 (iv) 2.3784

7. (i) 2.605 **8.** $m = 2$, 3.973 **9.** (i) 1.3247 (ii) 1.8393 (iii) 2.26 (iv) 3.14

10. (ii) 5.124, 1.6367, 0.2384, (iii) 5, 2.001, 0.9995 (iv) 3,2,1 (v) 2,1,1
(vi) 1.1892, 0.3377, $-0.5946 + 1.16149i$, $-0.5946 - 1.16149i$

11. (i) 3.162, 2.45 (iii) -1.9266, -1.8533

CHAPTER 3

1. (i) $x = 7, y = -9, z = 5$ (ii) $x = \dfrac{122}{109}, y = \dfrac{284}{327}, z = \dfrac{46}{327}$, (iii) $x = -12.75, y = 14.375, z = 8.75$

(iv) $x = -1, y = 4, z = 4$ (v) $x = 0.2,\ y = 0.4,\ z = 0.6,\ u = 0.8$

2. (i) $x=\frac{122}{109}, y=\frac{284}{327}, z=\frac{46}{327}$, (ii) $x=y=z=1$

(iii) $x=8.7, y=5.7, z=-1.3$, (iv) $x_1=2, x_2=\frac{1}{5}, x_3=0, x_4=\frac{4}{5}$

3. (i) $x=y=z=1$ (ii) $x=1, y=3, z=5$ (iii) $x=\frac{122}{109}, y=\frac{284}{327}, z=\frac{46}{327}$,

(iv) $x_1=-1, x_2=0, x_3=1, x_4=2$

4. (i) $\frac{1}{8}\begin{bmatrix} 1 & -1 & 3 \\ -1 & 1 & 5 \\ 3 & 5 & 23 \end{bmatrix}$ (ii) $\frac{1}{8}\begin{bmatrix} 2 & 2 & -2 \\ -9 & 11 & 5 \\ 5 & -7 & -1 \end{bmatrix}$ (iii) $\begin{bmatrix} -5/12 & 1/4 & 1/3 \\ 7/12 & 1/4 & -2/3 \\ 1/12 & -1/4 & 1/3 \end{bmatrix}$

5. (i) $x=1.013, y=-1.996, z=3.001$ (ii) $x=1.052, y=1.369, z=1.962$,

(iii) $x=2.556, y=1.722, z=-1.055$, (iv) $x=0.996, y=1.95, z=3.16$,

Actual value is $x=1, y=2, z=3$

(v) $x=1.926, y=3.573, z=2.425$ (vi) $x_1=1, x_2=2, x_3=3, \dot{x}_4=0$

6. (i) $x=1.35, y=2.103, z=2.845$, (ii) $x=y=z=1$ (iii) $x=32, y=26, z=21$

7. (i) $\begin{bmatrix} -0.5 & 0.2 & -1.6 \\ 0 & 0.2 & 0.4 \\ 0.5 & 0 & 1 \end{bmatrix}$ (ii) $\frac{1}{8}\begin{bmatrix} 1 & -1 & 3 \\ -1 & 1 & 5 \\ 3 & 5 & 23 \end{bmatrix}$,

(iii) $\frac{1}{8}\begin{bmatrix} 1.4 & 0.2 & -0.4 \\ -1.5 & 0 & 0.5 \\ 1.1 & -0.2 & -0.1 \end{bmatrix}$ (iv) $\frac{1}{9}\begin{bmatrix} -15 & -14 & 12 & 13 \\ 3 & 7 & -6 & -2 \\ 9 & -3 & 0 & -3 \\ -6 & -1 & 3 & 1 \end{bmatrix}$

8. (i) $9, \begin{bmatrix} 1 & 0 & -1 \end{bmatrix}'$ (ii) $11.66, \begin{bmatrix} 0.025 & 0.042 & .0-1 \end{bmatrix}'$ (iii) $7, \begin{bmatrix} \frac{37k}{66} & \frac{2k}{11} & k \end{bmatrix}$

9. $\begin{bmatrix} 1 & -5 & 0 \\ -5 & \frac{73}{25} & \frac{14}{25} \\ 0 & \frac{14}{25} & \frac{-23}{25} \end{bmatrix}$ (ii) $\lambda=0,3,3$ (iii) $\begin{bmatrix} 1 & -1.12 & 0 \\ -1.12 & 1.4 & -0.55 \\ 0 & -0.55 & 1.6 \end{bmatrix}$

10. (i) 4, -2, 6; $\begin{bmatrix} \frac{1}{\sqrt{2}} & 0 & \frac{-1}{\sqrt{2}} \end{bmatrix}$, $\begin{bmatrix} 0 & 1 & 0 \end{bmatrix}$, $\begin{bmatrix} \frac{1}{\sqrt{2}} & 0 & \frac{1}{\sqrt{2}} \end{bmatrix}$

(ii) 0.3856, -1.3126, 5.9269; $[0.5654 \quad -0.2949 \quad -0.8243]'$,

$[0.5457 \quad -0.736 \quad 0.4006]'$, $[-0.6185 \quad -0.6763 \quad -0.4001]'$

(iii) Largest eigenvalue is 5, corresponding eigenvector $[1, 1, 1]$

11. (i) $\begin{bmatrix} 0.76 & -0.65 \\ 0.60 & 0.74 \\ 0.26 & 0.19 \end{bmatrix} \begin{bmatrix} 27.78 & 0 \\ 0 & 9.85 \end{bmatrix} \begin{bmatrix} 0.87 & 0.49 \\ -0.49 & 0.87 \end{bmatrix}$

CHAPTER 4

1. $y = 13.6x$, **2.** 15.2 *thousand tons* **3.** $R = 70.052 + 0.29t$

4. $a = 0.5012$, $b = 1.9977$ **5.** $a = 99.86$, $b = 1.2$ **6.** $a = 146.3$, $k = -0.412$

7. $a = 10.5$, $b = -0.13$ **8.** $a = 6.32$, $b = 0.0095$ **9.** $y = 1.243 - 0.004x + 0.22x^2$

10. 31, **11.** $f(x) = 2x^3 - 7x^2 + 6x + 1$, $f(4) = 41$ **12.** 6.36, 11.02 **13.** 0.788

14. $y = \frac{2}{3}x^4 - 8x^3 + \frac{100}{3}x^2 - 56x + 31$ **15.** $u_2 = 42$, $u_4 = 49$ **16.** 755 **17.** 6.7531

18. 13.062, **19.** 0.3165, **20.** 111.874930, **21.** 0.19573, **22.** $x^3 - 3x^2 + 5x - 6$,

23. 12.45 **24.** (i) $\frac{5}{2(x-1)} - \frac{15}{x-2} + \frac{31}{2(x-3)}$, $\frac{1}{2(x-1)} + \frac{1}{x-2} - \frac{1}{2(x+1)}$,

(ii) $\frac{1}{5(x-1)} + \frac{3}{35(x+1)} - \frac{13}{10(x-4)} + \frac{71}{70(x-6)}$, **25.** 1.16314

27. $y(x) = \begin{cases} 3x^3 - 9x^2 + 11x - 11 & ; kx \le 2 \quad y(1.5) = -4.625 \\ -3x^3 + 27x^2 - 61x + 37 & ; 2 \le x \le 3 \quad y(2) = 11. \end{cases}$

CHAPTER 5

1. 4.75, 9 **2.** -27.9, 117.67 **3.** 4.054 **4.** 0.63, 6.6 **5.** 3.8140

6. (i) 0.5005, -0.2732 (ii) 0.4473, -0.1583 (iii) 0.4662, -0.2043

7. (i) 0.3950 (ii) 0.3341, -0.0256 (iii) 0.2719

8. -0.06, 0.5 **9.** -0.175 **10.** 0.5403 **11.** 1.58, 1.00 **12.** 0.26, **13.** 1.0987,

14. (i) 3.14 (ii) -0.747 **15.** 177.4830 **16.** (i) 0.7854 (ii) 0.7854 (iii) 0.78535

(iv) 0.78535, **17.** 53.87, 53.6 **18.** 0.9985, **19.** 1.1873, **20.** 29.0993, **22.** 0.7854,

23. 0.004999 **24.** 1.5 **25.** 1.00002 **26.** 0.6931 **27.** 1.8278

29. 0.876 **30.** 4.134 **31.** 0.49

CHAPTER 6

1. (i) $\bar{y}(x)=1+x+\frac{x^2}{2}+\frac{x^3}{3}+\frac{x^4}{8}+\ldots,\quad y(0.1)=1.1053$

(ii) $y(4.1)=4.005,\quad y(4.2)=4.0098$ **2.** $y(0.1)=3.005, y(0.2)=3.020$ *etc.*

5. $y(0.02)=1.0202, y(0.04)=1.0408, y(0.06)=1.0619$ **6.** $y(2)=5.0524$

7. $y(1)=1.4983$ **8.** 1.0101, 1.0207, 1.0318, 1.0438

9. 1.0569, 1.0713, 1.0871, 1.1048, 1.1244, 1.1464

10. 0.0918, 0.1487, 0.2221, 0.3138, 0.4255, 0.5596, 0.7183

11. $x(0.1)=1.1003,\ y(0.1)=1.1102,$ **12.** $y(1.1)=0.3707,\ z(1.1)=1.0361$

13. $y(0.2)=1.0204$, $y(0.4)=1.0$ **14.** *For* $n=2,\ y(0.5)=0.1389$

For $n=4,\ y(0.5)=0.1470,$ True value =0.1505

CHAPTER 7

1. (i) $u_1=7.9,\ u_2=13.7,\ u_3=17.9,\ u_4=6.6,\ u_5=11.9,\ u_6=16.3,$

$u_7=6.6,\ u_8=11.2,\ u_9=14.3,$

(ii) $u_1=1.999,\ u_2=2.999,\ u_3=3.999,\ u_4=2.999$

2. $u_1=26.66,\ u_2=33.33,\ u_3=43.33,\ u_4=46.66$

4. $u_1=1.57,\ u_2=3.71,\ u_3=6.57,\ u_4=2.06,\ u_5=4.69,\ u_6=8.06,$

$u_7=2.00,\ u_8=4.69,\ u_9=9.00$

5. $u_1=-3,\ u_2=-2,\ u_3=-2$

6.

j \ i	0	1	2	3	4
0	0	3	4	3	0
1	0	2	3	2	0
2	0	1.5	2	1.5	0
3	0	1	1.5	1	0
4	0	0.75	1	0.75	0
5	0	0.5	0.75	0.5	0

7.

j \ i	0	1	2	3	4	5	6	7	8	9	10
0	0	0.09	0.16	0.21	0.24	0.25	0.24	0.21	0.16	0.09	0
1	0	0.08	0.15	0.20	0.23	0.24	0.23	0.20	0.15	0.08	0
2	0	0.075	0.14	0.19	0.22	0.23	0.22	0.19	0.14	0.075	0
3	0	0.07	0.13	0.18	0.21	0.22	0.21	0.18	0.13	0.07	0

8.

j \ i	0	1	2	3	4	5
0	0	24	84	144	144	0
1	0	42	84	114	72	0
2	0	42	78	78	57	0
3	0	39	60	67.5	39	0
4	0	30	53.25	49.5	33.75	0
5	0	26.6	39.75	43.5	24.75	0
6	0	19.88	35.06	32.25	21.75	0

10.

t = 0.3, x =	0.1	0.2	0.3	0.4	0.5
u =	0.02	0.04	0.06	0.075	0.08
Exact u =	0.02	0.04	0.06	0.075	0.08

11.

j \ i	0	1	2	3	4
0	0	3	4	3	0
1	0	3	4	3	0
2	0	1	2	1	0
3	0	-1	-2	-1	0
4	0	-3	-4	-3	0

12.

t \ x	0	0.1	0.2	0.3	0.4	0.5
0.1	0	0.037	0.07	0.096	0.113	0.119
0.2	0	0.031	0.059	0.082	0.096	0.0101
0.3	0	0.023	0.043	0.059	0.07	0.074
0.4	0	0.012	0.023	0.031	0.037	0.039
0.5	0	0	0	0	0.05	0

14. $u_{11} = 0.148$, $u_{21} = 0.053$, $u_{12} = 210$, $u_{22} = 0.091$ etc.

15. $u_{11} = 0.1375$, $u_{21} = 0.1075$, $u_{31} = 0.0815$ etc.

16. $u_1 = 26.66$, $u_2 = 33.33$, $u_3 = 43.33$, $u_4 = 46.66$

CHAPTER 8

3. (i) $f(x) = \frac{\lambda\pi}{2} + \cot x$ (ii) $f(x) = \frac{2}{2-\lambda}\sin x, \quad \lambda \neq 2$

(iii) $f(x) = \frac{4\cos x + 2\lambda\pi \sin x}{4 + \lambda^2\pi^2}$

6. (i) $2x^2 + 2x$ (ii) $\frac{19}{8} - \frac{3}{2}x^2 + \frac{x^4}{2}$

8. $3x^2 + \frac{3}{2}x$

APPENDIX

SOME NUMERICAL METHODS TO SOLVE NONLINEAR COMPLEX EQUATIONS

In 1879, Cayley tried to use Newton-Raphson method to find complex roots of f(z)=0 where $f: C \to C$ is a complex function. If we take $z_0 \in C$ and iterate

$$z_{n+1} = z_n - f(z_n)/f'(z_n), n = 0,1,2,3... \quad \text{(A1)}$$

to look for conditions under which the sequence $\{z_n\}_{n=0}^{\infty}$ converges to a root. Cayley solved this problem when *f* is a quadratic polynomial and for cubic polynomials, after several years of trying he finally decline to continue. The Cayley's failure to make any real progress at that time was inevitable, because the Julia set, the set of points where Newton's method fails to converge, has fractional dimension. For instance in case of $f(z) = z^3 - 1$ it coincides with the frontier of the attraction basins of the three complex roots $e^{2k\pi i/3}, k = 0,1,2$ of unity. Here we outline some iterative methods $z_{n+1} = \phi(z_n)$ for solving $f(z) = 0$ for a complex function $f: C \to C$. In all these methods, we take a starting point $z_0 \in C$.

Newton's method

$$z_{n+1} = z_n - f(z_n)/f'(z_n), n = 0,1,2,3...$$

Its order of convergence is 2.

Newton's method for multiple roots

$$z_{n+1} = z_n - \frac{f(z_n)f'(z_n)}{(f'(z_n))^2 f(z_n) - f''(z_n)} \quad \text{(A2)}$$

Its order of convergence is 1.This method recovers the order 2 for multiple roots as deduced below. If f has a root of multiplicity $m \geq 1$ at ξ, then we only need to apply the ordinary Newton's method to the equation $g(z) = 0$.

Whittaker's method

$$z_{n+1} = z_n - \frac{f(z_n)}{2f'(z_n)}(2 - L_f(z_n)), \; L_f(z) = \frac{f(z)f''(z)}{(f'(z))^2} \quad \text{(A3)}$$

This method is also known as parallel-chord method, from its geometric interpretation for functions $f: R \to R$. It is a simplification of Newton's method in which to avoid computing the derivative, we make the approximation $f'(z) = 1/\lambda$, λ being constant. We try to choose λ in such a way that $F(z) = z - \lambda f(z)$ is a contractive function and so it will have a fixed point. Its order of convergence is 1 and convex acceleration is of order 2.

Hally's method

$$z_{n+1} = z_n - \frac{f(z_n)}{f'(z_n)} \frac{2}{2 - L_f(z_n)} = z_n - \frac{1}{\frac{f'(z_n)}{f(z_n)} - \frac{f''(z_n)}{2f'(z_n)}} \quad \text{(A4)}$$

This method was presented by Edmund Halley in about 1694, who is well known for first computing the orbit of the comet that carries his name. From its geometric interpretations for real functions, it is also known as method of tangent hyperbolas and is one of the most frequently rediscovered iterative functions in the literature. Alternatively, it can be interpreted as applying Newton's method to the equation g(z)=0 with $g(z) = f(z)/\sqrt{f'(z)}$. Its order of convergence is 3.

Chebyshev's method

$$z_{n+1} = z_n - \frac{f(z_n)}{f'(z_n)}\left(1 + \frac{L_f(z_n)}{2}\right) \quad \text{(A5)}$$

This is known as Euler-Chebyshev's method or method of tangent parabolas. It has order of convergence as 3.

Stirling's method

$$z_{n+1} = z_n - \frac{f(z_n)}{f'(z_n) - f(z_n)} \quad \text{(A6)}$$

Its order of convergence is 2.

Concept of speed of convergence

Let $\{z_n\}_{n=0}^{\infty}$ be a complex sequence. We say that $\alpha \in [1, \infty)$ is the order of convergence of the sequence if

$$\lim_{n \to \infty} \frac{|z_{n+1} - \xi|}{|z_n - \xi|^{\alpha}} = C \quad \text{(A7)}$$

where ξ is the complex number and C is nonzero constant. If $\alpha = 1$, then we assume as extra condition $|C| < 1$. The convergence of order α , if the constant C in (A7) is allowed to be zero or equivalently, if there exists a constant C and an index number such that $|z_{n+1} - \xi| \le C|z_n - \xi|^{\alpha}$, for any $n \ge n_0$.

The order of convergence is used to compare the speed of convergence of sequences. Here the speed we mean the number of iterations necessary to reach the limit with a required precision. Suppose that we have two sequences $\{z_n\}$ and $\{z'_n\}$ converging to the same limit ξ and assume that their respective order of convergence are α and α'. Then it is clear that asymptotically the sequence $\{z_n\}$ converges to its limits with little iteration for the some approximation (quickly) than $\{z'_n\}$. More refined measures for the speed of convergence are the concepts of information efficiency and efficiency index. If, each iteration require d new pieces of information (a "piece of information" typically is any evaluation of a function or one of its derivatives) then the informational efficiency is $\frac{\alpha}{d}$ and the efficiency index is $\alpha^{1/d}$, where the order of convergence is α.

The efficiency index is useful because it allows us to avoid artificial accelerations of an iterative method. For example, let us suppose that we have an iterative process $z_{n+1} = \phi(z_n)$ with order of convergence α. We take a new process $z_0^* = z_0, z_{n+1}^* = \phi(\phi(z_n^*))$. Then it is clear that the new sequence is merely $z_n^* = z_{2n}$ but $\{z_n'\}_{n=0}^{\infty}$ has order of convergence α^2. However both sequences $\{z_n\}_{n=0}^{\infty}$ and $\{z_n'\}_{n=0}^{\infty}$ has the same efficiency index. Therefore for the iterative method $z_{n+1} = \phi(z_n)$ the efficiency index is more suitable than the order of convergence to measure the computer time that a method uses to converge. But if ϕ involves a function f and its derivatives, the efficiency index still has a missing element namely, it does not take into count the computational work involved in computing $f, f', f'', \ldots$. To avoid this new concept of computational efficiency is given.

Suppose that in a method the cost of evaluating ϕ related with function f is $\theta(f)$ then the computational efficiency of ϕ relative to f is $E(\phi, f) = \alpha^{1/\theta(f)}, \alpha$ being the order of convergence. Because $\theta(f)$ may depend on computer, it is difficult to establish its value and hence computational efficiency is not very much used in practice. In the literature more commonly used measure is order of convergence however, this is the one that provides least information about the computer time necessary to find the root with a required precision. Finally, note that to ensure the convergence of an iterative method $z_{n+1} = \phi(z_n)$ for solving an equation f (z)=0, it is usually necessary to begin the method from a point z_0 close to the solution ξ.

Des Cartes method for complex quartic equations

Consider the equation

$$x^4 - Ax^3 + Bx^2 - Cx + D = 0, \tag{A8}$$

where A, B, C, D may be real or complex. In order to solve the complex equation (A8) we use DesCartes' algorithm as follows. Diminishing the roots of secular equation (A8) by a factor of $\frac{A}{4}$ to eliminate second term, we obtain

$$y^4 + Hy^2 + Gy + I = 0 \tag{A9}$$

where $y = x - \frac{A}{4}$, $H = B - \frac{3A^2}{8}$, $G = \frac{AB}{2} - \frac{A^3}{8} - C$, $I = D + \frac{A^2B}{16} - \frac{3A^4}{256} - \frac{AC}{4}$ (A10)

Factorizing the equation (A9) in to two quadratic factors, as per requirement of the Descartes procedure, we have

$$y^4 + H y^2 + Gy + I = (y^2 + ly + m)(y^2 - ly + n) \tag{A11}$$

Comparing the coefficients of various powers of y in (A11) on both sides, we get

$$m + n = l^2 + H \ , n - m = \frac{G}{l} \ , \ mn = I \tag{A12}$$

Eliminating m and n from equation (A12), we obtain

$$z^3 + 2Hz^2 + (H^2 - 4I)z - G^2 = 0 \tag{A13}$$

where $z = l^2$. The equation (14) being a cubic with complex coefficients will be solved by using irreducible case of Cardan's method with the help of De-Moivre's theorem. We again diminish the roots of equation (A13) by a factor of $\frac{-2H}{3}$ in order to obtain the standard cubic as

$$t^3 - 3H^* t - G^* = 0 \tag{A14}$$

$$\text{where } t = z + \frac{2H}{3}, H^* = \frac{(H^2 + 12I)}{9}, G^* = G^2 - \frac{8HI}{3} + \frac{2H^3}{27} \tag{A15}$$

Let the roots of equation (A14) be of the type

$$t = u + v \tag{A16}$$

so that $u^3 + v^3 = G^*$, $u^3 v^3 = H^{*3}$. From these equations it can be concluded that u^3 and v^3 are imaginary quantities and there is no general arithmetic or algebraic method of finding the exact value of the cube roots of imaginary quantities. This case is known as irreducible case of Cardan's method. However, we may find the cube roots with the help of De-Moivre's theorem as below:

Let $$u^3 = \frac{G^* + \sqrt{G^{*^2} - 4H^{*^3}}}{2} = L + iM, \quad L,\ M \in R \tag{A17}$$

Then the values of u are given by

$$u_1 = r^{\frac{1}{3}}\left[\cos\frac{\Phi}{3} + i\sin\frac{\Phi}{3}\right],$$

$$u_2 = r^{\frac{1}{3}}\left[\cos\left(\frac{2\pi+\Phi}{3}\right) + i\sin\left(\frac{2\pi+\Phi}{3}\right)\right],$$

$$u_3 = r^{\frac{1}{3}}\left[\cos\left(\frac{4\pi+\Phi}{3}\right) + i\sin\left(\frac{4\pi+\Phi}{3}\right)\right] \tag{A18}$$

where $r = \sqrt{L^2 + M^2}$ and $\Phi = \tan^{-1}(\frac{M}{L})$. Having determined u, the values of v can be obtained from the relation $uv = H^*$ which leads to the required values of t and hence to the values of $l^2 = z = t - \frac{2H}{3}$. One of the (convenient) value of l so obtained is then used to evaluate m and n from equation (A14). The values of m, n and l so obtained are used in the right hand side of equation (A11) to finally factorize the equation (A9) .The factored quadratic equations are then solved to find the four roots $y_i, i=1,2,3,4$ of the reduced secular equation (A9). The complex roots of secular equation (A8) are obtained from the relation $x_i = y_i + \frac{A}{4}, i=1,2,3,4$.

BIBLIOGRAPHY

1. Ahlberg, J. H., Nilson, E. N. and Walsh, J. L., The Theory of Splines and Their Applications, Academic Press, New York, 1967.

2. Ames, W. F., Numerical Methods for Partial Differential Equations, Thomas Nelson, London, 1969.

3. Anita, H. M., Numerical Methods for Scientists and Engineers, Tata McGraw Hill Publishing Company, New Delhi, 1991.

4. Atkinson, L. V., Harley, P. J. and Hudson, J. D., Numerical Methods with FORTRAN 77, Addison-Wesley Publishing Company, 1989.

5. Atkinson, K. E., An introduction to Numerical Analysis, John Wiley & Sons, New York, 1978.

6. Balagurusamy, E., Numerical Methods, Tata McGraw-Hill Publishing Company , New Delhi, 2002.

7. Booth, A. D., Numerical Methods, Academic Press, New York, 1958.

8. Buchanan, J. L. and Turner, P. R., Numerical Methods and Analysis, McGraw Hill Book Company, 1992.

9. Chopra, S. C. and Canale, R. P., Numerical Methods for Engineers, 2nd Edition, McGraw-Hill Book Company, 1989.

10. Constantinides, A., Applied Numerical Methods with personal Computers, McGraw Hill Book Company, 1987.

11. Conte, S. D. and Carl de Boor, Elementary Numerical Analysis, An Algorithmic Approach, Tata McGraw Hill, 3rd International Edition, New Delhi, 1981.

12. Crank, J. and Nicolson, P., A Practical Method for Numerical evaluation of partial differential equations of the heat conduction type, Proc. Camb. Phil. Soc. Vol. 43, 50- 67, 1947.

13. Davis, P. and Rabinowitz, P., Numerical Integration, Blaisdell, New York, 1967.

14. Dorn, W. S., Daniel, D., and Macracken, Numerical Methods with FORTRAN IV Case Studies, John Wiley, New York, 1972.

15. Froberg, C. E., Introduction to Numerical Analysis, Addison Wesley Reading, Massachusetts, 1965.

16. Gautschi, W., Numerical Analysis: An Introduction, Birkhauser, Boston, 1997.

17. Gerald, C. F. and Wheatly, P. O., Applied Numerical Analysis, 5th Edition, Addison Wesley Publishing Company, 1994.

18. Griffiths, D. V. and Smith, I. M., Numerical Methods for Engineers. Oxford University Press, 1991.

19. Hildebrand, F. B., Introduction to Numerical Analysis, Tata McGraw Hill, New Delhi, 1982.

20. Honig, G and Hirdes, U., A Method for the Numerical inversion of the Laplace Transform, J. Comp. Appl. Math., Vol. 10, 113-132,1984.

21. Ian Jacques and Colin Judd, Numerical Analysis, Champan and Hall, New York, 1987.

22. Isaacson, E. and Keller, H. B. , Analysis of Numerical Methods, John Wiley & Sons, New York, 1966.

23. Jain, M. K., Numerical Analysis for scientists and Engineers, S. B. W. Publishers, Delhi, 1971.

24. Kincaid, D., and Cheney, W., Numerical Analysis: Mathematics of scientific Computing, 2nd edition, Brooks/ Cole, Pacific Grove, CA, 1996.

25. Krishnamurthy, E. V. and Sen, S. K., Numerical Algorithms, East West Publishers, New Delhi, 1986.

26. Levy, H. and Baggott, E. A., Numerical Solutions of Differential Equations, Dover, New York, 1950.

27. Mathews, J. H., Numerical Methods for Mathematics, Science and Engineering, 2nd Edition, Prentice Hall of India, New Delhi, 1994.

28. McCormick, J. M. and Salvadori, M. G., Numerical Methods in FORTRAN, Prentice Hall of India, New Delhi, 1971.

29. Nielson, K. L., Methods in Numerical Analysis, Macmillan Company, New York, 1964.

30. Noble, B., Numerical Methods, Vol. 2, Oliver and Boyed, Edinburgh, 1964.

31. Nyhoff, L. and Leestma, S., FORTRAN 77 and Numerical Methods for engineers and Scientists, Prentice Hall, 1995.

32. Press, W. H., Teukolsky, S. K., Velleraing, W. T. and Flannery, B. P., Numerical Recipes in FORTRAN, 2nd Edition, Cambridge University Press, Cambridge, 1992.

33. Rao, K. S., Numerical Methods for Scientists and Engineers, Prentice Hall, India 2001.

34. Sastry, S. S., Introductory, Methods of Numerical Analysis, Prentice- Hall, India, 2001.

35. Scarborough, J. B., Numerical Mathematical Analysis, Johns Hopkins University Press, Baltimore, 1965.

36. Scheid, F., Theory and problems of Numerical Analysis, Schaum Series, McGraw Hill, New York, 1968.

37. Scheid, F., Numerical Analysis, Schaum Series, McGraw Hill Publishing Company, 1990.

38. Sharma J. N. and Singh, K., Partial Differential Equations for Engineers and Scientists, Narosa Publishing House, New Delhi India / Alpha Science International Ltd. UK, 2000.

39. Sharma J. N. Sharma, State space approach to generalized thermoelastic problems, Indian Journal of Pure and Applied Mathematics, Vol. 26, 1211- 1223, 1995.

40. Sharma J. N. and Kumar V., On the axisymmetric problems of thermoelasticity, Journal of Thermal Stresses, Vol. 19,781-794, 1996.

41. Sharma J. N. and Kumar V., Plane Strain problems of transversely isotropic thermoelastic media, Journal of Thermal Stresses, Vol. 20,463-476, 1997.

42. Sharma J. N. and Chauhan, R. S., Mechanical and thermal sources in a generalized thermoelastic half space, Journal of Thermal Stresses, Vol. 24, 651-675, 2001.

43. Smith G. D., Numerical Solutions of partial Differential Equations, Oxford University Press, London, 1965.

44. Stanton, R. G., Numerical Methods for Science and Engineering, Prentice Hall of India, New Delhi, 1967.

45. Traub, J. F., Iterative Methods for the solution of equations, Prentice- Hall, Englewood Cliffs, NJ, 1964.

46. Varono, J. L., Graphic and Numerical Comparison between iterative methods, The Mathematical Intelligencer, Vol. 24, 37-46, 2002.

47. Yakowitz, S. and Szidarovszky, F., An Introduction to Numerical Computations, 2nd Edition, Macmillan Publishing Company, 1990.

48. Yong, D. M. and Gregory, R. T., A Survey of Numerical Mathematics, Vol. 1 and 2, Addison Wesley Reading, Massachusetts, 1972.

INDEX